JAAatpl
JOINT AVIATION AUTHORITIES

Theoretical Training Manuals

Revised Edition

REFERENCE MATERIAL

OXFORD
Aviation Training
Succeed through our experience™

Cover photo by Derek Pedley: www.airteamimages.com

First published by Jeppesen GmbH, Frankfurt, Germany: 2001
Second edition: Jeppesen GmbH, Frankfurt, Germany: 2002
This edition published by Transair (UK) Ltd, Shoreham, England: 2004
Printed in Singapore by KHL Printing Co. Pte Ltd

Contact Details:

Ground Training Department
Oxford Aviation Services Ltd
Oxford Airport
Kidlington
Oxford OX5 1RA
England

Tel: +44 (0)1865 844299
E-mail: ddd@oxfordaviation.net

Transair Pilot Shop
Transair (UK) Limited
Shoreham Airport
Shoreham-by-Sea
West Sussex BN43 5PA
England

Tel: +44 (0)1273 466000
E-mail: info@transair.co.uk

For further information on products and services from Oxford Aviation Training and Transair visit our websites at: www.oxfordaviation.net and www.transair.co.uk

FOREWORD

Joint Aviation Authorities (JAA) pilot licences were first introduced in 1999, and have now been adopted by nearly all member states. A steadily increasing number of non-European countries have also expressed the intention of aligning their training with JAA requirements, and some have already begun this process. The syllabi and the regulations governing the award and the renewal of licences are currently defined by the JAA's licensing agency, known as "Joint Aviation Requirements-Flight Crew Licensing", or JAR-FCL. Over the next few years, JAA responsibilities, including licensing, will gradually be transferred to the new European Aviation Safety Agency (EASA).

The JAR-FCL ATPL theoretical training requirements and associated ground examinations, although possibly similar in scope to those previously used by many national authorities, are inevitably different in a number of respects from the syllabi and examinations previously used under national schemes. Consequently, students who wish to train for the JAA ATPL licence need access to study material which has been specifically designed to meet the requirements of the new licensing system. This series of text books, prepared by Oxford Aviation Training (OAT) and now published exclusively by Transair Pilot Shop, covers all JAR-FCL requirements and is specifically designed to help student pilots prepare for the ATPL theoretical knowledge examinations.

OAT is one of the world's leading professional pilot schools. Established for 40 years, Oxford has trained more than 14,000 professional pilots for over 80 airlines, world-wide. OAT was the first pilot school in the United Kingdom to be granted approval to train for the JAA ATPL, and has been the leading contributor within Europe to the process of defining and improving the training syllabus. OAT for example led and co-ordinated the joint-European effort to produce the ATPL Learning Objectives which are now published by the JAA as the definitive guide to the theoretical knowledge requirements of ATPL training.

Since JAA ATPL training started in 1999, OAT has achieved an unsurpassed success rate in the JAA ATPL examinations. At the start of 2004, for example, OAT students had successfully passed more than 20,000 individual JAR-FCL examinations and, currently, more than 300 students a year graduate from Oxford's theoretical training programmes. The text books, together with an increasing range of Computer Based Training (CBT) products, are also now used by other Flight Training Organizations both in Europe and, increasingly, throughout the world. Recognized by leading National Aviation Authorities as being fully compliant with JAR-FCL training requirements, the series has now effectively become the de-facto standard for JAR-FCL ATPL theoretical training. This achievement is the result of OAT's continued commitment to the development of the JAA licensing system. OAT's unrivalled experience and expertise make this series the best learning material available to any student who aspires to hold a JAA ATPL.

For those aspirant airline pilots who are not yet able to begin training, but hope to do so in the future, these text books provide high-quality study material to help them prepare thoroughly for their formal training. The books also make excellent reading for general aviation pilots or for aviation enthusiasts who simply wish to further their knowledge of aeronautical subjects. We trust that your study of these books will not only be enjoyable but, for those of you currently undergoing ATPL training, will also lead to success in the JAA ATPL ground examinations.

Whatever your aviation ambitions, we wish you every success and, above all, happy landings.

Oxford Aviation Training
March 2004

Textbook Series

Book	Title	JAR Ref. No.	Subject
1	010 Air Law		
2	020 Aircraft General Knowledge 1	021 01	Airframes & Systems
		021 01 01/04	Fuselage, Wings & Stabilising Surfaces
		021 01 07	Hydraulics
		021 01 05	Landing Gear
		021 01 06	Flight Controls
		021 01 08/09	Air Systems & Air Conditioning
		021 01 09/10	Anti-icing & De-icing
		021 04 00	Emergency Equipment
		021 01 11	Fuel Systems
3	020 Aircraft General Knowledge 2	021 02	Electrics – Electronics
		021 02 01	Direct Current
		021 02 02	Alternating Current
		021 02 05	Basic Radio Propagation.
4	020 Aircraft General Knowledge 3	021 00	Powerplant
		021 03 01	Piston Engines
		021 03 02	Gas Turbines
5	020 Aircraft General Knowledge 4	22	Instrumentation
		022 01	Flight Instruments
		022 03	Warning & Recording
		022 02	Automatic Flight Control
		022 04	Power Plant & System Monitoring Instruments
6	030 Flight Performance & Planning 1	031	Mass & Balance
		032	Performance
7	030 Flight Performance & Planning 2	033	Flight Planning & Monitoring
8	040 Human Performance & Limitations		
9	050 Meteorology		
10	060 Navigation 1	061	General Navigation
11	060 Navigation 2	062	Radio Navigation
12	070 Operational Procedures		
13	080 Principles of Flight		
14	090 Communications		
15	Reference Material		

REFERENCE MATERIAL

CONTENTS

jar.fcl.amc.1.470

AIRLINE TRANSPORT PILOT'S LICENCE SYLLABUS
Detailed Listing

010 00 00 00   **AIR LAW AND ATC PROCEDURES**

010 01 00 00   **INTERNATIONAL AGREEMENTS AND ORGANISATIONS**

010 01 01 00   The Convention of Chicago

010 01 01 01   Part I Air Navigation

- general principles and application: sovereignty, territory

- flight over territory of Contracting States: right of non-scheduled flight, scheduled air services, cabotage, landing at customs airports, applicability of air regulations, rules of the air, search of aircraft

- measures to facilitate air navigation: customs duty, conditions to be fulfilled with respect to aircraft: certificates of airworthiness, licences of personnel, recognition of certificates and licences, cargo restrictions, photographic apparatus: documents to be 01 carried in aircraft

- international standards and recommended practices: adoption of international standards and procedures, endorsement of certificates and licences, validity of endorsed certificates and licences: departure from international standards and procedures (notification of differences)

010 01 01 02   Part II The International Civil Aviation Organisation

- objectives and composition

010 01 01 03   Regional structure and offices

010 01 01 04   Duties in relation to:

- annexes to the convention
- standards and recommended practices
- procedures for air navigation services
- regional supplementary procedures
- regional air navigation
- manuals and circulars

010 01 02 00   Other international agreements

010 01 02 01   The International Air Transport Agreement:

- the five freedoms

010 01 02 02   The Convention of Tokyo, La Haye, Montreal

- jurisdiction
- authority of the pilot-in-command of the aircraft

010 01 02 03   European organisations name, composition, objectives and relevant documents relevant documents

- European Civil Aviation Conference (ECAC), including Joint Aviation Authorities (JAA)
- Eurocontrol
- European Commission (EC)

010 06 04 00   Approach procedures

- general criteria (except tables)
- approach procedure design: instrument approach areas, accuracy of fixes (only intersection fix tolerance factors, other fix tolerance factors, accuracy of facility providing track, approach area splays, descent gradient)
- arrival and approach segments: general, standard instrument arrival, initial approach segment (only general), intermediate approach segment, final approach segment (except tables), missed approach segment (only general)
- visual manoeuvring (circling) in the vicinity of the aerodrome: general, the visual manoeuvring (circling) area (except table), visual manoeuvring (circling) area not considered for obstacle clearance (except table), minimum descent altitude/height, visual flight manoeuvre, missed approach whilst circling
- simultaneous ILS operations on parallel or nearparallel runways
- area navigation (RNAV) approach procedures based on VOR/DME
- use of FMS/RNAV equipment to follow conventional nonprecision approach procedures

010 06 05 00   Holding procedures

- in flight procedures (except table), entry, holding
- obstacle clearance (except table)

010 06 06 00   Altimeter setting procedures (including ICAO Doe. 7030 regional supplementary procedures)

- basic requirements (except tables), procedures applicable to operators and pilots (except tables)

010 06 07 00   Secondary surveillance radar transponder operating procedures (including ICAO Doe. 7030-regional supplementary procedures)

- operation of transponders
- operation of ACAS equipment
- phraseology

010 07 00 00   AIR TRAFFIC SERVICES (based on ANNEX 11 and Doc. 4444)

010 07 01 00   Air Traffic Services Annex 11

- definitions (see general statements)

010 07 01 01   General

- objectives of ATS, divisions of ATS, designation of the portions of the airspace and controlled aerodromes where where ATS will be provided, classification of airspaces (appendix 4 of annex 11), required navigation performance (RNP), establishment and designation of the units providing ATS, specifications for flight information regions, control areas and control zones, minimum flight altitudes, priority in the event of an aircraft in emergency, inflight contingencies, time in ATS

010 07 01 02   Air Traffic Control

- application
- provision of air traffic control service, operation of air traffic control service, separation minima, contents of clearances, coordination of clearances, control of persons and vehicles at aerodromes

010 07 01 03   Flight Information Service

- application
- scope of flight information service
- operational flight information service broadcasts

010 07 01 04   Alerting Service

- application, notification of rescue coordination centres (only INCERFA, ALERFA, DETRESFA), information to aircraft operating in the vicinity of an aircraft in a state of emergency

010 07 01 055   Principles governing the identification of RNP types and the identification of ATS routes other than standard departure and arrival routes (Appendix 1)

010 07 03 00   Rules of the air and air traffic services (ICAO Doc. 4444RAC/501/11 and ICAO Doc. 7030Regional supplementary procedures)

- definitions (see general statements)
- relationship to other document

010 07 03 01   General provisions

- general air traffic services operating practices: submission of a flight plan, change from IFR to VFR flight, clearances and information, control of air traffic flow, altimeter setting procedures, indication of heavy wake turbulence category and MLS capacity, position reporting, air traffic incident report, procedures in regard to aircraft equipped with airborne collision avoidance systems (ACAS)
- Appendix 1

010 07 03 02   Area Control Service

- general provisions for the separation of controlled traffic
- vertical separation: vertical separation application, vertical separation minimum, minimum cruising level, assignment of cruising level, vertical separation during ascent or descent
- horizontal separation: lateral separation application, lateral separation application, longitudinal separation application (except between supersonic aircraft)
- reduction in separation minima
- air traffic control clearances: contents, description of air traffic control clearances, clearance to fly maintaining own separations while in visual meteorological conditions, essential traffic information, clearance of a requested change in flight plan
- emergency and communication failure: emergency procedures (only general priority, emergency descent, action by pilotincommand), airground communication failure (only concerning the actions by pilotincommand), interception of civil aircraft

010 07 03 03   Approach Control Service

- departing aircraft: general procedures for departing aircraft, clearances for departing aircraft to climb maintaining own separation while in visual meteorological conditions, information   for departing aircraft
- arriving aircraft: general procedures for arriving aircraft, clearance to descend subject to maintaining own separation in visual meteorological conditions, visual approach, instrument approach, holding, approach sequence, expected approach time, information for arriving aircraft

010 07 03 04   Aerodrome Control Service

- functions of aerodrome control towers: general, alerting service provided by aerodrome control towers, suspension of VFR operations by aerodrome control towers
- traffic and taxi circuits: selection of runway-in-use
- information to aircraft by aerodrome control towers: information related to the operation of the aircraft, information on aerodrome conditions
- control of aerodrome traffic: order of priority for arriving and departing aircraft, control of departing and arriving aircraft, wake turbulence categorisation of aircraft and increased longitudinal separation minima, authorisation of special VFR flights

010 07 03 05   Flight Information Service and Alerting Service

-      flight information service
-      alerting service

010 07 03 06   Use of radar in Air Traffic Services

-      general provisions: limitations in the use of radar, identification procedures (only establishment of radar identity), position information, radar vectoring use of radar in the air traffic control service

010 08 00 00   **AERONAUTICAL INFORMATION SERVICE (based on ANNEX 15)**

010 08 01 00   Annex 15

-      essential definitions
-      applicability

010 09 00 00   **AERODROMES (based on ANNEX 14, VOL 1 & 2)**

010 09 01 00   Annex 14

-      definitions

010 09 01 01   Aerodrome data:

-      conditions of the movement area and related facilities

010 09 01 02   Visual aids for navigation

-      indicators and signalling devices markings
-      lights
-      signs
-      markers

010 09 01 03   Visual aids for denoting obstacles

-      marking of objects
-      lighting of objects

010 09 01 04   Visual aids for denoting restricted use of areas

010 09 01 05   Emergency and other services

-      rescue and fire fighting
-      apron management service
-      ground servicing of aircraft

010 00 00 00   **AIR LAW AND ATC PROCEDURES**

010 09 01 06   Attachment A to Annex 14

-      calculation of declared distances
-      radio altimeter operating areas
-      approach lighting systems

010 10 00 00   **FACILITATION (based on ANNEX 9)**

-      definitions

| 010 10 01 00 | Entry and departure of aircraft |
| - | description, purpose and use of aircraft documents: general declaration |

| 010 10 02 00 | Entry and departure of persons and their baggage |
| - | entry requirement and procedures crew and other operator's personnel |

010 11 00 00   **SEARCH AND RESCUE (based on ANNEX 12)**

| 010 11 01 00 | Annex 12 |
| - | definitions |

010 11 01 01   Organisation

- establishment and provision of SAR service
- establishment of SAR regions
- establishment and designation of SAR services units

010 11 01 02   Cooperation

- cooperation between States
- cooperation with other services

010 11 01 03   Operating procedures

- procedures for pilots-in-command at the scene of an accident
- procedures for pilots-in-command
- intercepting a distress transmission
- search and rescue signals

010 11 01 04   Search and rescue signals:

- signals with surface craft
- ground/air visual signal code
- air/ground signals

010 12 00 00   **SECURITY (based on ANNEX 17)**

010 12 01 00   Annex 17

| 010 12 01 01 | General: |
| | aims and objectives |

| 010 12 01 02 | Organisation |
| - | cooperation and coordination |

010 12 01 03   Operators: operators security programme

010 13 00 00   **AIRCRAFT ACCIDENT INVESTIGATION (based on ANNEX 13)**

010 13 01 00	Annex 13
-	definitions
-	applicability

010 14 00 00   JARFCL

010 15 00 00   **NATIONAL LAW**

010 15 01 00   National law and differences to relevant

CIVIL AVIATION
AUTHORITY

CAP 696

CIVIL AVIATION AUTHORITY
JAR FCL EXAMINATIONS
LOADING MANUAL

CIVIL AVIATION AUTHORITY, LONDON

CAP 696

CIVIL AVIATION AUTHORITY
JAR FCL EXAMINATIONS
LOADING MANUAL

CIVIL AVIATION AUTHORITY, LONDON, AUGUST 1999

ISBN 0 86039 769 6

Printed and distributed by
Westward Digital Limited, 37 Windsor Street, Cheltenham, England

Contents

MASS AND BALANCE

SECTION 1 - GENERAL NOTES

INTRODUCTION

IMPORTANT NOTICE

These data sheets are intended for the use of candidates for the European Professional Pilot's Licence Examinations.

The data contained within these sheets is for **examination purposes only**. The data must not be used for any other purpose and, specifically, **are not to be used for the purpose of planning activities associated with the operation of any aircraft in use now or in the future**.

AIRCRAFT DESCRIPTION

The aircraft used in these data sheets are of generic types related to the classes of aircraft on which the appropriate examinations are based.

Candidates must select the correct class of aircraft for the question being attempted. To assist in this, the data for each class is presented on different coloured paper.

Generic Aircraft

Single engine piston	not certified under JAR 25 (Light Aeroplanes) Performance Class B	**SEP1**
Multi engine piston	not certified under JAR 25 (Light Aeroplanes) Performance Class B	**MEP1**
Medium range jet transport	certified under JAR 25 Performance Class A	**MRJT**

The same set of generic aircraft will be utilised in the following subjects:
- 031 - Mass and Balance - Aeroplanes
- 032 - Performance - Aeroplanes
- 033 - Flight Planning and Monitoring - Aeroplanes

LAYOUT OF DATA SHEETS

Each set of data sheets will consist of an introduction that will contain some pertinent information relating to the aircraft and the subject being examined. This data will include (but not be limited to) a list of abbreviations and some conversion factors.

This will be followed by a selection of graphs and/or tables that will provide coverage suitable for the syllabus to be examined. A worked example will accompany each graph/table and will demonstrate typical usage.

Data sheets for each type will appear on different colour paper as follows:-

- SEP1   green paper
- MEP1   blue paper
- MRJT   white paper

DEFINITIONS

Definitions given in italics are not given in ICAO or JAA documentation but are in common use.

MASS DEFINITIONS:-

Basic Empty Mass (Basic Mass)	*is the mass of an aeroplane plus standard items such as: unusable fuel and other unusable fluids; lubricating oil in engine and auxiliary units; fire extinguishers; pyrotechnics; emergency oxygen equipment; supplementary electronic equipment.*
Dry Operating Mass (D.O.M.)	is the total mass of the aeroplane ready for a specific type of operation excluding all usable fuel and traffic load. The mass includes items such as:- (i) Crew and crew baggage (ii) Catering and removable Passenger service equipment (iii) Potable water and lavatory chemicals (iv) Food & beverages
Operating Mass (OM)	*is the DOM plus fuel but without traffic load.*
Traffic Load	The total mass of passengers, baggage and cargo, including any 'non-revenue' load

Zero Fuel Mass	*is D.O.M. plus traffic load but excluding fuel*
Maximum Zero Fuel Mass (MZFM)	The maximum permissible mass of an aeroplane with no useable fuel.
Taxi Mass	*is the mass of the aircraft at the start of the taxi (at departure from the loading gate).*
Maximum Structural Taxi Mass	*is the structural limitation on the mass of the aeroplane at commencement of taxi.*
Take-Off Mass (TOM)	is the mass of an aeroplane including everything and everyone contained within it at the start of the take-off run.
Performance Limited Take-Off Mass	*is the take-off mass subject to departure airfield limitations. It must never exceed the maximum structural limit.*
Regulated T.O.M.	*is the lowest of 'performance limited' & 'structural limited' T.O.M.*
Maximum Structural Take-Off Mass	the maximum permissible total aeroplane mass at the start of the take-off run.
Performance Limited Landing Mass	*is the mass subject to the destination airfield limitations, It must never exceed the structural limit.*
Maximum Structural Landing Mass	the maximum permissible total aeroplane mass on landing under normal circumstances.
Regulated Landing Mass	*is the lowest of 'performance limited' and 'structural limited' landing mass.*

OTHER DEFINITIONS

Centre of Gravity (CG)	*is that point through which the force of gravity is said to act on a mass.*
Datum	(relative to an aeroplane) is that plane from which the centres of gravities of all masses are referenced.
Balance Arm (BA)	*is the distance from the datum to the centre of gravity of a mass.*

Moment	*is the product of the mass and the balance arm*
Loading Index (LI)	*is a non-dimensional figure that is a scaled down value of a moment. It is used to simplify mass and balance calculations.*
Dry Operating Index (DOI)	*is the index for the position of the centre of gravity at Dry Operating Mass.*

CONVERSIONS

All conversions are taken from ICAO Annex

Mass conversions

Pounds (LB) to Kilograms (KG)	LB x 0.45359237 KG
Kilograms (KG) to Pounds (LB)	KG x 2.20462262 LB

Volumes (Liquid)

Imperial Gallons to Litres (L)	Imp. Gall x 4.546092
US Gallons to Litres (L)	US Gall x 3.785412

Lengths

Feet (ft) to Metres (m)	Feet x 0.3048

Distances

Nautical mile (NM) to metres (m)	NM x 1852.0

MASS AND BALANCE

SECTION II- DATA FOR SINGLE ENGINE PISTON/PROPELLER (SEP1) AEROPLANE

1.   AEROPLANE DESCRIPTION AND DATA

- monoplane
- single reciprocating engine
- propeller - constant speed
- retractable undercarriage
- Performance Class B.

Figure 2.1

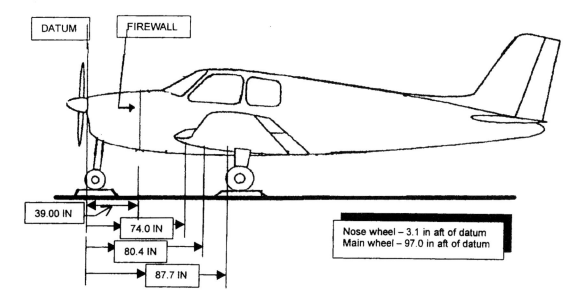

Reference datum	39.00 inches forward of firewall
Centre of Gravity (CG) limits	forward limit 74.00 - 80.4 inches aft limit 87.7 inches.
Maximum T.O.M.	3650 lb.
Maximum Landing Mass	3650 lb.
Basic Empty Mass (BEM)	2415 lb.
CG @ BEM	77.7 inches
Moment (x100) =	1876.46 lb./inches
Landing Gear retraction/extension	does not significantly affect CG position
Floor structure load limit	50 lb. per square foot between front and rear spars (includes Baggage Zone A) 100 lb. per square foot elsewhere. (Baggage Zones B & C

Figure.2.2   SEATING AND BAGGAGE ARRANGEMENTS

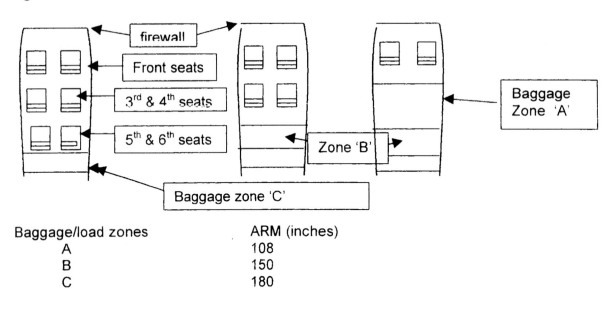

Baggage/load zones	ARM (inches)
A	108
B	150
C	180

Figure 2.3   USEFUL LOAD WEIGHTS AND MOMENTS

USABLE FUEL

LEADING EDGE TANKS ARM 75					
GALLONS	WEIGHT	MOM/100	GALLONS	WEIGHT	MOM/100
5	30	23	44	264	198
10	60	45	50	300	225
15	90	68	55	330	248
20	120	90	60	360	270
25	150	113	65	390	293
30	180	135	70	420	315
35	210	158	74	444	333
40	240	180			

2.   PROCEDURE FOR MASS AND BALANCE CALCULATION (FIG 2.4)

2.1   Record the Basic Empty Mass and Moment under the Basic Empty Condition block, he moment must be divided by 100 to correspond to 'Useful Load' Mass and Moments tables.

2.2   Record the Mass and corresponding moment for each of the useful load items (except fuel.) to be carried in the aeroplane (occupants, baggage).

2.3   Total the Mass column and moment column. The SUB-TOTAL is the Zero Fuel Condition.

2.4   Determine the Mass and corresponding moment for the fuel loading to be used. This fuel loading includes fuel for the flight, plus that required for start, taxi and take-off. Add the Fuel to Zero Fuel Condition to obtain the SUB-TOTAL Ramp Condition.

2.5   Subtract the fuel to be used for start, taxi and take-off to arrive at the SUB-TOTAL Take-off Condition.

2.6   Subtract the Mass and moment of the fuel in the incremental sequence in which it is to be used from the take-off weight and moment. The Zero Fuel Condition, the Take-off Condition and the Landing Condition moment must be within the minimum and maximum moments shown on the Moment Limit vs Mass graph for that mass. If the total moment is less than the minimum moment allowed, useful load items must be shifted aft or forward load items reduced. If the total moment is greater than the maximum moment allowed, useful load items must be shifted forward or aft load items reduced. If the quantity or location of load items is changed, the calculations must be revised and the moments rechecked.

Figure.2.4   LOADING MANIFEST   SEP1

ITEM	MASS	ARM (IN)	MOMENT x **100**
1. BASIC EMPTY CONDITION			
2. FRONT SEAT OCCUPANTS		79	
3. THIRD & FOURTH SEAT PAX		117	
4. BAGGAGE ZONE 'A'		108	
5. FIFTH & SIXTH SEAT PAX		152	
6. BAGGAGE ZONE 'B'		150	
7. BAGGAGE ZONE 'C'		180	
SUB – TOTAL = ZERO FUEL MASS			
8. FUEL LOADING			
SUB – TOTAL = RAMP MASS			
9. SUBTRACT FUEL FOR START, TAXI & RUN UP. *(SEE NOTE)*			
SUB- TOTAL = TAKE OFF MASS			
10. TRIP FUEL			
SUB – TOTAL = LANDING MASS			

NB. FUEL FOR START' TAXI AND RUN UP IS NORMALLY 13 LBS AT AN AVERAGE ENTRY OF 10 IN THE COLUMN HEADED **MOMENT (X 100)**

Figure 2.5   CENTRE OF GRAVITY ENVELOPE

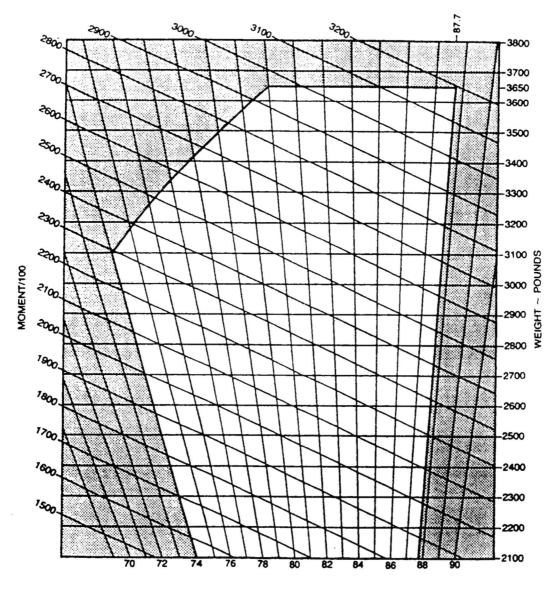

MASS AND BALANCE

SECTION III - DATA FOR LIGHT TWIN ENGINE PISTON/PROPELLER AEROPLANE

CONTENTS

1.   Aeroplane Description and data
      -    monoplane
      -    twin reciprocating supercharged engines
      -    counter- rotating, constant speed propellers
      -    retractable undercarriage
      -    Performance Class B.

Figure 3.1   LOCATIONS DIAGRAM

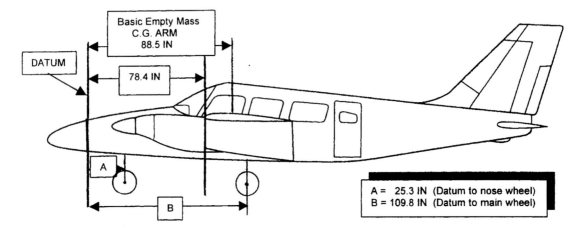

Reference datum	78.4 inches forward wing leading edge at inboard edge of inboard fuel tank
CG limits fwd aft	82.0" to 90.8" (subject to aeroplane mass) 94.6".
Max T.O. Mass	4750 LB Max Landing Mass 4513 lb.
Max Zero Fuel Mass	4470 LB
Basic Empty Mass	3210 LB arm 88.5 inches

Gear retraction/extension does not significantly affect CG position

Structural Floor Loading Limit   120 LB/square foot.

CONFIGURATION OPTIONS

BAGGAGE/FREIGHT ZONES

	Max Mass	Arm	
Zone 1	100 LB	22.5	
Zone 2	360 LB	118.5	available only with centre seats removed
Zone 3	400 LB	157.6	available only with rear seats removed
Zone 4	100 LB	178.7	

STANDARD ALLOWANCES

Fuel relative density - an average mass of 6 LB per US gallon should be used.
Passenger and pilot mass - actual mass values should be used.

2.   PROCEDURE FOR MASS AND BALANCE CALCULATIONS

See example at figures 3.2 and 3.3. Figures 3.4 and 3.5 are provided for your use.

a)   Enter all mass values in correct locations on table (Figure 3.2/3.4)
b)   Calculate moments for each entry
c)   Total mass values to obtain zero fuel mass
d)   Total moments for zero fuel mass condition
e)   Determine arm at zero fuel mass
f)   Add total fuel mass and arm
g)   Obtain moment for fuel load
h)   Add fuel mass and moment to determine ramp mass and moment
i)   Deduct start-up, taxi and run-up fuel allowance and correct moment to obtain take-off conditions.
j)   Check CG position lies within envelope (chart at Figure 3.3/3.5)
k)   Deduct estimated fuel burn to destination
l)   Obtain estimated landing mass and moment
m)   Check CG position at landing to ensure that it lies within envelope (chart at figure 3.3/3.5)

Figure 3.2   LOADING MANIFEST (Example)   MEP1

ITEM	Mass (Lbs.)	Arm Aft Of Datum (IN)	Moment (IN/Lbs.)
Basic Empty Mass	3210	88.5	284085
Pilot and Front Passenger	340	85.5	29070
Passengers (Centre Seats) or Baggage Zone 2 (360 LB Max.)	236	118.5	29766
Passengers (Rear Seats) or Baggage Zone 3 (400 LB Max.)	340	157.6	53585
Baggage Zone 1 (100 LB Max.)	100	22.5	2250
Baggage Zone 4 (100 LB Max.).		178.7	
Zero Fuel Mass (4470 LB Max - Std)	4228	93.9	396956
Fuel (123 Gal. Max.).	545	93.6	51012
Ramp Mass (4773 LB Max)	4773	93.9	447968
Fuel Allowance for Start, Taxi, Run-up	-23	93.6	-2153
Take-off Mass (4750 LB Max.)	4750	93.9	446134
Minus Estimated Fuel Burn-off	-450	93.6	-42120
Landing Mass (4513 LB Max.)	4300	93.6	404014

* N.B. Maximum mass values given in this table are for **structural limits only**

Figure 3.3   CG ENVELOPE

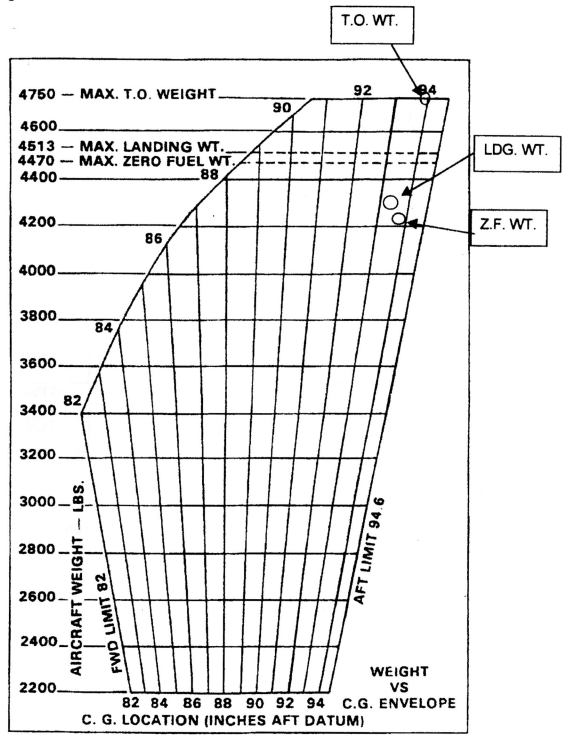

Figure 3.4   LOADING MANIFEST

ITEM	Mass (Lbs.)	Arm Aft Of Datum (IN)	Moment (IN/Lbs.)
Basic Empty Mass	3210	88.5	
Pilot and Front Passenger		85.5	
Passengers (Centre Seats) or Baggage Zone 2 (360 LB Max.)		118.5	
Passengers (Rear Seats) or Baggage Zone 3 (400 LB Max.)		157.6	
Baggage Zone 1 (100 LB Max.)		22.5	
Baggage Zone 4 (100 LB Max.)		178.7	
Zero Fuel Mass (4470 LB Max - Std)			
Fuel (123 Gal. Max.).		93.6	
Ramp Mass (4773 LB Max)			
Fuel Allowance for Start, Taxi, Run-up		93.6	
Take-off Mass (4750 LB Max.)			
Minus Estimated Fuel Burn-off		93.6	
Landing Mass (4513 LB Max.)			

* N.B. Maximum mass values given in this table are for **structural limits only**

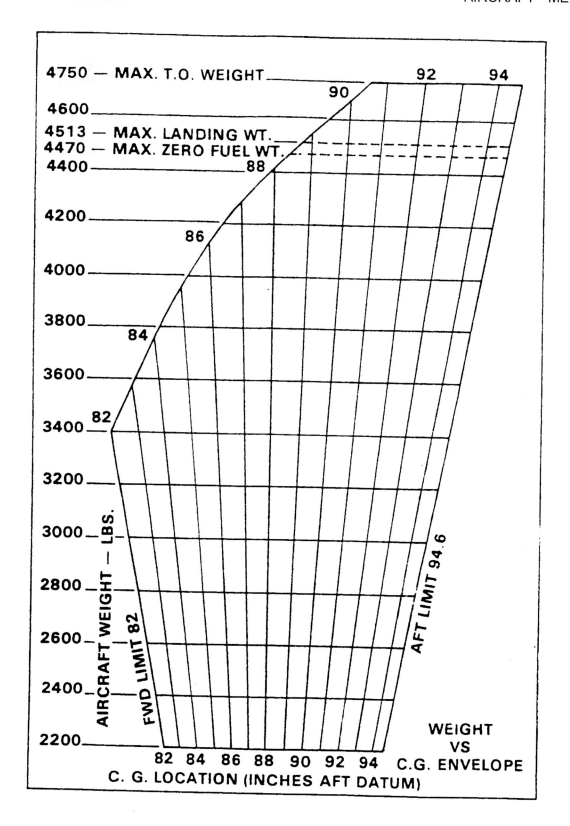

MASS AND BALANCE

SECTION IV- DATA FOR MEDIUM RANGE TWIN JET (MRJT.1)

CONTENTS

1.   AIRCRAFT DESCRIPTION

- monoplane
- twin high-bypass gas turbine engines
- retractable undercarriage
- certified under FAA/JAR - 25
- Performance Class A

2.   AIRCRAFT DATA CONSTANTS

Figure 4.1   Locations Diagram

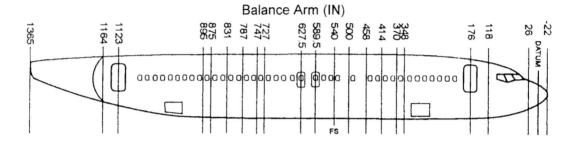

Figure 4.2   TABLE TO CONVERT BODY STATION TO BALANCE ARM

BODY STATION	CONVERSION	BALANCE ARM - IN
130 to 500	B.S. – 152 IN	-22 to 348
500A	348 + 22 IN	370
500B	348 + 44 IN	392
500C	348 + 66 IN	414
500D	348 + 88 IN	436
500E	348 +110 IN	458
500F	348 +132 IN	480
500G	348 +152 IN	500
540 to 727	B.S. + 0 IN	540 to 727
727A	727 + 20 IN	747
727B	727 + 40 IN	767
727C	727 + 60 IN	787
727D	727 + 82 IN	809
727E	727 +104 IN	831
727F	727 +126 IN	853
727G	727 +148 IN	875
747 to 1217	B.S. +148 IN	895 to 1365

2.1   Datum point   540 inches forward of front spar (FS)

2.2   Landing Gear Retraction/extension   negligible effect from operation of
                                                  landing gear

1.3   Flap Retraction

Figure 4.3   Effect of flap retraction

From	To	Moment Change (Kg – inches x 1000)
5	0	-11
15	0	-14
30	0	-15
40	0	-16

1.4 Take-off Horizontal Stabiliser Trim Setting

Figure 4.4   Graph of trim units for C.G. position

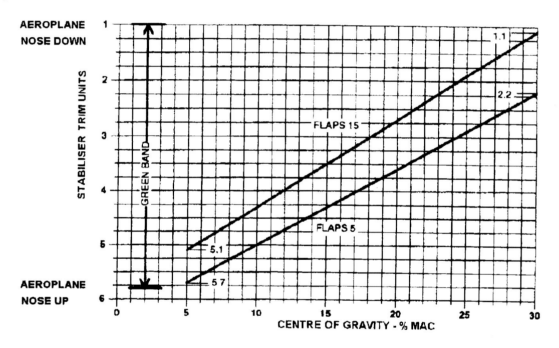

2.5   Mean Aerodynamic Chord   134.5 inches
                                                  Leading edge 625.6 inches aft of
                                                  datum

3.   MASS AND BALANCE LIMITATIONS

3.1   Mass Limits

        Maximum Structural Taxi Mass   63060
        Maximum Structural Take-off Mass   62800
        Maximum Structural Landing Mass   54900
        Maximum Structural Zero Fuel Mass   51300

3.2   Centre of Gravity Limits

The centre of gravity for this aeroplane must at all times be within the limits prescribed by the CG envelope shown in fig. 4.11 on page 9.

4.   FUEL

4.1

Figure 4.5   Fuel Tank Location and Maximum Volume

	BA (full tanks)	Volume (US Gallons)	Mass (Kg)
Left Wing Main Tank 1	650.7	1499	4542
Right Wing Main Tank 2	650.7	1499	4542
Centre Tank	600.4	2313	7008
Max. Total Fuel (**assumes 3.03 Kg/US Gall.**)	628.8	5311	16092

Caution - If centre tank contains more than 450 Kg the wing tanks **must** be full.

4.2

Figure 4.6   Unusable Fuel Quantities

Location	Volume (US Galls)	Mass (Kg)	BA
Wing Tank 1	4.6	14.0	599.0
Wing Tank 2	4.6	14.0	599.0
Centre Tank	7.9	24.0	600.9

<u>FUEL TANK LOCATION DIAGRAM</u>

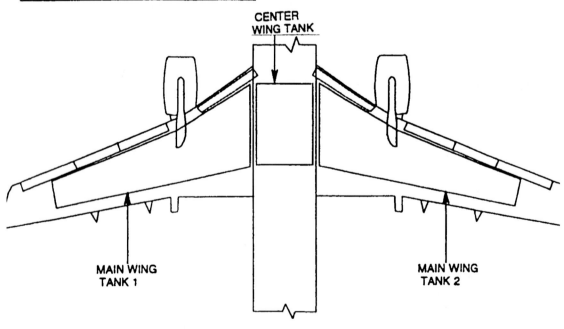

CENTER WING TANK

MAIN WING TANK 1

MAIN WING TANK 2

5.   PASSENGERS (PAX) AND PERSONNEL

5.1   Maximum Passenger Load   141
       Club/Business   33
       Economy   108

5.2   Passenger Distribution

       Figure 4.7 shows the balance arms (in inches) for the distribution of passengers. If the pax load is low, zones B, C and D are the preferred seating areas.

Figure 4.7   BALANCE ARMS (IN)

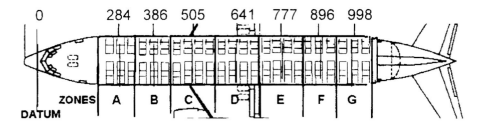

Figure 4.8   Table of pax. Zones /Balance Arms

ZONE	NO. PAX	B.A.
A	15	284
B	18	386
C	24	505
D	24	641
E	24	777
F	18	896
G	18	998

5.3   Passenger Mass

       Unless otherwise stated passenger mass is assumed to be 84 Kg. (this includes a 6 Kg. allowance for hand baggage)

5.4   Passenger Baggage

       Unless otherwise stated a baggage allowance of 13 Kg may be made per passenger.

5.5   Personnel

       Standard Crewing

	No.	BA	Standard Mass (Kg) each
Flight Deck	2	78.0	90
Cabin Staff Forward	2	162.0	90
Cabin Staff Aft	1	1107.0	90

6. CARGO

Figure 4.9   Cargo Compartment Limitations

FORWARD CARGO COMPARTMENT

BA – IN	228	286	343	500
MAXIMUM COMPARTMENT RUNNING LOAD (Kg. per IN.)	13.15	8.47	13.12	
MAXIMUM DISTRIBUTION LOAD INTENSITY (Kg. per Ft.^{2})	68			
MAXIMUM COMPARTMENT LOAD (Kg)	762	483	2059	
COMPARTMENT CENTROID (BA – IN)	257	314.5	421.5	
MAXIMUM TOTAL LOAD (Kg.)	3305			
FWD HOLD CENTROID (BA – IN)	367.9			
FWD HOLD VOLUME (CU. Ft.)	607			

AFT CARGO COMPARTMENT

BA – IN	731	940	997	1096
MAXIMUM COMPARTMENT RUNNING LOAD (Kg. per IN)	14.65	7.26	7.18	
MAXIMUM DISTRIBUTION LOAD INTENSITY (Kg. per Ft.^{2})	68			
MAXIMUM COMPARTMENT LOAD (Kg)	3062	414	711	
COMPARTMENT CENTROID (BA – IN)	835.5	968.5	1046.5	
MAXIMUM TOTAL LOAD (Kg.)	4187			
FWD HOLD CENTROID (BA – IN)	884.5			
FWD HOLD VOLUME (CU. Ft.)	766			

7.   MASS AND BALANCE CALCULATIONS

7.1   Using Loading Manifest (Figure 4.10) and CG limits envelope (Figure 4.11)

(a)   Enter DOM and balance arm.
(b)   Enter all details of passenger loads and distribution.
(c)   Enter all details of cargo loads and distribution.
(d)   Calculate all moments.
(e)   Total for Zero Fuel Mass and ZFM moment.
(f)   Check ZFM does not exceed max. ZFM.
(g)   Add total fuel load and distribution.
(h)   Calculate fuel load moment.
(i)   Determine total ramp mass and moment.
(j)   Check ramp mass does not exceed structural max.
(k)   Deduct for taxi.
(l)   Determine Take-off Mass and moment.
(m)   Check TOM does not exceed relevant limit.
(n)   Determine Take-off CG and check envelope.
(o)   Determine Take-off stabiliser trim setting.
(p)   Deduct estimated fuel burn-off to destination.
(q)   Determine landing mass and ensure that max. landing mass limit is not exceeded.
(r)   Determine moment at landing mass.
(s)   Determine CG at landing and ensure that it lies within envelope.

Figure 4.10   Loading Manifest- MRJT 1

<u>Max Permissible Aeroplane Mass Values:</u> -

TAXI MASS - _____   ZERO FUEL MASS - _____

TAKE OFF MASS - _____   LANDING MASS - _____

ITEM	MASS (KG.)	B.A. I.N	MOMENT KG-IN/1000	C.G. %MAC
1. D.O.M				
2. PAX Zone A		284		-
3. PAX Zone B		386		-
4. PAX Zone C		505		-
5. PAX Zone D		641		-
6. PAX Zone E		777		-
7. PAX Zone F		896		-
8. PAX Zone G		998		-
9. CARGO HOLD 1		367.9		-
10. CARGO HOLD 4		884.5		-
11. ADDITIONAL ITEMS				-
ZERO FUEL MASS				
12. FUEL TANKS 1 & 2				-
13. CENTRE TANK				-
TAXI MASS				-
LESS TAXI FUEL				-
TAKE OFF MASS				
LESS FLIGHT FUEL				-
EST. LANDING MASS				

Figure 4.11   C.G. ENVELOPE   (MRJT1)

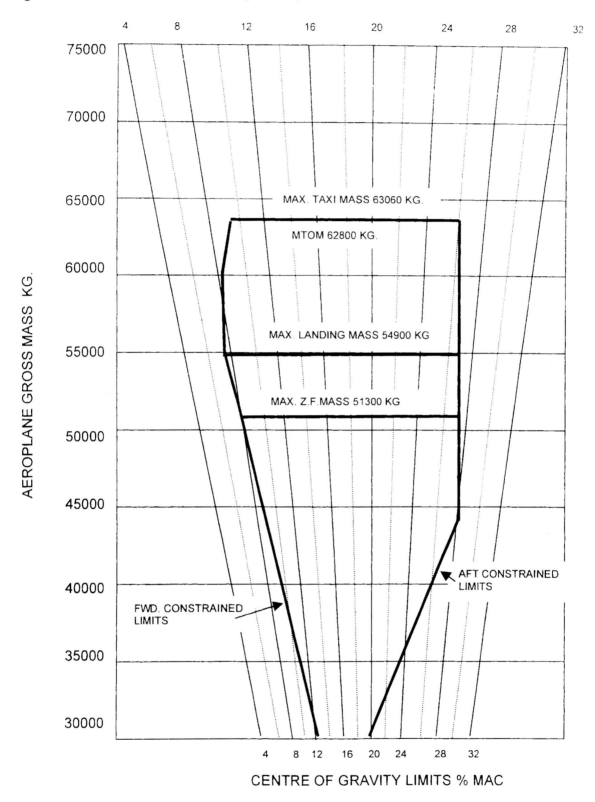

7.2   Using Load and Trim Sheet

The load/trim sheet (as shown in the example at fig 4.12) is in two parts.

Part A (to the left) is a loading summary which should be completed as follows: -

Section 1   is used to establish the limiting take-off mass; maximum allowable traffic load; underload before last minute changes (LMC).

Section 2   shows the distribution of the traffic load.
                   In this section the following abbreviations are used:-

TR	Transit
B	Baggage
C	Cargo
M	Mail
Pax	Passengers
Pax F	First Class
Pax C	Club/Business
Pax M	Economy

Section 3   is used to summarise load and cross check that limits have not been exceeded.

The example shown uses the following data:-

DOM 34300 Kg.	DOI	45.0		Max. TOM	62800
MZFM 51300				MLDGM	54900

Passengers	130	Aver. Mass 84 Kg.	
Baggage	130 @ 14 Kg per piece		
Cargo 630 Kg.			
Fuel Total	14500 Kg	Flight Fuel	8500 Kg.

Part B is the trim portion

Using data from the loading summary, start by entering the index for the DOM.
Move the index in turn (for the mass in each cargo hold) then in accordance with the passenger distribution.
Establish the CG % MAC at ZFM and ensure that it lies within the envelope.
Add fuel index correction (from figure 4.13) to obtain the TOM index and ensure that the CG lies within the envelope.
Extract the % MAC value for the TOM/CG position.

Figure 4.12   Load and Trim Sheet (Example)

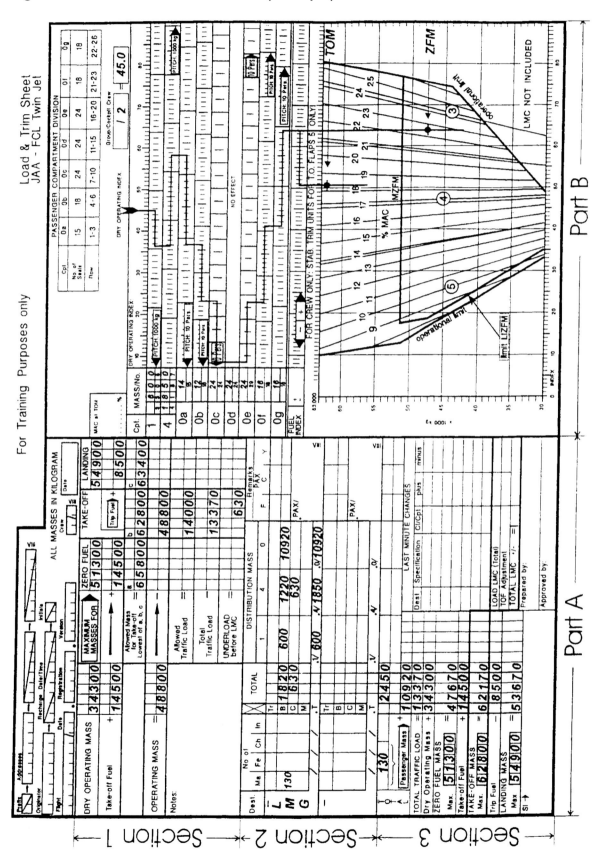

Figure 4.13   FUEL INDEX CORRECTION TABLE

Fuel Mass (Kg)	Index Units		Fuel Mass (Kg)	Index Units
500	-1.0		9330	-0.3
750	-1.5		9580	-0.9
1000	-1.9		9830	-1.5
1250	-2.3		10080	-2.1
1500	-2.6		10330	-2.7
1750	-3.0		10580	-3.3
2000	-3.3		10830	-3.9
2500	-3.7		11080	-4.5
3000	-4.3		11330	-5.1
3500	-4.7		11580	-5.7
4000	-5.1		11830	-6.3
4500	-5.4		12080	-6.9
5000	-5.7		12330	-7.5
5500	-5.9		12580	-8.1
6000	-6.0		12830	-8.7
6500	-6.1		13080	-9.3
7000	-5.9		13330	-9.9
7500	-5.0		13580	-10.5
7670	-4.6		13830	-11.1
7830	-4.1		14080	-11.7
8000	-3.7		14330	-12.3
8170	-3.2		14580	-12.9
8330	-2.6		14830	-13.5
8500	-2.1		15080	-14.1
8630	-1.6		15330	-14.8
8750	-1.1		15580	-15.4
8880	-0.6		15830	-16.3
9000	-0.1		16080	-17.1
tanks 1 & 2 full 9080	+0.3		centre tank full 16140	-17.3

Useable fuel quantities in lines = 20 Kg. (included in the tables).
Interpolation not necessary!

For mass figures not printed in these tables the index of the next higher mass is applicable.

Figure 4.14   Load and Trim Sheet (Blank)

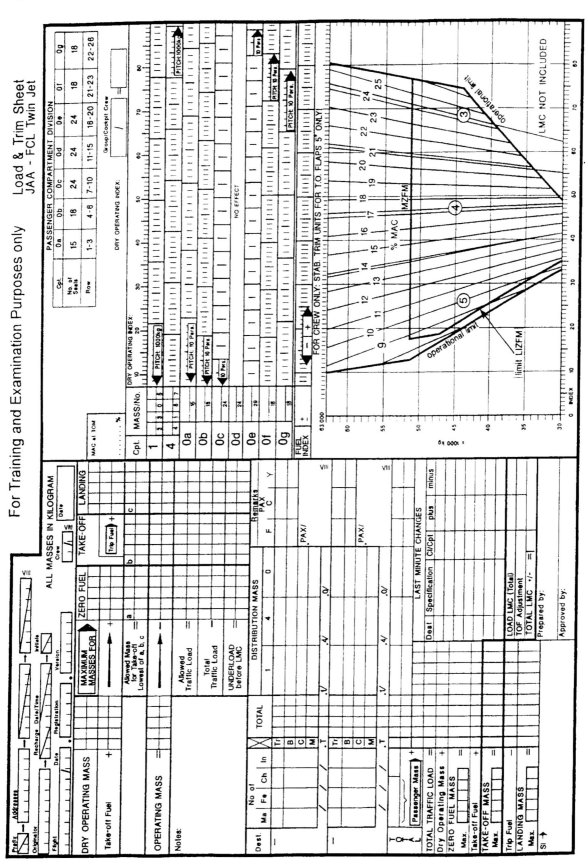

CAP 697

**CIVIL AVIATION AUTHORITY
JAR FCL EXAMINATIONS
FLIGHT PLANNING MANUAL**

CIVIL AVIATION AUTHORITY, LONDON

CAP 697

CIVIL AVIATION AUTHORITY
JAR FCL EXAMINATIONS
FLIGHT PLANNING MANUAL

CIVIL AVIATION AUTHORITY, LONDON, AUGUST 1999

ISBN 0 86039 770 X

Printed and distributed by
Westward Digital Limited, 37 Windsor Street, Cheltenham, England

General Contents

INTENTIONALLY BLANK

FLIGHT PLANNING & MONITORING

SECTION I - GENERAL NOTES

INTRODUCTION

IMPORTANT NOTICE

These data sheets are intended for the use of candidates for the European Professional Pilot's Licence Examinations.

The data contained within these sheets is for **examination purposes only**. The data must not be used for any other purpose and, specifically, **are not to be used for the purpose of planning activities associated with the operation of any aircraft in use now or in the future.**

AIRCRAFT DESCRIPTION

The aircraft used in these data sheets are of generic types related to the classes of aircraft on which the appropriate examinations are based.

Candidates must select the correct class of aircraft for the question being attempted. To assist in this, the data for each class is presented on different coloured paper.

Generic Aircraft

Single engine piston	not certified under JAR 25 (Light Aeroplanes) Performance Class B	**SEP1**
Multi engine piston	not certified under JAR 25 (Light. Aeroplanes) Performance Class B	**MEP1**
Medium range jet transport	certified under JAR 25 Performance Class A	**MRJT**

The same set of generic aircraft will be utilised in the following subjects:
- 031 - Mass and Balance - Aeroplanes
- 032 - Performance - Aeroplanes
- 033 - Flight Planning and Monitoring - Aeroplanes

LAYOUT OF DATA SHEETS

Each set of data sheets will consist of an introduction that will contain some pertinent information relating to the aircraft and the subject being examined. This data will include (but not be limited to) a list of abbreviations and some conversion factors.

This will be followed by a selection of graphs and/or tables that will provide coverage suitable for the syllabus to be examined. A worked example will accompany each graph/table and will demonstrate typical usage.

Data sheets for each type will appear on different colour paper as follows:-
- SEP1   green paper
- MEP1   blue paper
- MRJT   white paper

DEFINITIONS

Definitions given in italics are not given in ICAO or JAA documentation but are in common use.

MASS DEFINITIONS:-

Basic Empty Mass (Basic Mass)	*is the mass of an aeroplane plus standard items such as: unusable fuel and other unusable fluids; lubricating oil in engine and auxiliary units; fire extinguishers; pyrotechnics; emergency oxygen equipment; supplementary electronic equipment.*
Dry Operating Mass (D.O.M.)	is the total mass of the aeroplane ready for a specific type of operation excluding all usable fuel and traffic load. The mass includes items such as:- (i) Crew and crew baggage (ii) Catering and removable Passenger service equipment (iii) Potable water and lavatory chemicals (iv) Food & beverages
Operating Mass (OM)	*is the DOM plus fuel but without traffic load.*
Traffic Load	The total mass of passengers, baggage and cargo, including any 'non-revenue' load

Zero Fuel Mass	*is D.O.M. plus traffic load but excluding fuel*
Maximum Zero Fuel Mass (MZFM)	The maximum permissible mass of an aeroplane with no useable fuel.
Taxi Mass	*is the mass of the aircraft at the start of the taxi (at departure from the loading gate).*
Maximum Structural Taxi Mass	*is the structural limitation on the mass of the aeroplane at commencement of taxi.*
Take-Off Mass (TOM)	is the mass of an aeroplane including everything and everyone contained within it at the start of the take-off run.
Performance Limited Take-Off Mass	*is the take-off mass subject to departure airfield limitations. It must never exceed the maximum structural limit.*
Regulated T.O.M.	*is the lowest of 'performance limited' & 'structural limited' T.O.M.*
Maximum Structural Take-Off Mass	the maximum permissible total aeroplane mass at the start of the take-off run.
Performance Limited Landing Mass	*is the mass subject to the destination airfield limitations. It must never exceed the structural limit.*
Maximum Structural Landing Mass	the maximum permissible total aeroplane mass on landing under normal circumstances.
Regulated Landing Mass	*is the lowest of 'performance limited' and 'structural limited' landing mass.*

N.B.
Within these data sheets the term 'weight' should be considered to have the same meaning as 'mass.

CONVERSIONS

All conversions are taken from ICAO Annex

Mass conversions

Pounds (LB) to Kilograms (KG)	LB x 0.45359237 KG
Kilograms (KG) to Pounds (LB)	KG x 2.20462262 LB

Volumes (Liquid)

Imperial Gallons to Litres (L)	Imp. Gall x 4.546092
US Gallons to Litres (L)	US Gall x 3.785412

Lengths

Feet. (ft) to Metres (m)	Feet x 0.3048

Distances

Nautical mile (NM) to metres (m)	NM x 1852.0

FLIGHT PLANNING AND MONITORING

SECTION II - SINGLE ENGINE PISTON AEROPLANE (SEP1)

CONTENTS

1.   DATA FOR SINGLE ENGINE PISTON AEROPLANE - (SEP)

Aeroplane
-    monoplane
-    single reciprocating engine
-    constant speed propeller
-    retractable undercarriage.

MTOM   3650 lbs.
MLM   3650 lbs.
Max fuel load   74 US gallons
Fuel density   6 lbs. per US gallon* (unless advised otherwise)

2.   FUEL, TIME AND DISTANCE TO CLIMB

Enter graph as follows:
a.   From OAT at take off move vertically to airport or start of climb pressure
      altitude
b.   Move horizontally to aeroplane mass
c.   Move vertically down and read time, fuel and distance respectively
d.   Enter with OAT at cruise altitude. From there move vertically to cruise
      altitude.
e.   Move horizontally to aeroplane mass
f.   Move vertically downwards and read time, fuel and distance respectively
g.   Subtract c from f to obtain climb time, fuel and distance respectively.

EXAMPLE

OAT AT TAKE-OFF   *15°C*
OAT AT CRUISE   *-5°C*
AIRPORT PRESSURE ALTITUDE   *5653 FEET*
CRUISE PRESSURE ALTITUDE   *11500 feet*
INITIAL CLIMB WEIGHT   *3650 LBS*

_____

FROM GRAPH (FIGURE FP 2-1)
TIME TO CLIMB   *11.5 MINS*
FUEL TO CLIMB   *3.5 GAL*
DISTANCE TO CLIMB   *23.5 N.A.M.*

FIGURE 2.1   TIME FUEL AND DISTANCE TO CLIMB

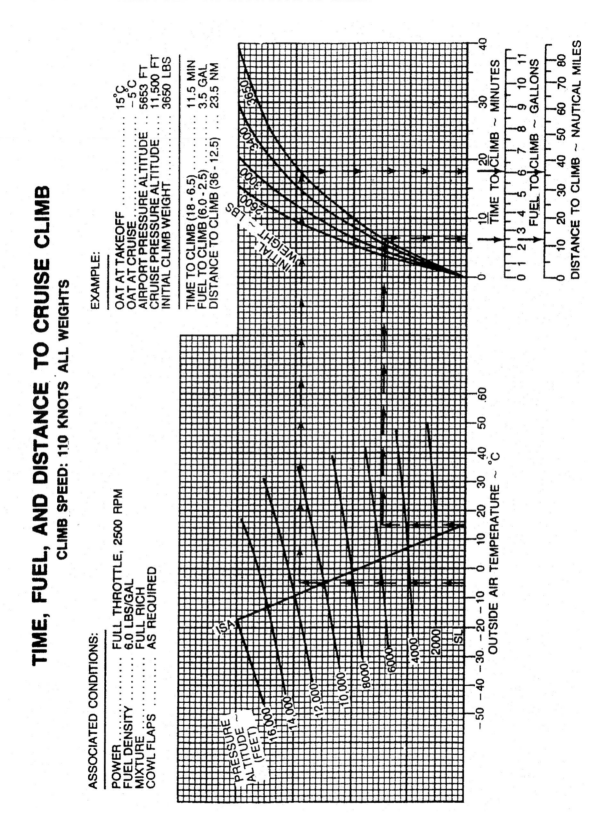

3.   Recommended and economy Cruise Power Settings

The following tables (Figures 2.2 and 2.3) cover cruise with lean mixture

Table	2.2.1	25.0 in. HG (or full throttle)	2500 RPM
	2.2.2	25.0 in. HG (or full throttle)	2100 RPM
	2.2.3	23.0 in. HG (or full throttle)	2300 RPM
	2.3.1	21.0 in. HG (or full throttle)	2100 RPM

Method of use:
    a.   Select the correct table
    b.   Select the appropriate temperature deviation condition
    c.   Enter with cruising level and read off required data,

Figure 2.2   RECOMMENDED CRUISE POWER SETTINGS

Table 2.2.1

20°C LEAN   25.0 IN. HG (or full throttle) @ 2500 RPM

Of Peak EGT   Cruise Lean Mixture
3400 lbs.

	Press. Alt.	IOAT		Man. Press.	Fuel Flow		Air Speed	
	Feet	°C	°F	IN. HG	PPH	GPH	KIAS	KTAS
ISA –20°C (ISA –36°F)	0	-3	27	25.0	86.3	14.4	168	159
	2000	-6	20	25.0	89.3	14.9	168	164
	4000	-10	13	25.0	92.3	15.4	168	169
	6000	-14	6	24.1	89.8	15.0	164	170
	8000	-18	-1	22.3	82.6	13.8	157	168
	10,000	-22	-8	20.6	76.0	12.7	150	165
	12,000	-26	-15	19.1	70.2	11.7	143	162
	14,000	-30	-23	17.7	65.5	10.9	135	158
	16,000	-35	-30	16.3	60.8	10.1	126	152
Standard Day (ISA)	0	17	63	25.0	82.9	13.8	163	160
	2000	14	56	25.0	85.6	14.3	163	165
	4000	10	50	25.0	88.5	14.8	163	170
	6000	6	42	24.1	86.1	14.4	159	171
	8000	2	35	22.3	79.3	13.2	152	169
	10,000	-2	28	20.6	73.3	12.2	145	166
	12,000	-6	21	19.1	67.8	11.3	137	162
	14,000	-10	13	17.7	63.5	10.6	129	157
	16,000	-15	6	16.3	59.1	9.9	120	150
ISA + 20°C (ISA + 36°F)	0	37	99	25.0	79.5	13.3	158	161
	2000	34	92	25.0	82.1	13.7	158	166
	4000	30	86	25.0	84.7	14.1	158	171
	6000	26	79	24.1	82.5	13.8	154	172
	8000	22	71	22.3	76.2	12.7	147	169
	10,000	18	64	20.6	70.5	11.8	140	165
	12,000	14	57	19.1	65.5	10.9	132	161
	14,000	10	49	17.7	61.5	10.3	123	155
	16,000	5	42	16.3	57.5	9.6	113	146

Notes: 1.   Full throttle manifold pressure settings are approximate
        2.   Shaded area represents operation with full throttle
        3.   Fuel flows are to be used for flight planning only and will vary from
              aeroplane to aeroplane. Lean using the EGT.

Figure 2.2 RECOMMENDED CRUISE POWER SETTINGS

TABLE 2.2.2

20°C LEAN   25.0 IN. HG (or full throttle) @ 2100 RPM

Of Peak EGT

CRUISE LEAN MIXTURE
3400 Lbs.

	Press. Alt.	IOAT		Man. Pres	Fuel Flow		Air Speed	
	Feet	°C	°F	IN. HG	PPH	GPH	KIAS	KTAS
ISA – 20° C (ISA – 36° F)	0	-3	26	25.0	63.8	10.6	148	140
	2000	-7	19	25.0	66.4	11.1	149	145
	4000	-11	12	25.0	68.9	11.5	149	150
	6000	-15	5	24.3	68.3	11.4	147	152
	8000	-19	-2	22.5	63.9	10.7	139	148
	10,000	-23	-9	20.8	60.1	10.0	132	144
	12,000	-27	-17	19.3	56.7	9.5	123	139
	14,000	-31	-24	17.9	54.5	9.1	113	132
	16,000	-35	-32	16.5	52.2	8.7	95	114
Standard Day (ISA)	0	17	62	25.0	61.9	10.3	143	140
	2000	13	55	25.0	64.2	10.7	143	145
	4000	9	48	25.0	66.6	11.1	144	150
	6000	5	41	24.3	66.1	11.0	141	152
	8000	1	34	22.5	61.9	10.3	134	148
	10,000	-3	27	20.8	58.5	9.8	126	143
	12,000	-7	19	19.3	55.6	9.3	116	136
	14,000	-11	12	17.9	53.5	8.9	103	125
	16,000	-	-	-	-	-	-	-
ISA + 20° C (ISA + 36° F)	0	37	98	25.0	60.1	10.0	138	140
	2000	33	91	25.0	62.1	10.4	138	145
	4000	29	84	25.0	64.4	10.7	139	150
	6000	25	77	24.3	63.9	10.7	136	151
	8000	21	70	22.5	60.2	10.0	128	147
	10,000	17	63	20.8	56.8	9.5	119	141
	12,000	13	55	19.3	54.5	9.1	108	131
	14,000	-	-	-	-	-	-	-
	16,000	-	-	-	-	-	-	-

Notes: 1. Full throttle manifold pressure settings are approximate
2. Shaded area represents operation with full throttle
3. Fuel flows are to be used for flight planning only and will vary from aeroplane to aeroplane. Lean using the EGT.

Figure 2.2   RECOMMENDED CRUISE POWER SETTINGS

TABLE   2.2.3

20°C LEAN

Of Peak EGT

23.0 IN. HG (OR FULL THROTTLE) @ 2300 RPM

CRUISE LEAN MIXTURE
3400 lbs.

	Press. Alt.	IOAT		Man. Press.	Fuel Flow		Air Speed	
	Feet	°C	°F	IN. HG.	PPH	GPH	KIAS	KTAS
ISA –20° C (ISA –36° F)	0	-3	26	23.0	67.6	11.3	152	144
	2000	-7	20	23.0	69.7	11.6	152	149
	4000	-11	13	23.0	72.1	12.0	153	154
	6000	-15	6	23.0	74.4	12.4	153	158
	8000	-18	-1	22.4	73.8	12.3	150	160
	10,000	-23	-9	20.7	68.4	11.4	143	157
	12,000	-17	-16	19.2	63.8	10.6	135	153
	14,000	-31	-23	17.8	60.0	10.0	127	148
	16,000	-35	-31	16.4	56.3	9.4	117	141
Standard Day (ISA)	0	17	62	23.0	65.4	10.9	147	145
	2000	13	56	23.0	67.4	11.2	147	149
	4000	9	49	23.0	69.4	11.6	148	154
	6000	5	42	23.0	71.7	12.0	148	159
	8000	2	35	22.4	71.1	11.9	145	160
	10,000	-3	27	20.7	66.2	11.0	137	157
	12,000	-7	20	19.2	61.8	10.3	129	152
	14,000	-11	13	17.8	58.5	9.8	120	146
	16,000	-15	5	16.4	55.3	9.2	109	137
ISA + 20° C (ISA + 36° F)	0	37	98	23.0	63.2	10.5	142	145
	2000	33	92	23.0	65.1	10.9	143	149
	4000	29	85	23.0	67.1	11.2	143	154
	6000	25	78	23.0	69.0	11.5	142	158
	8000	22	71	22.4	68.5	11.4	140	160
	10,000	17	63	20.7	64.0	10.7	132	156
	12,000	13	56	19.2	60.0	10.0	123	151
	14,000	9	48	17.8	57.1	9.5	113	142
	16,000	-	-	-	-	-	-	-

Notes: 1.   Full throttle manifold pressure settings are approximate
2.   Shaded area represents operation with full throttle
3.   Fuel flows are to be used for flight planning only and will vary from aeroplane to aeroplane. Lean using the EGT.

Figure 2.3   ECONOMY CRUISE POWER SETTINGS

TABLE   2.3.1

| 20°C LEAN |

21.0 IN. HG. (OR FULL THROTTLE) @ 2100 RPM

Of Peak EGT

CRUISE LEAN MIXTURE
3400 lbs.

	Press. Alt.	IOAT		Man. Press.	Fuel Flow		Air Speed	
	Feet	°C	°F	IN. HG.	PPH	GPH	KIAS	KTAS
ISA – 20° C (ISA – 36°F)	0	-4	25	21.0	52.7	8.8	126	120
	2000	-8	18	21.0	54.0	9.0	128	125
	4000	-11	12	21.0	55.4	9.2	130	130
	6000	-15	5	21.0	56.9	9.5	131	136
	8000	-19	-2	21.0	58.9	9.8	132	141
	10,000	-23	-9	20.8	60.1	10.0	132	144
	12,000	-27	-17	19.3	56.7	9.5	123	139
	14,000	-31	-24	17.9	54.5	9.1	113	132
	16,000	-35	-32	16.5	52.2	8.7	95	114
Standard Day (ISA)	0	16	61	21.0	51.8	8.6	120	118
	2000	12	54	21.0	53.1	8.9	123	124
	4000	9	48	21.0	54.4	9.1	124	129
	6000	5	41	21.0	55.7	9.3	125	134
	8000	1	34	21.0	57.3	9.6	126	140
	10,000	-3	27	20.8	58.5	9.8	126	143
	12,000	-7	19	19.3	55.6	9.3	116	137
	14,000	-11	12	17.9	53.5	8.9	103	125
	16,000	-	-	-	-	-	-	-
ISA + 20° C (ISA + 36°F)	0	36	97	21.0	50.8	8.5	114	115
	2000	32	90	21.0	52.1	8.7	116	121
	4000	29	83	21.0	53.4	8.9	118	127
	6000	25	77	21.0	54.7	9.1	119	132
	8000	21	70	21.0	55.9	9.3	120	137
	10,000	17	63	20.8	56.8	9.5	119	141
	12,000	13	55	19.3	54.5	9.1	108	131
	14,000	-	-	-	-	-	-	-
	16,000	-	-	-	-	-	-	-

Notes: 1.   Full throttle manifold pressure settings are approximate
2.   Shaded area represents operation with full throttle
3.   Fuel flows are to be used for flight planning only and will vary from aeroplane to aeroplane. Lean using the EGT.

4. RANGE PROFILE

The graph at Figure 2.4 (page 9) provides a simple and rapid means of determining the still air range (nautical air miles) for the sample aeroplane. An example of the use of the graph is shown.
Note that the figures make allowance for the taxi, run-up and 45 minutes reserve fuel.

Figure 2.4 **RANGE**

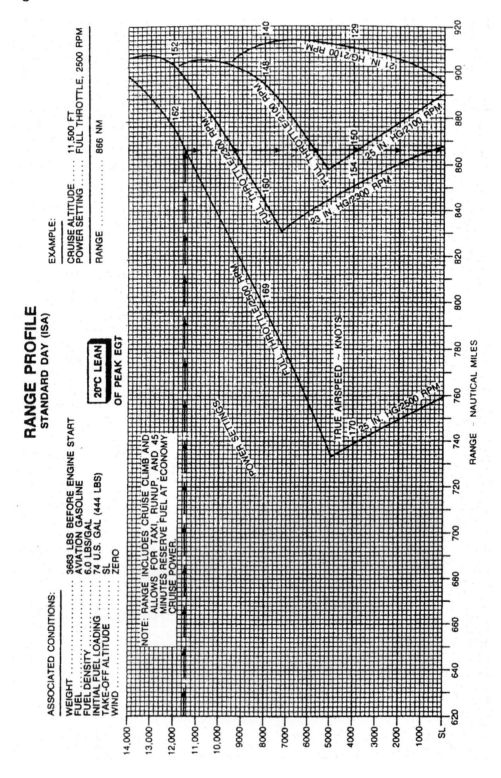

5.   ENDURANCE PROFILE

The graph at Figure 2.5 (page 1) provides a rapid method for determination of endurance for the sample aeroplane. An example is shown on the graph.

Figure 2.5   ENDURANCE

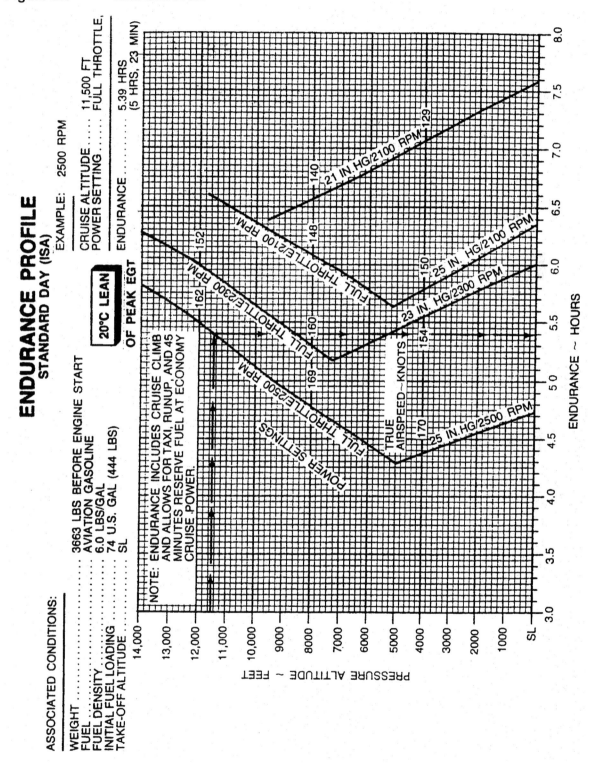

INTENTIONALLY BLANK

FLIGHT PLANNING AND MONITORING

SECTION III - MULTI-ENGINE PISTON AEROPLANE   (MEP1)

Contents

1.   AEROPLANE DATA

- Monoplane
- Twin reciprocating engine
- Twin counter-rotating, constant speed propellers
- Retractable undercarriage.

MTOM   4750 lb.
MZFM   4470 lb.
MLM   4513 lb.
Maximum fuel load -   123 US Gallons
Fuel Density -   6 lb. Per US Gallon (unless otherwise advised)

2.   FUEL, TIME AND DISTANCE TO CLIMB

Method of Use

1. Enter graph (fig. 3.1) at departure airfield temperature, move vertically to airfield pressure altitude
2. Move horizontally to intersect fuel, time and distance curves respectively.
3. From each intersection move vertically down to establish corrections of fuel, time and distance for a non sea level take-off.
4. Repeat steps 1, 2 and 3 for cruise altitude and temperature.
5. Subtract results of 3 from values of fuel, time and distance obtained at 4 to obtain fuel, time and distance for climb.
   Remember range figures are for still air.

Example from graph (Figure 3.1)

1. Departure airport altitude   2000 ft
2. Departure airport OAT   21°C
3. Cruise altitude   16500 ft
4. Cruise OAT   -13°C

From graph
Fuel to climb   = 15-2 = 13 gal
Time to climb   = 27-3 = 24 min
Distance   = 50-5 = 45 NAM

Figure 3.1   **CLIMB**

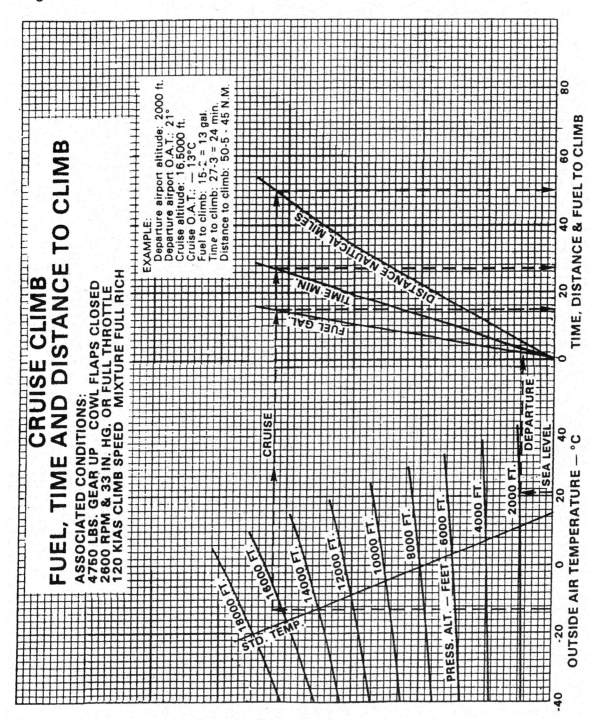

3.   RANGE AT STANDARD TEMPERATURES

Method of Use

1.   Enter graph (Fig.3.2) with cruise altitude
2.   Move horizontally to power selected intersection (with or without reserve)
3.   Move vertically to read range in nautical miles still air distance.

Figure 3.2   RANGE

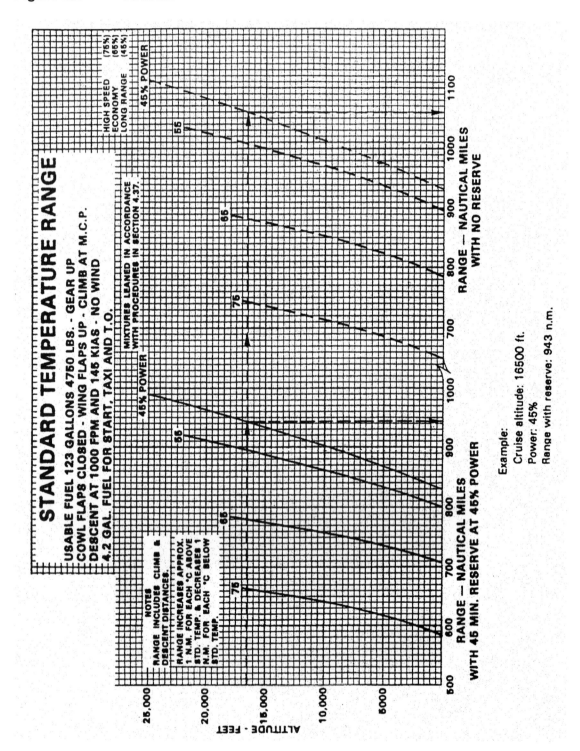

4. POWER SETTING, FUEL FLOW AND TAS

Enter the power setting table (fig. 3.3) with required % power to obtain fuel flow in US gallons per hour.

Manifold Pressure is read off against pressure altitude and RPM in the correct % power column.

Figure 3.3 POWER SETTING TABLE

POWER		75%		65%			55%					
FUEL FLOW		29.0 GPH		23.3 GPH			18.7 GPH					
RPM		2500	2600	2400	2500	2600	2100	2200	2300	2400	2500	2600
Pres. Alt (ft)	Temp °C ISA	MANIFOLD PRESSURE										
SL	15	34.0	33.0	33.8	32.0	31.0	31.2	30.3	29.4	28.2	27.2	26.3
2000	11	33.8	32.7	33.2	31.7	30.7	30.5	29.7	28.8	27.8	26.8	26.0
4000	7	33.6	32.4	32.8	31.5	30.5	30.0	29.2	28.3	27.4	26.4	25.6
6000	3	33.4	32.2	32.5	31.2	30.3	29.7	28.8	28.0	27.0	26.2	25.3
8000	-1	33.1	32.0	32.3	31.0	30.1	29.4	28.4	27.7	26.8	25.7	25.0
10000	-5	33.0	31.9	32.0	30.9	30.0	-	28.3	27.5	26.5	25.5	24.7
12000	-9	32.5	31.8	31.8	30.7	29.8		28.3	27.2	26.3	25.3	24.6
14000	-13	-	31.7	-	30.5	29.7		-	27.1	26.1	25.2	24.4
16000	-17		31.6		30.4	29.5			-	25.9	25.0	24.3
18000	-21		-		-	29.4				-	25.0	24.2
20000	-25					29.3					-	24.2
22000	-28					-						24.1
		MAX EGT 1525°F					MAX EGT 1650°F					
24000	-33											-
25000	-34											

POWER		45%					
FUEL FLOW		16.0 GPH					
RPM		2100	2200	2300	2400	2500	2600
Pres. Alt (ft)	Temp °C ISA	MANIFOLD PRESSURE					
SL	15	27.1	26.4	25.5	24.3	23.3	22.5
2000	11	26.4	25.8	24.6	23.7	22.8	22.1
4000	7	25.8	25.0	24.0	23.2	22.3	21.8
6000	3	25.3	24.5	23.5	22.8	21.9	21.5
8000	-1	24.8	24.0	23.0	22.4	21.6	21.2
10000	-5	24.4	23.7	22.8	22.0	21.4	21.0
12000	-9	24.0	23.3	22.5	21.7	21.2	20.9
14000	-13	-	23.0	22.3	21.4	21.1	20.8
16000	-17		-	22.0	21.3	21.0	20.6
18000	-21			-	21.2	20.9	20.5
20000	-25			-	21.2	20.8	20.4
22000	-28			-	-	-	20.4
		MAX EGT 1650°F					
24000	-33	-	-	-	-	-	20.4
25000	-34	-	-	-	-	-	20.4

To maintain constant power, add 1 % for each 6° C above standard. Subtract 1 % for each 6° C below standard.
Do not exceed 34 inches MAP in cruise.

TRUE AIR SPEED

The graph at figure 3.4 should be used to determine the true airspeed for the various power setting/ altitude/ temperature combinations in the cruise configuration.

The example on the graph illustrates the method of use

Figure 3.4   SPEED POWER

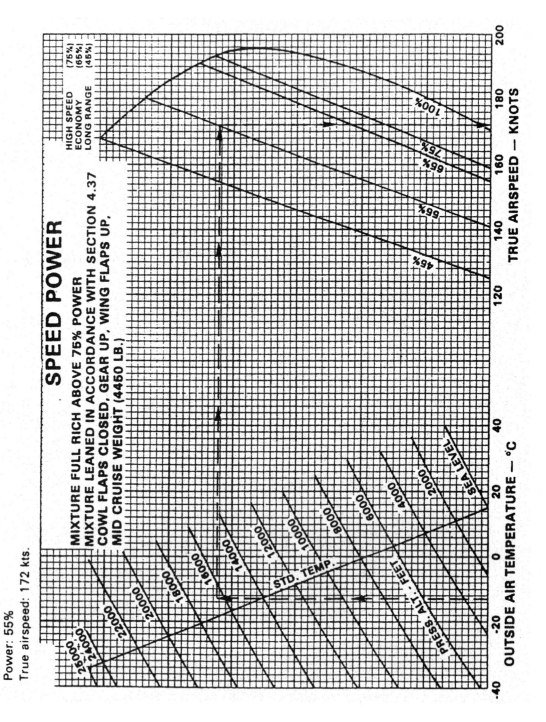

5.   ENDURANCE

Method of Use

1.   Enter graph with cruise altitude.
2.   Move horizontally to selected power line (with or without reserves as directed).
3.   Move vertically down to read endurance in hours.

Figure 3.5   ENDURANCE

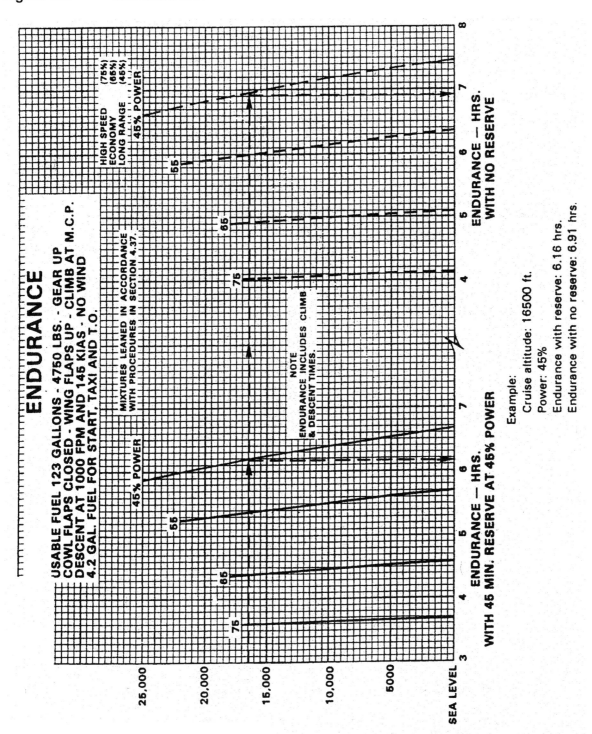

6. DESCENT

Method of Use.

1. Enter graph with OAT at cruise altitude and move vertically to cruise altitude.
2. From there move horizontally to fuel, time and distance lines.
3. Move vertically down respectively from each and read values for fuel (gallons), time (minutes) and distance (nautical miles).
4. Repeat 1,2 and 3 for altitude of airfield
5. Subtract results of 4 from 3 and derive fuel, time and distance for descent profile.

Figure 3.6 DESCENT

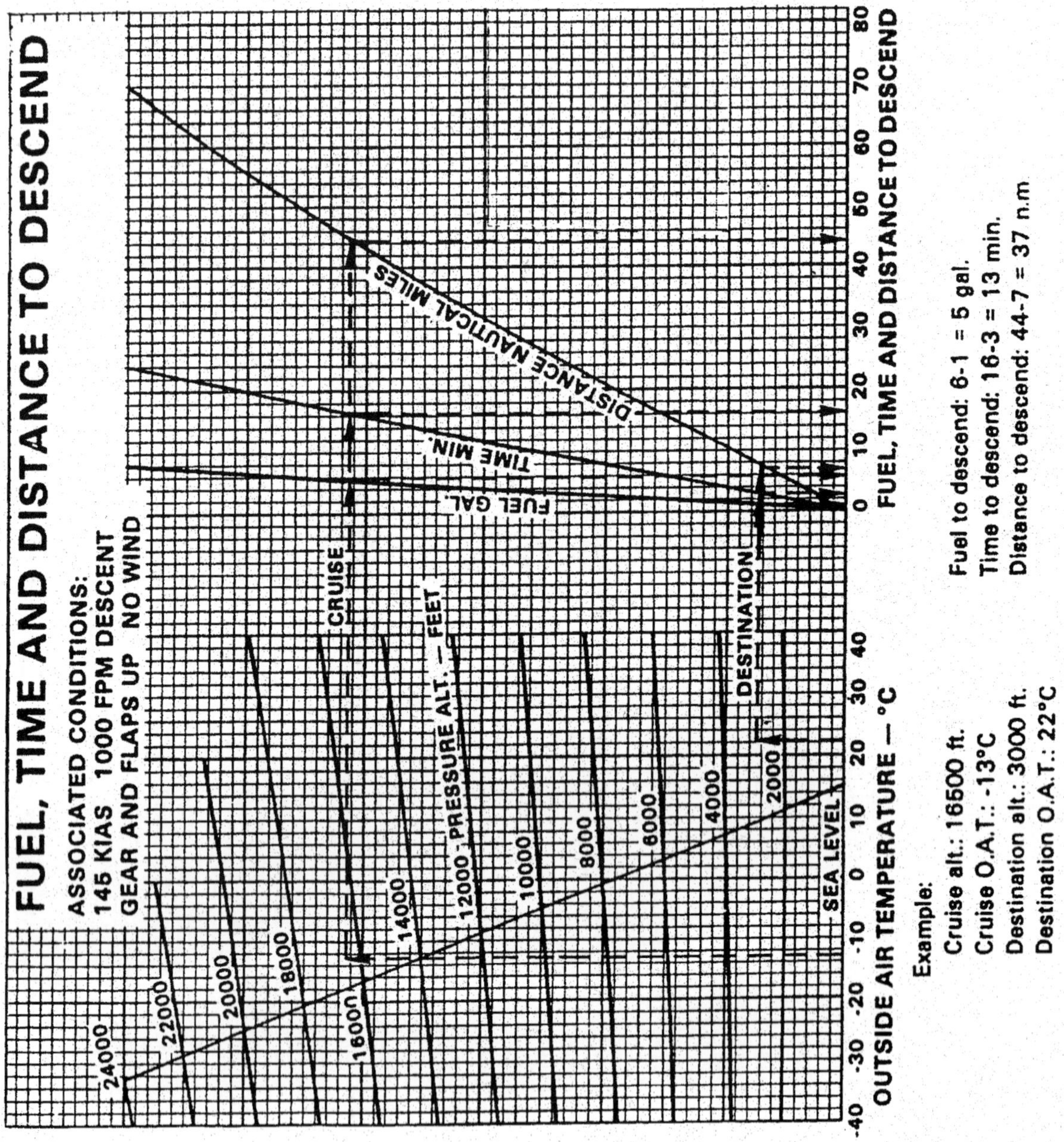
FUEL, TIME AND DISTANCE TO DESCEND
ASSOCIATED CONDITIONS:
145 KIAS 1000 FPM DESCENT
GEAR AND FLAPS UP NO WIND
DISTANCE NAUTICAL MILES
TIME MIN.
FUEL GAL.
CRUISE
DESTINATION
PRESSURE ALT. — FEET
24000
22000
20000
18000
16000
14000
12000
10000
8000
6000
4000
2000
SEA LEVEL
-40 -30 -20 -10 0 10 20 30 40
OUTSIDE AIR TEMPERATURE — °C
0 10 20 30 40 50 60 70 80
FUEL, TIME AND DISTANCE TO DESCEND
Example:
Cruise alt.: 16500 ft.
Cruise O.A.T.: -13°C
Destination alt.: 3000 ft.
Destination O.A.T.: 22°C
Fuel to descend: 6-1 = 5 gal.
Time to descend: 16-3 = 13 min.
Distance to descend: 44-7 = 37 n.m

FUEL PLANNING AND MONITORING

SECTION IV   MEDIUM RANGE JET TRANSPORT

CONTENTS

1   AEROPLANE DATA AND CONSTANTS

1.1   Aeroplane Data
•   Monoplane
•   Twin turbo-jet engines
•   Retractable undercarriage

Structural Limits: -
Maximum Taxi (Ramp) Mass   63060 Kg.
Maximum Take Off Mass   62800 Kg.
Maximum Landing Mass   54900 Kg.
Maximum Zero Fuel Mass   51300 Kg.

Maximum Fuel Load   5311 U.S. Gallons
                                  16145 Kg. (@3.04 Kg./Gal.)

1 .2   Constants
Fuel Density (unless otherwise notified)
            3.04 Kg./US Gallon
            6.7 lbs./US Gallon

2.   OPTIMUM ALTITUDES

2.1   Optimum Cruise Altitude (Fig. 4.2.1)

Enter graph with cruise mass (56800 Kg.)
Move vertically to selected cruise profile (LRC)
Move horizontally to read optimum altitude (33500 ft.)

NB.   Fuel Penalties will be incurred by operating "off optimum" altitude as shown in table.

OFF - OPTIMUM	FUEL MILEAGE PENALTY %	
CONDITION	LRC	0.74
2000 ft. above	1	1
Optimum	0	0
2000 ft below	1	2
4000 ft. below	4	4
8000 ft. below	10	11
12,000 ft. below	15	20

2.2   Short Distance Cruise Altitude (Fig. 4.2.2)

Enter with trip distance (Nautical Air Miles).
Move to temperature deviation.
Move horizontally to reference line.
Follow the trade lines to intersect with vertical through brake release weight.
Move horizontally to read maximum pressure altitude.

Figure 4.2.1   OPTIMUM ALTITUDE

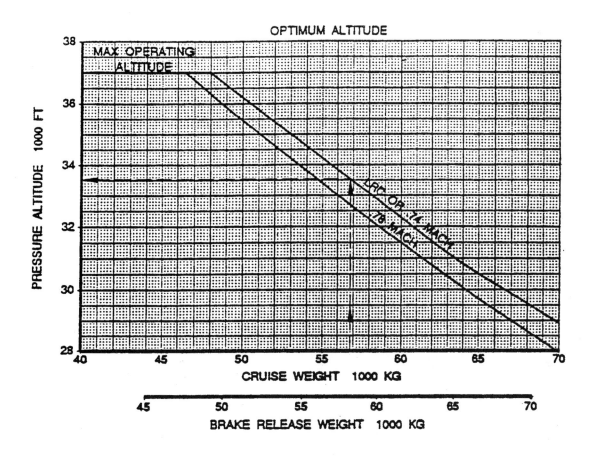

Figure 4.2.2   SHORT DISTANCE CRUISE ALTITUDE

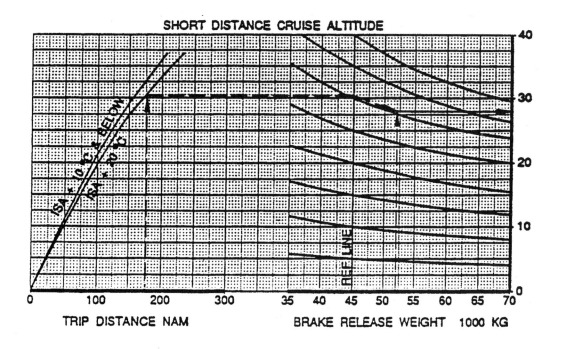

3.   SIMPLIFIED FUEL PLANNING

Graphs are provided, for the various cruise options, as follows: -
Fig. 4.3.1   Long Range Cruise (LRC)
Fig. 4 3.2   0 74 Mach Cruise
Fig. 4.3.3   0.78 Mach Cruise
Fig. 4.3.4   Step Climb
Fig. 4.3.5   Alternate Planning - LRC

These graphs are similar in layout and use.
An example is shown in Fig. 4.3.1A
Trip Distance   350 nautical ground miles
Cruise Altitude   29000 ft.
Estimated Landing Weight   30000 Kg.
Average Wind Component   50 Kts. - headwind
Temperature Deviation   ISA + 20°C

Method
Enter with trip distance in nautical ground miles.
Correct for wind component.
Move vertically to cruise altitude intersections.
From lower scale intersection move right to reference line, correct for landing
weight **(interpolating for altitude between the two sets of trade lines)** then move
horizontally to read fuel.
From upper scale intersection move left to reference line, correct for temperature
deviation then move horizontally to read trip time.
Apply corrections in accordance with paragraph 3.1 as required.

NB.   If actual wind component is greater than range given on chart, convert trip
       distance nautical ground miles (NGM) to nautical air miles (NAM).

$$NAM = NGM \times \frac{TAS\ (av.)}{TAS +/- WC}$$

3.1   Flight Planning Allowances - Simplified Flight Planning Charts.
The "Simplified Flight Planning" charts (Figs. 4.3.1 to 4.3.3) allow rapid determination
of estimated trip time and fuel from brake release to landing.

Additional allowances will be required if the climb, cruise and descent schedules
differ from those stated.

(1)   Cost Index Adjustment
If the flight is planned to operate with the FMS in the 'ECON' mode,
adjustments to the LRC trip fuel and time are necessary in order to account
for the different speed profiles flown. These are given in the following table:

COST INDEX	FUEL ADJUSTMENT %	TIME ADJUSTMENT %
0	-1	+4
20	+1	+1
40	+2	-1
60	+4	-2
80	+5	-3
100	+7	-4
150	+10	-5
200	+14	-7

(2)   Ground Operations
APU fuel flow   115 Kg per hour
Taxi fuel   11 Kg per minute

(3)   Altitude Selection
Operation "off optimum" altitude will result in fuel penalties (see table in para. 2.1 page 2).

(4)   Cruise
Increase trip fuel by 1% for operation with A.C. packs at high flow.
Increase fuel for anti-ice: -
        Engine anti-ice only   70 Kg/hour
        Engine and wing anti-ice   180 Kg/hour

(5)   Descent.
Simplified Charts assume a descent at 0.74M/250 KIAS and a straight in approach.
For every additional minute of flaps down manoeuvre add 75 Kg. fuel.
For engine anti-ice during descent add 50 Kg.

(6)   Holding Fuel
Determine from the table at Fig. 4.4

Figure 4.3.1A SIMPLIFIED FLIGHT PLANNING

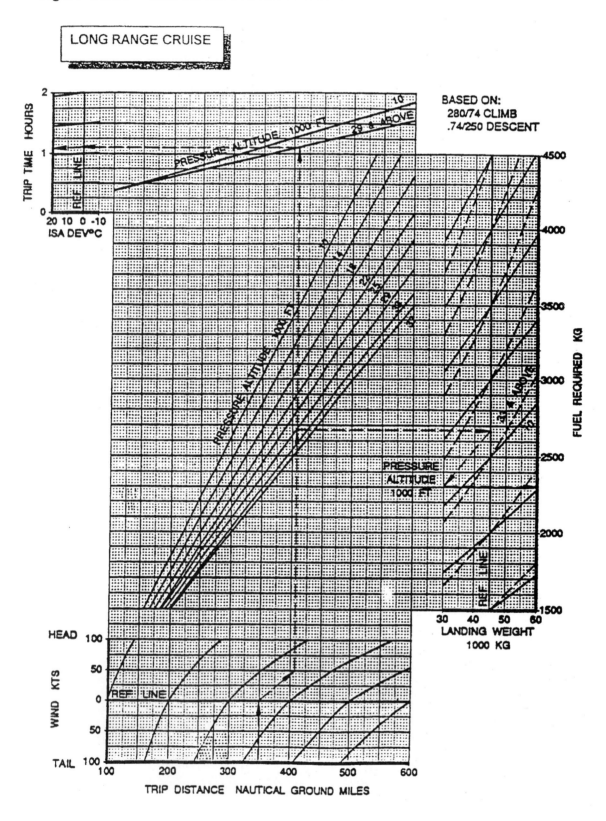

LONG RANGE CRUISE

BASED ON:
280/74 CLIMB
.74/250 DESCENT

Figure. 4.3.1B SIMPLIFIED FLIGHT PLANNING

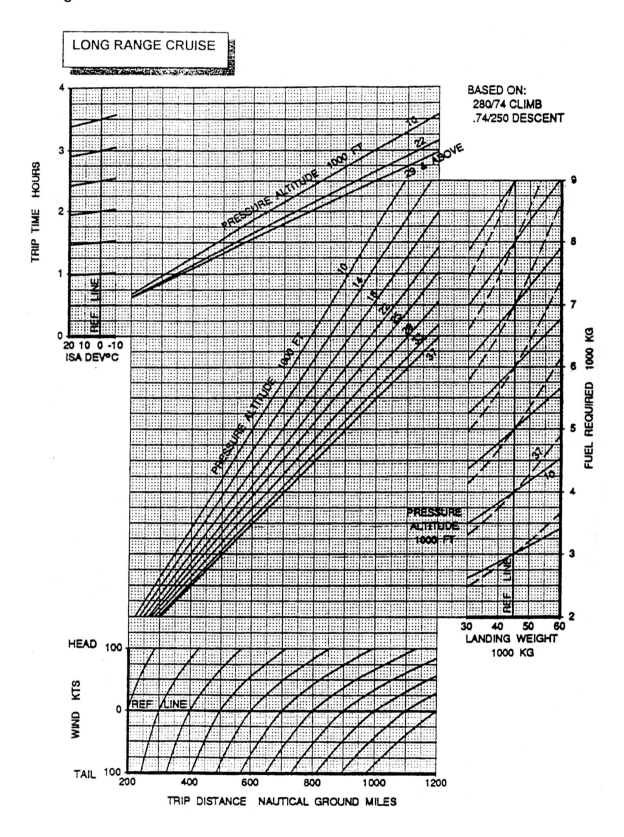

LONG RANGE CRUISE

BASED ON:
280/74 CLIMB
.74/250 DESCENT

Figure. 4.3.1C   SIMPLIFIED FLIGHT PLANNING

LONG RANGE CRUISE

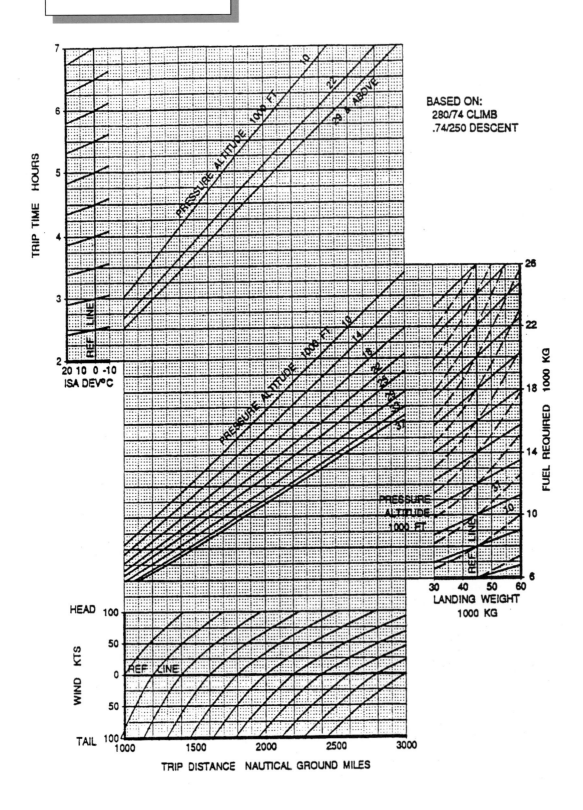

BASED ON:
280/74 CLIMB
.74/250 DESCENT

Figure 4.3.2A   SIMPLIFIED FLIGHT PLANNING

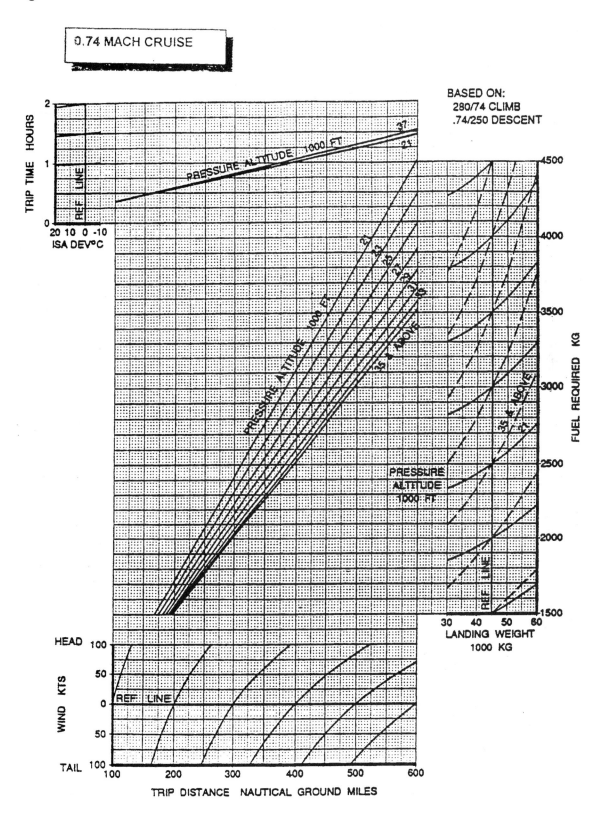

Figure 4.3.2B SIMPLIFIED FLIGHT PLANNING

0.74 MACH CRUISE

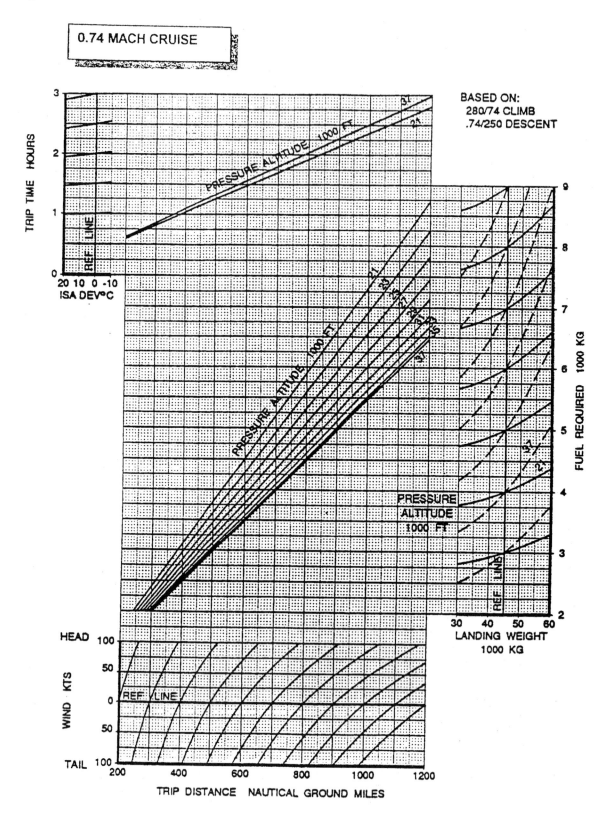

BASED ON:
280/74 CLIMB
.74/250 DESCENT

Figure 4.3.2C   SIMPLIFIED FLIGHT PLANNING

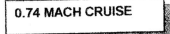

0.74 MACH CRUISE

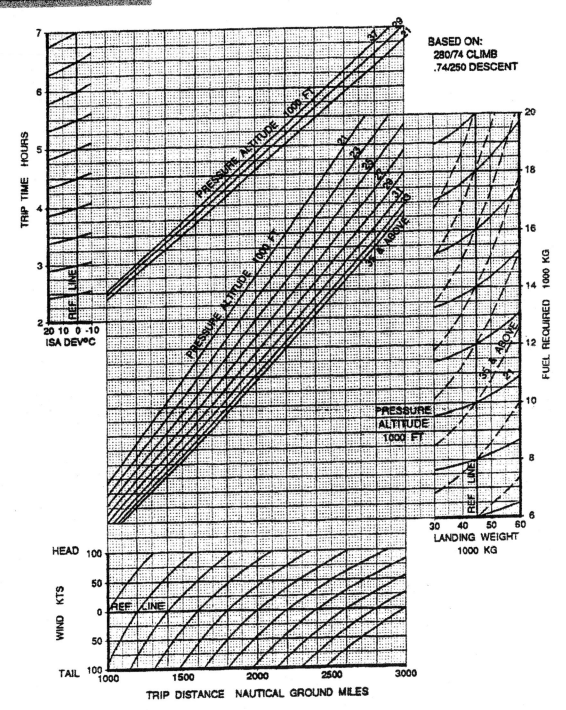

Figure. 4.3.3A   SIMPLIFIED FLIGHT PLANNING

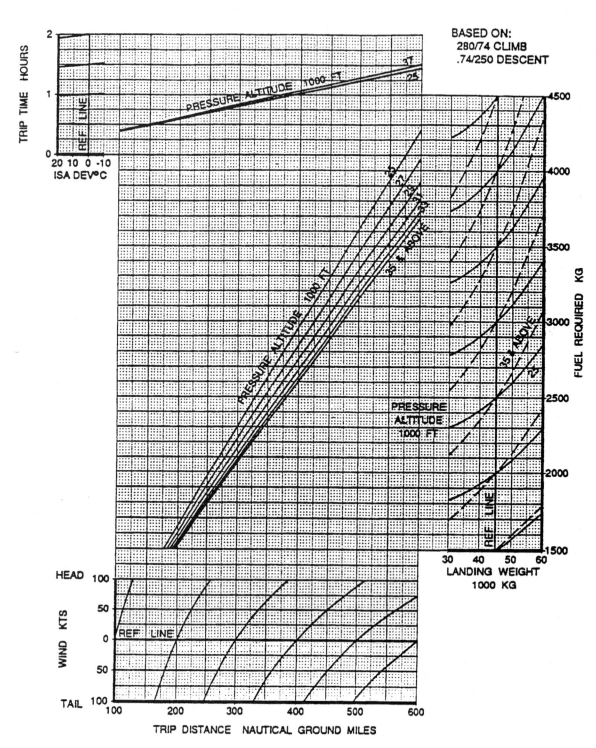

Figure. 4.3.3B   SIMPLIFIED FLIGHT PLANNING

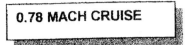

0.78 MACH CRUISE

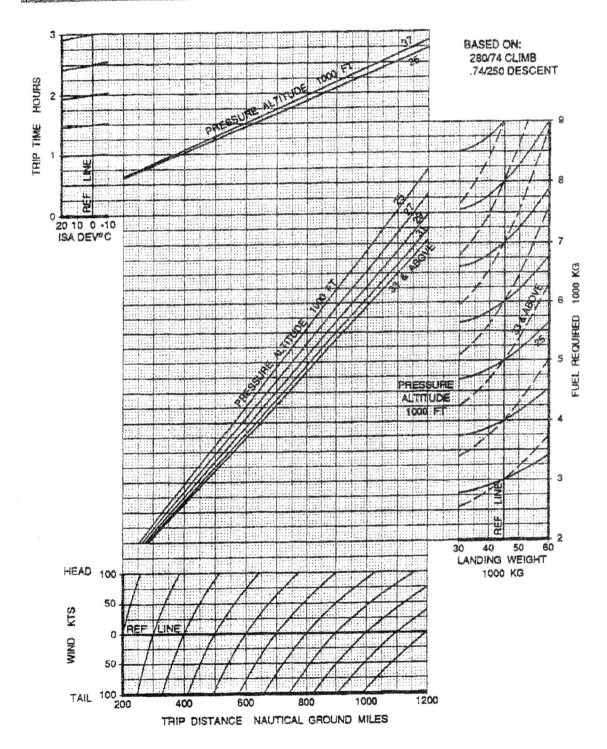

Figure. 4.3.3C SIMPLIFIED FLIGHT PLANNING

0.78 MACH CRUISE

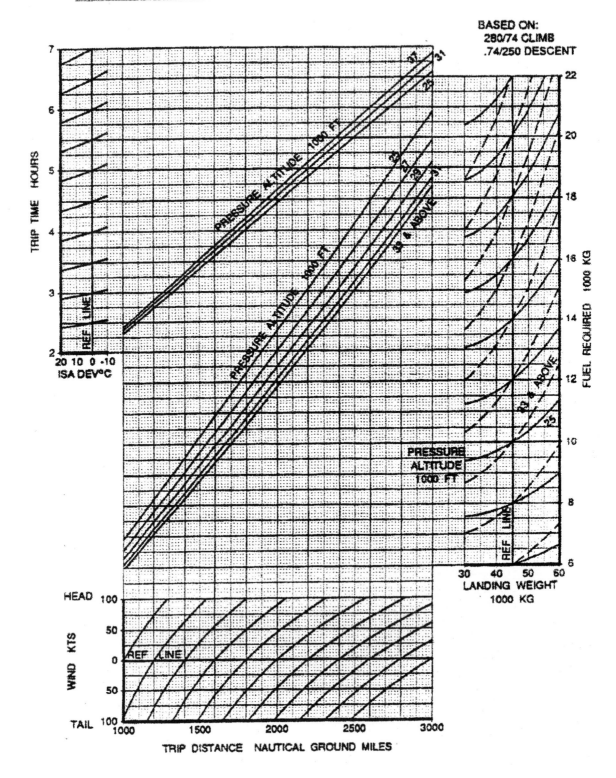

BASED ON:
280/74 CLIMB
.74/250 DESCENT

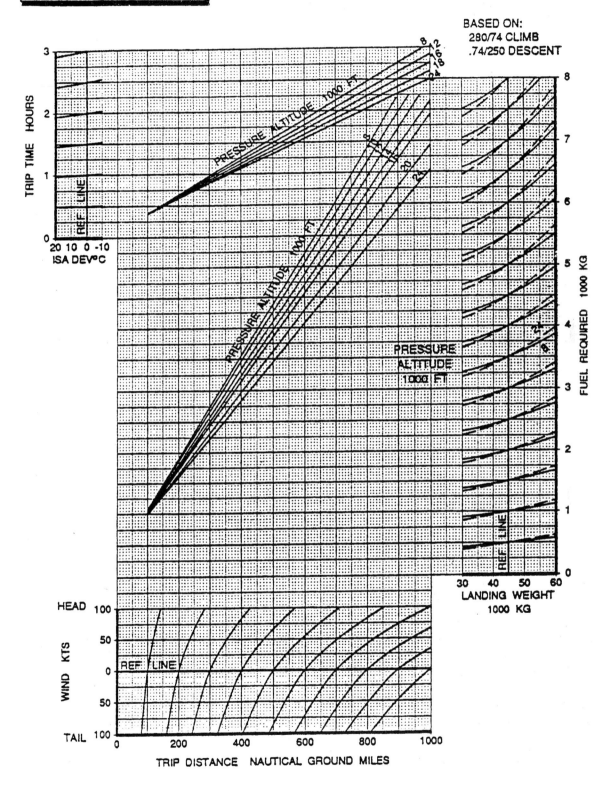

SIMPLIFIED FLIGHT PLANNING
300 KIAS CRUISE

BASED ON:
280/74 CLIMB
.74/250 DESCENT

3.2   Step Climb Simplified Fuel Planning (Fig. 4.3.5)

This chart allows the planner to optimise aeroplane performance by increasing the cruise altitude in 4000 ft. steps in order to allow for the increase in optimum altitude as aeroplane weight decreases.
The graph is valid for altitudes with 'Step Climb' of 4000 ft. to 2000 ft above optimum altitude.
The graph provides trip fuel and time, at LRC or 0.74M, from brake release to touchdown.
The method of use is the same as that for the constant altitude charts except that the argument of 'brake release weight' is used in place of cruise 'pressure altitude'- see example on chart

Figure 4.3.5   SIMPLIFIED FLIGHT PLANNING

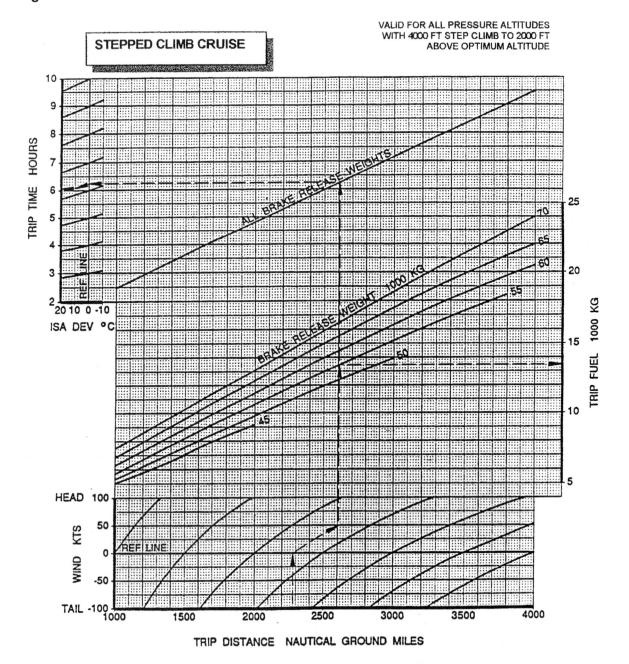

3.3   Alternate Planning (Fig. 4.3.6)

The fuel and time figures extracted from this chart include the following:
- Missed approach
- Climb to cruise altitude
- Cruise at LRC
- Descent and straight on approach.

Method of use is similar to previous graphs.
For distances greater than 500 NM use the LRC Simplified Flight Planning Charts.

Figure 4.3.6   SIMPLIFIED FLIGHT PLANNING

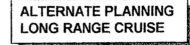

ALTERNATE PLANNING
LONG RANGE CRUISE

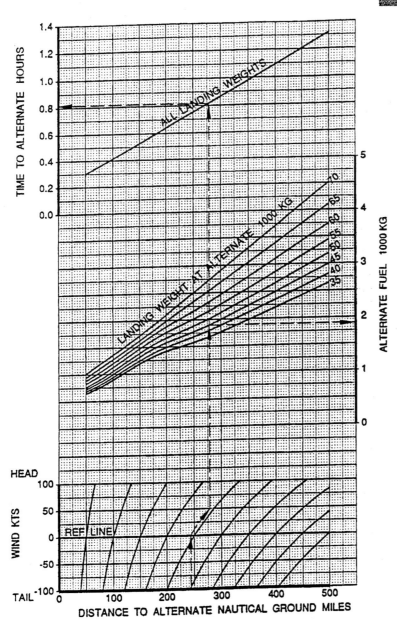

4.   HOLDING FUEL PLANNING

The table below provides fuel flow information necessary for planning holding reserve fuel requirements.

Chart is based on racetrack pattern at minimum drag airspeed - minimum speed 210 KIAS.

For holding in straight and level reduce table values by 5%

Figure 4.4

FLAPS UP

Press. Alt. Ft.	WEIGHT 1000 Kg														
	66	64	62	60	58	56	54	52	50	48	46	44	42	40	38
	TOTAL FUEL FLOW KG/HR														
37000					2740	2540	2400	2260	2160	2080	1980	1900	1800	1740	1680
35000		3020	2820	2660	2520	2420	2320	2220	2140	2060	1960	1880	1800	1720	1660
30000	2840	2740	2660	2560	2480	2400	2300	2220	2140	2060	1960	1880	1800	1740	1680
25000	2840	2760	2660	2580	2500	2420	2320	2240	2160	2080	2000	1920	1840	1780	1720
20000	2840	2760	2680	2580	2500	2420	2340	2260	2180	2100	2020	1940	1860	1800	1760
15000	2880	2800	2700	2620	2540	2460	2380	2300	2220	2140	2060	1980	1920	1860	1800
10000	2920	2820	2740	2660	2580	2500	2420	2340	2260	2180	2100	2020	1980	1920	1880
5000	2960	2860	2780	2700	2620	2540	2460	2380	2300	2220	2140	2080	2020	1960	1920
1500	3000	2900	2820	2740	2660	2580	2520	2440	2360	2280	2220	2140	2080	2020	1980

Fuel flow is based on a racetrack pattern.
For holding in straight and level flight reduce fuel values by 5%

5.   DETAILED FUEL PLANNING

5.1   En-route Climb (Figures 4.5.1)

Tables are provided for a range of temperature deviations from ISA -15°C to ISA +25°C
Fuel and time given in these tables are from brake release and distance from 1500 ft. with a climb airspeed schedule 280 KIAS/0.74 Mach.
The stated TAS is the average for the climb and should be used to correct the still air distance shown.

$$NGM = NAM \times \frac{\{TAS +/-. WIND\}}{TAS}$$

Figure 4.5.1   EN-ROUTE CLIMB   280/.74

ISA -6°C TO -15°C

Press. Alt. Ft.	Units Min/Kg. NAM/Kts	68000	66000	64000	62000	60000	58000	56000	52000	48000	44000	40000
37000	Time/Fuel				30/ 2100	25/ 1800	22/ 1650	20/ 1550	17/ 1350	15/ 1200	13/ 1050	12/ 950
	Dist/TAS				184/391	148/387	130/385	117/383	98/381	85/379	73/378	64/377
36000	Time/Fuel			28/ 2050	24/ 1800	22/ 1650	20/ 1550	19/ 1450	16/ 1300	14/ 1150	13/ 1000	11/ 900
	Dist/TAS			166/388	142/385	127/383	115/381	106/380	91/378	79/377	69/376	60/375
35000	Time/Fuel	32/ 2350	27/2000	24/1850	22/ 1700	20/ 1600	19/ 1500	17/ 1400	15/ 1250	13/ 1100	12/ 1000	11/ 900
	Dist/TAS	195/390	158/385	139/383	125/381	114/380	105/378	97/377	85/376	74/375	65/374	57/373
34000	Time/Fuel	26/ 2000	23/ 1850	21/ 1700	20/ 1600	19/ 1500	17/ 1400	16/ 1350	14/ 1200	13/ 1100	11/ 950	10/ 850
	Dist/TAS	152/383	136/381	123/379	113/378	105/376	97/375	90/375	79/373	70/372	61/371	54/371
33000	Time/Fuel	23/ 1850	21/ 1750	20/ 1650	19/ 1550	17/ 1450	16/ 1350	15/ 1300	14/ 1150	12/ 1050	11/ 950	10/ 850
	Dist/TAS	133/378	121/376	112/375	104/374	97/373	90/372	84/372	74/371	66/370	58/369	51/368
32000	Time/Fuel	21/ 1750	20/ 1650	19/ 1550	17/ 1500	16/ 1400	16/ 1300	15/ 1250	13/ 1100	11/ 1000	11/ 900	9/ 800
	Dist/TAS	120/374	111/373	103/372	96/371	90/370	84/369	79/359	70/368	62/367	55/366	48/366
31000	Time/Fuel	20/ 1700	19/ 1600	18/ 1500	17/ 1400	16/ 1350	15/ 1300	14/ 1200	13/ 1100	11/ 1000	10/ 900	9/ 800
	Dist/TAS	110/370	102/369	95/368	89/367	84/367	79/366	74/366	66/365	58/364	52/364	46/363
30000	Time/Fuel	19/ 1600	18/ 1550	17/ 1450	16/ 1350	15/ 1300	14/ 1250	13/ 1200	12/ 1050	11/ 950	10/ 850	9/ 800
	Dist/TAS	101/366	95/365	89/364	83/364	78/363	74/ 363	70/362	62/362	55/361	49/ 361	43/360
29000	Time/Fuel	17/ 1550	16/ 1450	16/ 1400	15/ 1300	14/ 1250	13/ 1200	12/ 1100	11/ 1050	10/ 950	9/ 850	8/ 750
	Dist/TAS	92/361	87/360	81/360	77/359	72/359	68/358	64/358	57/357	51/357	46/357	41/356
28000	Time/Fuel	16/ 1450	15/ 1400	15/ 1300	14/ 1250	13/ 1200	13/ 1150	12/ 1100	11/ 1000	10/ 900	9/ 800	8/ 750
	Dist/TAS	84/356	79/356	75/355	70/355	67/355	63/354	59/354	53/353	48/353	42/353	38/352
27000	Time/Fuel	15/ 1400	14/ 1350	14/ 1250	13/ 1200	12/ 1150	12/ 1100	11/ 1050	10/ 950	9/ 850	6/ 800	8/ 700
	Dist/TAS	77/352	73/351	69/351	65/351	61/350	58/350	55/350	49/349	44/349	39/349	35/348
26000	Time/Fuel	14/ 1350	14/ 1250	13/ 1200	12/ 1150	12/ 1100	11/ 1050	11/ 1000	10/ 900	9/ 850	8/ 750	7/ 700
	Dist/TAS	71/348	67/347	63/347	60/347	57/347	54/346	51/346	46/346	41/345	37/345	33/345
25000	Time/Fuel	13/ 1300	13/ 1200	12/ 1150	12/ 1100	11/ 1050	11/ 1000	10/ 950	9/ 900	8/ 800	8/ 750	7/ 650
	Dist/TAS	65/344	61/343	58/343	55/343	52/343	49/343	47/342	42/342	38/342	34/342	30/341
24000	Time/Fuel	13/ 1200	12/ 1150	11/ 1100	11/ 1050	10/ 1000	10/ 950	10/ 950	9/ 850	8/ 750	7/ 700	6/ 650
	Dist/TAS	60/340	56/340	54/340	51/339	48/339	46/339	43/339	39/339	35/338	32/338	28/338
23000	Time/Fuel	12/ 1150	11/ 1100	11/ 1000	10/ 1000	10/ 1000	9/ 950	9/ 900	8/ 800	7/ 750	7/ 700	6/ 600
	Dist/TAS	55/336	52/336	49/336	47/336	44/336	42/335	40/335	36/335	33/335	29/335	26/335
22000	Time/Fuel	11/ 1100	11/ 1000	10/ 1000	10/ 1000	9/ 950	9/ 900	9/ 900	8/ 800	7/ 700	6/ 650	6/ 600
	Dist/TAS	50/333	48/333	45/333	43/332	41/332	39/332	37/332	33/332	30/332	27/332	24/331
21000	Time/Fuel	10/ 1050	10/ 1000	10/ 1000	9/ 950	9/ 900	8/ 850	8/ 800	7/ 750	7/ 700	6/ 650	6/ 550
	Dist/TAS	46/330	44/329	42/329	40/329	38/329	36/329	34/329	31/329	28/328	25/328	22/328
20000	Time/Fuel	10/ 1000	9/ 950	9/ 950	9/ 900	8/ 850	8/ 800	8/ 800	7/ 700	6/ 650	6/ 600	5/ 550
	Dist/TAS	42/326	40/326	38/326	36/326	35/326	33/326	31/326	28/326	26/325	23/325	21/325
19000	Time/Fuel	9/ 950	9/ 950	8/ 900	8/ 850	8/ 800	7/ 800	7/ 750	7/ 700	6/ 600	6/ 600	5/ 500
	Dist/TAS	39/323	37/323	35/323	33/323	32/323	30/323	29/323	26/323	24/322	21/322	19/322
18000	Time/Fuel	9/ 900	8/ 900	8/ 850	8/ 800	7/ 800	7/ 750	7/ 700	6/ 650	6/ 600	5/ 550	5/ 500
	Dist/TAS	35/320	34/320	32/320	31/320	29/320	28/320	26/320	24/320	22/320	19/319	17/319
17000	Time/Fuel	8/ 900	8/ 850	8/ 800	7/ 800	7/ 750	7/ 700	6/ 700	6/ 650	5/ 600	5/ 550	5/ 500
	Dist/TAS	32/317	31/317	29/317	28/317	27/317	25/317	24/317	22/317	20/317	18/317	16/317
16000	Time/Fuel	8/ 850	7/ 800	7/ 750	7/ 750	7/ 700	6/ 700	6/ 650	6/ 600	5/ 550	5/ 500	4/ 450
	Dist/TAS	29/314	28/314	27/314	25/314	24/314	23/314	22/314	20/314	18/314	16/314	15/314
15000	Time/Fuel	7/ 800	7/ 750	7/ 750	6/ 700	6/ 700	6/ 650	6/ 650	5/ 600	5/ 550	4/ 500	4/ 450
	Dist/TAS	26/312	25/312	24/312	23/311	22/311	22/311	20/311	18/311	16/311	15/311	13/311
14000	Time/Fuel	7/ 750	6/ 700	6/ 700	6/ 650	6/ 650	6/ 600	5/ 600	5/ 550	5/ 500	4/ 450	4/ 400
	Dist/TAS	24/309	23/309	22/309	21/309	20 309	19/309	18/309	16/309	15/309	13/309	12/309
13000	Time/Fuel	6/ 700	6/ 700	6/ 650	6/ 650	5/ 600	5/ 600	5/ 550	5/ 500	4/ 500	4/ 450	4/ 400
	Dist/TAS	21/306	20/306	19/306	19/306	18/306	17/306	16/306	15/306	13/306	12/306	11/306
12000	Time/Fuel	6/ 650	6/ 650	5/ 600	5/ 600	5/ 600	5/ 550	5/ 550	4/ 500	4/ 450	4/ 400	3/ 400
	Dist/TAS	19/304	18/304	17/304	17/304	16/304	15/304	14/304	13/304	12/304	11/304	10/304
11000	Time/Fuel	5/ 600	5/ 600	5/ 600	5/ 550	5/ 550	5/ 500	4/ 500	4/ 450	4/ 450	3/ 400	3/ 350
	Dist/TAS	17/301	16/301	15/301	15/301	14/301	13/301	13/301	12/301	11/301	10/301	9/301
10000	Time/Fuel	5/ 600	5/ 550	5/ 550	5/ 550	4/ 500	4/ 500	4/ 500	4/ 400	4/ 400	3/ 350	3/ 350
	Dist/TAS	15/299	14/299	13/299	13/299	12/299	12/299	11/299	10/299	9/299	8/299	7/299
8000	Time/Fuel	4/ 500	4/ 500	4/ 500	4/ 450	4/ 450	4/ 450	3/ 400	3/ 400	3/ 350	3/ 350	3/ 300
	Dist/TAS	11/294	10/294	10/294	9/294	9/294	9/294	8/294	7/294	7/294	6/294	6/294
6000	Time/Fuel	4/ 450	3/ 400	3/ 400	3/ 400	3/ 400	3/ 350	3/ 350	3/ 350	3/ 300	2/ 300	2/ 250
	Dist/TAS	7/290	7/290	6/290	6/290	6/290	6/290	5/290	5/290	5/290	4/290	4/290
1500	Time/Fuel	2/ 250	2/ 250	2/ 250	2/ 250	2/ 250	2/ 250	2/ 250	2/ 200	2/ 200	2/ 200	1/ 150

Fuel Adjustment for high elevation airports	Airport Elevation	2000	4000	6000	8000	10000	12000
Effect on time and distance is negligible	Fuel Adjustment	-50	-100	-150	-250	-300	-350

Figure 4.5.1 EN-ROUTE CLIMB 280/.74

ISA -5°C TO +5°C

Press. Alt. Ft.	Units Min/Kg. NAM/Kts	BRAKE RELEASE WEIGHT KG.										
		68000	66000	64000	62000	60000	58000	56000	52000	48000	44000	40000
37000	Time/Fuel				32/ 2250	26/ 1900	23/ 1750	21/ 1600	18/ 1400	16/ 1250	14/ 1100	12/ 1000
	Dist/TAS				197/400	158/396	138/393	124/392	105/389	90/388	78/386	68/385
36000	Time/Fuel			29/ 2150	25/ 1900	23/ 1750	21/ 1650	19/ 1550	17/ 1350	15/ 1200	13/ 1050	11/ 950
	Dist/TAS			177/397	151/393	135/391	122/390	112/388	96/386	84/385	73/384	64/383
35000	Time/Fuel	33/ 2450	28/ 2150	25/ 1950	22/ 1800	21/ 1650	19/ 1550	18/ 1450	16/ 1300	14/ 1150	12/ 1050	11/ 950
	Dist/TAS	209/399	169/394	148/391	133/389	121/388	112/387	104/386	90/384	78/383	69/382	60/381
34000	Time/Fuel	27/ 2150	24/ 1950	22/ 1800	21/ 1700	19/ 1600	18/ 1500	17/ 1400	15/ 1250	13/ 1150	12/ 1000	11/ 900
	Dist/TAS	162/391	144/389	131/387	120/386	111/385	103/384	96/383	84/381	74/380	65/379	57/379
33000	Time/Fuel	24/ 1950	22/ 1850	21/ 1700	19/ 1600	18/ 1550	17/ 1450	16/ 1350	14/ 1200	13/ 1100	11/ 1000	10/ 900
	Dist/TAS	142/386	129/385	119/383	110/382	103/381	96/380	90/380	79/379	70/378	62/377	54/376
32000	Time/Fuel	22/ 1850	21/ 1750	19/ 1650	18/ 1550	17/ 1450	16/ 1400	15/ 1300	14/ 1200	12/ 1050	11/ 950	10/ 850
	Dist/TAS	128/382	118/381	110/380	102/379	96/378	90/377	84/377	74/376	66/375	58/374	51/374
31000	Time/Fuel	21/ 1800	19/ 1650	18/ 1600	17/ 1500	16/ 1400	15/ 1350	14/ 1300	13/ 1150	12/ 1050	10/ 950	9/ 850
	Dist/TAS	117/378	109/377	102/376	95/375	89/375	84/374	79/374	70/373	62/372	55/371	49/371
30000	Time/Fuel	19/ 1700	18/ 1600	17/ 1500	16/ 1450	15/ 1350	15/ 1300	14/ 1250	12/ 1100	11/ 1000	10/ 900	9/ 800
	Dist/TAS	108/374	101/373	95/372	89/372	84/371	79/371	74/370	66/369	59/369	52/368	46/368
29000	Time/Fuel	18/ 1600	17/ 1550	16/ 1450	15/ 1400	14/ 1300	14/ 1250	13/ 1200	12/ 1100	11/ 950	10/ 900	9/ 800
	Dist/TAS	98/369	92/368	87/367	82/367	77/367	73/366	69/366	61/365	55/365	49/364	43/364
28000	Time/Fuel	17/ 1550	16/ 1450	15/ 1400	14/ 1300	14/ 1250	13/ 1200	12/ 1150	11/ 1050	10/ 950	9/ 850	8/ 750
	Dist/TAS	90/364	84/363	79/363	75/362	71/362	67/362	63/361	57/361	51/360	45/360	40/360
27000	Time/Fuel	16/ 1450	15/ 1400	14/ 1350	13/ 1250	13/ 1200	12/ 1150	12/ 1100	11/ 1000	10/ 900	9/ 800	8/ 750
	Dist/TAS	82/359	77/359	73/358	69/358	65/358	62/357	58/357	52/357	47/356	42/356	37/356
26000	Time/Fuel	15/ 1400	14/ 1350	13/ 1300	13/ 1250	12/ 1150	12/ 1100	11/ 1050	10/ 950	9/ 850	8/ 800	7/ 700
	Dist/TAS	75/355	71/355	67/354	63/354	60/354	57/353	54/353	48/353	43/352	39/352	35/352
25000	Time/Fuel	14/ 1350	13/ 1250	13/ 1200	12/ 1150	11/ 1050	11/ 1100	10/ 1000	9/ 900	9/ 850	8/ 750	7/ 700
	Dist/TAS	69/351	65/351	62/350	58/350	55/350	53/350	50/349	45/349	40/349	36/348	32/348
24000	Time/Fuel	13/ 1300	12/ 1200	12/ 1150	11/ 1100	11/ 1050	10/ 1000	10/ 950	9/ 900	8/ 800	7/ 750	7/ 650
	Dist/TAS	63/347	60/347	57/346	54/346	51/346	48/346	46/346	41/345	37/345	33/345	30/345
23000	Time/Fuel	12/ 1200	12/ 1150	11/ 1100	11/ 1050	10/ 1000	10/ 950	9/ 950	8/ 850	8/ 750	7/ 700	6/ 650
	Dist/TAS	58/343	55/343	52/343	50/343	47/342	45/342	42/342	38/342	34/342	31/341	28/341
22000	Time/Fuel	11/ 1150	11/ 1100	10/ 1050	10/ 1000	10/ 1000	9/ 950	9/ 900	8/ 800	7/ 750	7/ 700	6/ 600
	Dist/TAS	53/340	51/339	48/339	46/339	43/339	41/339	39/339	35/338	32/338	29/338	26/338
21000	Time/Fuel	11/ 1100	10/ 1050	10/ 1000	9/ 1000	9/ 950	9/ 900	8/ 850	8/ 800	7/ 700	6/ 650	6/ 600
	Dist/TAS	49/336	46/336	44/336	42/336	40/336	38/335	36/335	33/335	29/335	26/335	24/335
20000	Time/Fuel	10/ 1050	10/ 1000	9/ 950	9/ 950	9/ 900	8/ 850	8/ 800	7/ 750	7/ 700	6/ 600	5/ 550
	Dist/TAS	45/333	42/333	40/332	39/332	37/332	35/332	33/332	30/332	27/332	24/332	22/332
19000	Time/Fuel	10/ 1000	9/ 950	9/ 950	8/ 900	8/ 850	8/ 800	7/ 800	7/ 700	6/ 650	6/ 600	5/ 550
	Dist/TAS	41/329	39/329	37/329	35/329	34/329	32/329	30/329	28/329	25/329	22/329	20/329
18000	Time/Fuel	9/ 950	9/ 900	8/ 900	8/ 850	8/ 800	7/ 800	7/ 750	6/ 700	6/ 650	5/ 550	5/ 550
	Dist/TAS	37/326	36/326	34/326	32/326	31/326	29/326	28/326	25/326	23/326	21/326	18/326
17000	Time/Fuel	8/ 900	8/ 900	7/ 800	7/ 800	7/ 800	7/ 750	7/ 700	6/ 650	6/ 600	5/ 550	5/ 500
	Dist/TAS	34/323	32/323	31/323	29/323	28/323	27/323	26/323	23/323	21/323	19/323	17/323
16000	Time/Fuel	8/ 850	8/ 850	7/ 800	7/ 750	7/ 750	6/ 700	6/ 700	6/ 650	5/ 600	5/ 550	4/ 450
	Dist/TAS	31/320	29/320	28/320	27/320	26/320	24/320	23/320	21/320	19/320	17/320	15/320
15000	Time/Fuel	7/ 800	7/ 800	7/ 750	7/ 750	6/ 700	6/ 700	6/ 650	5/ 600	5/ 550	5/ 550	4/ 450
	Dist/TAS	28/318	27/318	25/317	24/317	23/317	22/317	21/317	19/317	17/317	16/317	14/317
14000	Time/Fuel	7/ 800	7/ 750	6/ 700	6/ 700	6/ 650	6/ 650	5/ 600	5/ 550	5/ 500	4/ 500	4/ 450
	Dist/TAS	25/315	24/315	23/315	22/315	21/315	20/315	19/315	17/315	16/315	15/315	13/314
13000	Time/Fuel	6/ 750	6/ 700	6/ 700	6/ 650	6/ 650	5/ 600	5/ 600	5/ 550	4/ 500	4/ 450	4/ 400
	Dist/TAS	22/312	21/312	21/312	20/312	19/312	18/312	17/312	16/312	14/312	13/312	11/312
12000	Time/Fuel	6/ 700	6/ 650	6/ 650	5/ 600	5/ 600	5/ 550	5/ 550	4/ 500	4/ 450	4/ 450	3/ 400
	Dist/TAS	20/310	19/310	18/310	17/309	17/309	16/309	15/309	14/309	13/309	11/309	10/309
11000	Time/Fuel	6/ 650	5/ 650	5/ 600	5/ 600	5/ 550	5/ 550	4/ 500	4/ 500	4/ 450	4/ 400	3/ 350
	Dist/TAS	18/307	17/307	16/307	15/307	15/307	14/307	13/307	12/307	11/307	10/307	9/307
10000	Time/Fuel	5/ 600	5/ 600	5/ 550	5/ 550	4/ 550	4/ 500	4/ 500	4/ 450	4/ 400	3/ 400	3/ 350
	Dist/TAS	15/305	15/305	14/305	13/305	13/305	12/305	12/305	11/304	10/304	9/304	8/304
8000	Time/Fuel	4/ 500	4/ 500	4/ 500	4/ 450	4/ 450	4/ 450	4/ 450	3/ 400	3/ 350	3/ 350	3/ 300
	Dist/TAS	11/300	11/300	10/300	10/300	9/300	9/300	9/300	8/300	7/300	6/300	6/300
6000	Time/Fuel	4/ 450	3/ 450	3/ 400	3/ 400	3/ 400	3/ 400	3/ 350	3/ 350	3/ 300	2/ 300	2/ 250
	Dist/TAS	7/295	7/295	7/295	7/295	6/295	6/295	6/295	5/295	5/295	4/295	4/295
1500	Time/Fuel	2/ 250	2/ 250	2/ 250	2/ 250	2/ 250	2/ 250	2/ 250	2/ 200	2/ 200	2/ 200	1/ 150

Fuel Adjustment for high elevation airports	Airport Elevation	2000	4000	6000	8000	10000	12000
Effect on time and distance is negligible	Fuel Adjustment	-50	-100	-200	-250	-300	-350

Figure 4.5.1   EN-ROUTE CLIMB   280/.74

ISA +6°C TO +15°C

Press. Alt. Ft.	Units Min/Kg. NAM/Kts	BRAKE RELEASE WEIGHT KG.										
		68000	66000	64000	62000	60000	58000	56000	52000	48000	44000	40000
37000	Time/Fuel				33/ 2350	27/ 2000	24/ 1850	22/ 1700	18/ 1500	16/ 1300	14/ 1150	12/ 1000
	Dist/TAS				212/409	169/404	147/402	132/400	111/397	95/396	82/394	72/393
36000	Time/Fuel			30/ 2250	26/ 2000	23/ 1850	21/ 1700	20/ 1600	17/ 1400	15/ 1250	13/ 1100	12/ 1000
	Dist/TAS			189/405	161/402	143/400	130/398	119/397	102/395	89/393	77/392	68/391
35000	Time/Fuel	35/ 2600	29/ 2250	26/ 2050	23/ 1900	21/ 1750	20/ 1650	19/ 1550	16/ 1350	14/ 1200	13/ 1100	11/ 950
	Dist/TAS	224/407	180/402	157/399	141/397	129/396	119/395	110/394	95/392	83/391	73/390	64/389
34000	Time/Fuel	28/ 2250	25/ 2050	23/ 1900	21/ 1800	20/ 1650	19/ 1550	18/ 1500	16/ 1300	14/ 1200	12/ 1050	11/ 950
	Dist/TAS	173/400	154/397	140/395	128/394	118/393	110/392	102/391	89/389	78/388	69/387	61/386
33000	Time/Fuel	25/ 2100	23/ 1950	21/ 1800	20/ 1700	19/ 1600	18/ 1500	17/ 1450	15/ 1300	13/ 1150	12/ 1050	10/ 900
	Dist/TAS	151/394	138/393	127/391	118/390	109/389	102/388	95/388	84/386	74/385	65/385	58/384
32000	Time/Fuel	23/ 1950	21/ 1850	20/ 1750	19/ 1650	18/ 1550	17/ 1450	16/ 1400	14/ 1250	13/ 1100	11/ 1000	10/ 900
	Dist/TAS	136/390	126/389	117/388	109/387	102/386	95/385	89/384	79/383	70/383	62/382	55/381
31000	Time/Fuel	22/ 1850	20/ 1750	19/ 1650	18/ 1550	17/ 1500	16/ 1400	15/ 1350	13/ 1200	12/ 1100	11/ 1000	10/ 900
	Dist/TAS	125/386	116/385	108/384	101/383	95/382	89/382	84/381	74/380	66/380	59/379	52/378
30000	Time/Fuel	20/ 1800	19/ 1700	18/ 1600	17/ 1500	16/ 1450	15/ 1350	14/ 1300	13/ 1150	12/ 1050	10/ 950	9/ 850
	Dist/TAS	115/382	107/381	100/381	95/379	89/379	84/378	79/378	70/377	62/376	56/376	49/375
29000	Time/Fuel	19/ 1700	18/ 1600	17/ 1550	16/ 1450	15/ 1400	14/ 1300	14/ 1250	12/ 1150	11/ 1000	10/ 900	9/ 850
	Dist/TAS	105/376	98/376	92/375	87/374	82/374	77/374	73/373	65/373	58/372	52/372	46/371
28000	Time/Fuel	17/ 1600	17/ 1550	16/ 1450	15/ 1400	14/ 1300	13/ 1250	13/ 1200	12/ 1100	10/1000	9/ 900	8/ 800
	Dist/TAS	95/371	90/371	84/370	80/370	75/369	71/369	67/369	60/368	54/368	48/367	42/367
27000	Time/Fuel	16/ 1550	15/ 1450	15/ 1400	14/ 1350	13/ 1250	13/ 1200	12/ 1150	11/ 1050	10/ 950	9/ 850	8/ 750
	Dist/TAS	87/366	82/366	77/366	73/365	69/365	66/365	62/364	56/364	50/363	44/363	39/363
26000	Time/Fuel	15/ 1450	15/ 1400	14/ 1350	13/ 1250	13/ 1200	12/ 1150	11/ 1100	10/ 1000	9/ 900	8/ 800	8/ 750
	Dist/TAS	80/362	75/362	71/361	67/361	64/361	60/360	57/360	51/360	46/359	41/359	37/359
25000	Time/Fuel	14/ 1400	14/ 1350	13/ 1250	12/ 1200	12/ 1150	11/ 1100	11/ 1050	10/ 950	9/ 850	8/ 800	7/ 700
	Dist/TAS	73/358	69/357	65/357	62/357	59/357	56/356	53/356	47/356	43/356	38/355	34/355
24000	Time/Fuel	13/ 1350	13/ 1250	12/ 1200	12/ 1150	11/ 1100	11/ 1050	10/ 1000	9/ 900	8/ 850	8/ 750	7/ 700
	Dist/TAS	67/354	63/353	60/353	57/353	54/353	51/353	49/352	44/352	39/352	35/352	32/351
23000	Time/Fuel	13/ 1250	12/ 1200	11/ 1150	11/ 1100	10/ 1050	10/ 1000	10/ 950	9/ 900	8/ 800	7/ 750	7/ 650
	Dist/TAS	61/350	58/350	55/349	53/349	50/349	47/349	45/349	41/348	37/348	33/348	29/348
22000	Time/Fuel	12/ 1200	11/ 1150	11/ 1100	10/ 1050	10/ 1000	9/ 950	9/ 950	8/ 850	8/ 750	7/ 700	6/ 650
	Dist/TAS	56/346	54/346	51/346	48/346	46/345	44/345	42/345	37/345	34/345	30/345	27/344
21000	Time/Fuel	11/ 1150	11/ 1100	10/ 1050	10/ 1000	9/ 950	9/ 950	9/ 900	8/ 800	7/ 750	6/ 700	6/ 600
	Dist/TAS	52/343	49/342	47/342	44/342	42/342	40/342	38/342	35/342	31/341	28/341	25/341
20000	Time/Fuel	10/ 1100	10/ 1050	10/ 1000	9/ 950	9/ 950	8/ 900	8/ 850	7/ 800	7/ 700	6/ 650	6/ 600
	Dist/TAS	47/339	45/339	43/339	41/339	39/339	37/338	35/338	32/338	29/338	26/338	23/338
19000	Time/Fuel	10/ 1050	9/ 1000	9/ 950	9/ 950	8/ 900	8/ 850	8/ 800	7/ 750	6/ 700	6/ 600	5/ 550
	Dist/TAS	43/336	41/336	39/336	37/335	36/335	34/335	32/335	29/335	26/335	24/335	21/335
18000	Time/Fuel	9/ 1000	9/ 950	8/ 900	8/ 900	8/ 850	7/ 800	7/ 800	7/ 700	6/ 650	6/ 600	5/ 550
	Dist/TAS	39/332	38/332	36/332	34/332	33/332	31/332	30/332	27/332	24/332	22/332	19/332
17000	Time/Fuel	9/ 950	8/ 900	8/ 900	8/ 850	7/ 800	7/ 750	7/ 750	6/ 700	6/ 600	5/ 550	5/ 500
	Dist/TAS	36/329	34/329	33/329	31/329	30/329	28/329	27/329	24/329	22/329	20/329	18/329
16000	Time/Fuel	8/ 900	8/ 850	7/ 850	7/ 800	7/ 750	7/ 750	6/ 700	6/ 650	5/ 600	5/ 550	4/ 500
	Dist/TAS	33/326	31/326	30/326	28/326	27/326	26/326	25/326	22/326	20/326	18/326	16/326
15000	Time/Fuel	8/ 850	7/ 800	7/ 800	7/ 750	6/ 750	6/ 700	6/ 650	6/ 650	5/ 600	5/ 500	4/ 450
	Dist/TAS	29/323	28/323	27/323	26/323	24/323	23/323	22/323	20/323	18/323	16/323	15/323
14000	Time/Fuel	7/ 800	7/ 800	7/ 750	6/ 700	6/ 700	6/ 650	6/ 650	5/ 600	5/ 550	4/ 500	4/ 450
	Dist/TAS	26/321	25/321	24/321	23/320	22/320	21/320	20/320	18/320	17/320	15/320	13/320
13000	Time/Fuel	7/ 750	6/ 750	6/ 700	6/ 700	6/ 650	5/ 650	5/ 600	5/ 550	4/ 500	4/ 450	4/ 450
	Dist/TAS	24/318	23/318	22/318	21/318	20/318	19/318	18/318	16/318	15/318	13/318	12/318
12000	Time/Fuel	6/ 700	6/ 700	6/ 650	5/ 650	5/ 600	5/ 600	5/ 550	5/ 500	4/ 500	4/ 450	4/ 400
	Dist/TAS	21/315	20/315	19/315	18/315	18/315	17/315	16/315	15/315	13/315	12/315	11/315
11000	Time/Fuel	6/ 650	5/ 650	5/ 600	5/ 600	5/ 600	5/ 550	5/ 550	4/ 500	4/ 450	4/ 400	3/ 400
	Dist/TAS	19/313	18/313	17/313	16/313	16/313	15/312	14/312	13/312	12/312	11/312	9/312
10000	Time/Fuel	5/ 600	5/ 600	5/ 600	5/ 550	5/ 550	4/ 500	4/ 500	4/ 450	4/ 450	3/ 400	3/ 350
	Dist/TAS	16/310	16/310	15/310	14/310	14/310	13/310	12/310	11/310	10/310	9/310	8/310
8000	Time/Fuel	4/ 550	4/ 500	4/ 500	4/ 500	4/ 450	4/ 450	4/ 450	3/ 400	3/ 350	3/ 350	3/ 300
	Dist/TAS	12/305	11/305	11/305	10/305	10/305	10/305	9/305	8/305	8/305	7/305	6/305
6000	Time/Fuel	4/ 450	4/ 450	3/ 400	3/ 400	3/ 400	3/ 400	3/ 350	3/ 350	3/ 300	2/ 300	2/ 250
	Dist/TAS	8/301	8/301	7/301	7/301	7/301	6/301	6/301	6/301	5/301	5/301	4/301
1500	Time/Fuel	2/ 250	2/ 250	2/ 250	2/ 250	2/ 250	2/ 250	2/ 250	2/ 200	2/ 200	2/ 200	1/ 150

Fuel Adjustment for high elevation airports	Airport Elevation	2000	4000	6000	8000	10000	12000
Effect on time and distance is negligible	Fuel Adjustment	-50	-100	-200	-250	-300	-400

Figure 4.5.1   **EN-ROUTE CLIMB**   280/.74M

ISA +16°C TO + 25°C

Press. Alt. Ft.	Units Min/Kg. NAM/Kts	BRAKE RELEASE WEIGHT KG.										
		68000	66000	64000	62000	60000	58000	56000	52000	48000	44000	40000
37000	Time/Fuel					37/ 2550	31/ 2150	27/ 1950	22/ 1650	19/ 1450	17/ 1300	15/ 1150
	Dist/TAS					246/417	198/413	172/410	140/407	118/405	101/403	88/402
36000	Time/Fuel				35/ 2450	30/ 2200	27/ 2000	24/ 1850	21/ 1600	18/ 1400	16/ 1250	14/ 1100
	Dist/TAS				227/414	192/411	170/408	153/406	128/404	110/402	95/400	82/399
35000	Time/Fuel		42/ 2950	34/ 2500	30/ 2200	27/ 2050	25/ 1900	23/ 1750	20/ 1550	17/ 1350	15/ 1200	13/ 1050
	Dist/TAS		281/418	220/412	190/409	169/406	153/405	140/403	119/401	103/399	90/398	78/397
34000	Time/Fuel	40/ 2850	34/ 2500	30/ 2250	27/ 2100	25/ 1950	23/ 1800	21/ 1700	19/ 1500	16/ 1300	14/ 1150	13/ 1050
	Dist/TAS	260/414	215/409	188/406	169/404	153/403	141/401	130/400	112/398	97/397	85/396	74/395
33000	Time/Fuel	33/ 2500	30/ 2300	27/ 2100	25/ 1950	23/ 1850	21/ 1700	20/ 1600	18/ 1450	16/ 1300	14/ 1150	12/ 1000
	Dist/TAS	210/407	186/404	168/402	153/400	141/399	130/398	121/397	105/395	92/394	80/393	70/392
32000	Time/Fuel	30/ 2350	27/ 2150	25/ 2000	23/ 1900	22/ 1750	20/ 1650	19/ 1550	17/ 1400	15/ 1250	13/ 1100	12/ 1000
	Dist/TAS	185/401	167/399	153/398	141/396	130/395	121/394	113/394	98/392	86/391	76/390	67/389
31000	Time/Fuel	27/ 2200	25/ 2050	23/ 1900	22/ 1800	20/ 1700	19/ 1600	18/ 1500	16/ 1350	14/ 1200	13/ 1100	11/ 950
	Dist/TAS	166/396	152/395	141/394	130/393	121/392	113/391	106/390	93/389	82/388	72/387	63/387
30000	Time/Fuel	25/ 2100	24/ 1950	22/ 1850	21/ 1750	19/1650	18/1550	17/ 1450	15/ 1300	14/ 1150	12/ 1050	11/ 950
	Dist/TAS	152/392	140/391	130/389	121/389	113/388	106/387	99/387	87/385	77/385	68/384	60/383
29000	Time/Fuel	23/ 1950	22/ 1850	20 /1750	19/ 1650	18/ 1550	17/ 1450	16/ 1400	14/ 1250	13/ 1100	12/ 1000	10/ 900
	Dist/TAS	136/386	126/385	118/383	110/383	103/383	97/382	91/382	80/381	71/380	63/379	56/379
28000	Time/Fuel	21/ 1850	20/ 1750	19/ 1650	18/ 1550	17/ 1500	16/ 1400	15/ 1350	14/ 1200	12/ 1100	11/ 950	10/ 850
	Dist/TAS	123/380	114/379	107/379	100/378	94/378	89/377	83/377	74/376	66/375	58/375	52/375
27000	Time/Fuel	20/ 1750	19/ 1650	18/ 1550	17/ 1500	16/ 1400	15/ 1350	14/ 1250	13/ 1150	11/ 1050	10/ 950	9/ 850
	Dist/TAS	111/375	104/374	98/374	92/373	86/373	81/372	77/372	68/371	61/371	54/371	48/370
26000	Time/Fuel	18/ 1650	17/ 1550	16/ 1500	16/ 1400	15/ 1350	14/ 1300	13/ 1200	12/ 1100	11/ 1000	10/ 900	9/ 800
	Dist/TAS	101/370	95/370	89/369	84/369	79/368	75/368	70/368	63/367	56/367	50/366	44/366
25000	Time/Fuel	17/ 1550	16/ 1500	15/ 1400	15/ 1350	14/ 1300	13/ 1200	13/ 1150	11/ 1050	10/ 950	9/ 850	8/ 750
	Dist/TAS	92/365	86/365	81/365	77/364	73/364	69/364	65/363	58/363	52/363	46/362	41/362
24000	Time/Fuel	16/ 1500	15/ 1400	14/ 1350	14/ 1350	13/ 1300	12/ 1200	12/ 1150	11/ 1000	10/ 900	9/ 850	8/ 750
	Dist/TAS	84/361	79/361	75/360	70/360	67/360	63/360	60/359	53/359	48/359	43/358	38/358
23000	Time/Fuel	15/ 1400	14/ 1350	13/ 1300	13/ 1250	12/ 1150	12/ 1100	11/ 1050	10/ 950	9/ 900	8/ 800	7/ 700
	Dist/TAS	77/357	72/357	68/356	65/356	61/356	58/356	55/356	49/355	44/355	39/355	35/355
22000	Time/Fuel	14/ 1350	13/ 1300	13/ 1250	12/ 1150	11/ 1100	11/ 1050	10/ 1000	9/ 900	9/ 850	8/ 750	7/ 700
	Dist/TAS	70/353	66/353	63/352	59/352	56/352	53/352	50/352	45/351	41/351	36/351	32/351
21000	Time/Fuel	13/ 1300	12/ 1200	12/ 1150	11/ 1100	11/ 1050	10/ 1000	10/ 950	9/ 900	8/ 800	7/ 750	7/ 650
	Dist/TAS	64/349	60/349	57/349	54/349	51/348	49/348	46/348	42/348	37/348	34/348	30/347
20000	Time/Fuel	12/ 1200	12/ 1150	11/ 1100	11/ 1050	10/ 1000	10/ 950	9/ 950	8/ 850	8/ 750	7/ 700	6/ 650
	Dist/TAS	58/345	55/345	52/345	50/345	47/345	45/345	43/345	38/344	34/344	31/344	28/344
19000	Time/Fuel	11/ 1150	11/ 1100	10/ 1050	10/ 1000	9/ 950	9/ 900	9/ 900	8/ 800	7/ 750	7/ 650	6/ 600
	Dist/TAS	53/342	50/342	48/342	45/342	43/342	41/341	39/341	35/341	32/341	28/341	25/341
18000	Time/Fuel	11/ 1100	10/ 1050	10/ 1000	9/ 950	9/ 900	9/ 900	8/ 850	7/ 750	7/ 700	6/ 650	6/ 600
	Dist/TAS	48/339	46/339	44/338	42/338	39/338	38/338	36/338	32/338	29/338	26/338	23/338
17000	Time/Fuel	10/ 1050	10/ 1000	9/ 950	9/ 900	8/ 850	8/ 850	8/ 800	7/ 750	6/ 650	6/ 600	5/ 550
	Dist/TAS	44/335	42/335	40/335	38/335	36/335	34/335	33/335	29/335	27/335	24/335	21/335
16000	Time/Fuel	9/ 1000	9/ 950	9/ 900	8/ 850	8/ 850	7/ 800	7/ 750	7/ 700	6/ 650	5/ 600	5/ 550
	Dist/TAS	40/332	38/332	36/332	34/332	33/332	31/332	30/332	27/332	24/332	22/332	19/332
15000	Time/Fuel	9/ 950	8/ 900	8/ 850	8/ 800	7/ 800	7/ 750	7/ 700	6/ 650	6/ 600	5/ 550	5/ 500
	Dist/TAS	36/329	34/329	33/329	31/329	30/329	28/329	27/329	24/329	22/329	20/329	18/329
14000	Time/Fuel	8/ 850	8/ 850	7/ 800	7/ 750	7/ 750	7/ 700	6/ 700	6/ 650	5/ 550	5/ 500	4/ 500
	Dist/TAS	32/326	31/326	29/326	28/326	27/326	25/326	24/326	22/326	20/326	18/326	16/326
13000	Time/Fuel	7/ 800	7/ 800	7/ 750	7/ 750	6/ 700	6/ 650	6/ 650	5/ 600	5/ 550	5/ 500	4/ 450
	Dist/TAS	29/323	28/323	26/323	25/323	24/323	23/323	22/323	20/323	18/323	16/323	14/323
12000	Time/Fuel	7/ 750	7/ 750	6/ 700	6/ 700	6/ 650	6/ 650	5/ 600	5/ 550	5/ 500	4/ 450	4/ 450
	Dist/TAS	26/321	25/321	23/321	22/321	21/321	20/321	19/321	18/321	16/320	14/320	13/320
11000	Time/Fuel	6/ 700	6/ 700	6/ 650	6/ 650	5/ 600	5/ 600	5/ 550	5/ 500	4/ 500	4/ 450	4/ 400
	Dist/TAS	23/318	22/318	21/318	20/318	19/318	18/318	17/318	16/318	14/318	13/318	11/318
10000	Time/Fuel	6/ 650	6/ 650	5/ 600	5/ 600	5/ 550	5/ 550	5/ 550	4/ 500	4/ 450	4/ 400	3/ 400
	Dist/TAS	20/315	19/315	18/315	17/315	16/315	16/315	15/315	14/315	12/315	11/315	10/315
8000	Time/Fuel	5/ 550	5/ 550	5/ 500	5/ 500	4/ 450	4/ 450	4/ 450	4/ 450	3/ 400	3/ 350	3/ 350
	Dist/TAS	14/310	14/310	13/310	13/310	12/310	11/310	11/310	10/310	9/310	8/310	7/310
6000	Time/Fuel	4/ 450	4/ 450	4/ 450	4/ 450	3/ 400	3/ 400	3/ 400	3/ 350	3/ 350	3/ 300	2/ 300
	Dist/TAS	10/306	9/306	9/306	8/306	8/306	8/306	8/306	7/306	6/306	5/306	5/306
1500	Time/Fuel	2/ 250	2/ 250	2/ 250	2/ 250	2/ 250	2/ 250	2/ 250	2/200	2/200	2/200	1/150

Fuel Adjustment for high elevation airports	Airport Elevation	2000	4000	6000	8000	10000	12000
Effect on time and distance is negligible	Fuel Adjustment	-50	-150	-200	-300	-350	-400

5.2 Wind Range Correction (Fig. 4.5.2)

This graph is used for conversion of nautical ground miles to nautical air miles. (This is intended for use in conjunction with the 'integrated range' tables).
Enter graph with average TAS. Correct for wind component.
Move to ground distance at the right then vertically down to read corresponding air distance.
For longer distances than shown on the graph apply a factor of 10 to the tabulated values

Figure 4.5.2 WIND RANGE CORRECTION GRAPH

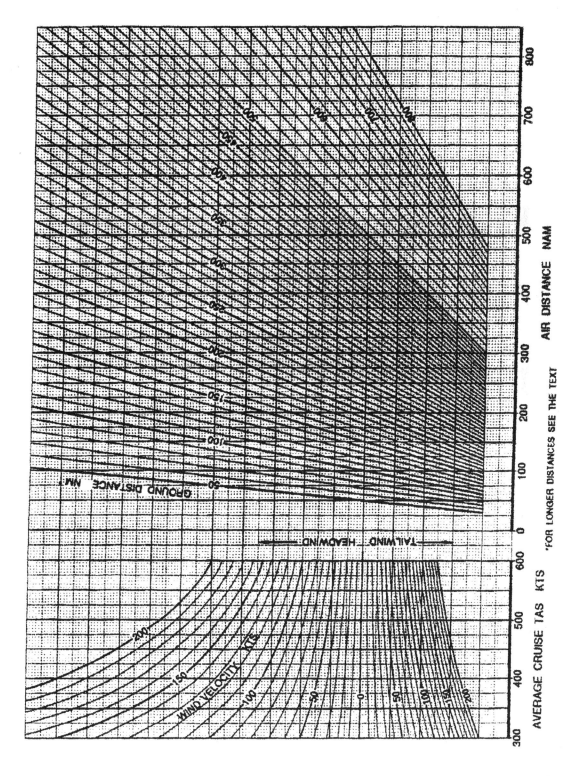

5.3   Integrated Range

This section allows for detailed flight planning for the cruise.

Tables are given as follows:

Figures 4.5.3.1	Long Range Cruise FL 270 to FL 370	(pages 25- 35)
Figures 4.5.3.2	0.74 Mach Cruise FL 210 to FL 370	(pages 36- 52)
Figures 4.5.3.3	0.78 Mach Cruise FL 290,300,310, 330,350,370.	(pages 53- 58)
Figures 4.5.3.4	Low level (300 KIAS) Cruise FL 140 to FL 210.	(pages 59 -66)

The tables in this section are identical in use.
The tables are based on a "differences,, principle, the difference between two gross weights representing a weight of fuel used. The corresponding difference in tabulated distance represents the still air (zero wind) distance available for that weight of fuel used.

Example:

An aircraft commences a cruise at 0.74 Mach at FL 330 where the temperature is ISA.
The follbwing gives the relevant data for the first two sectors:

	NGM	W.C (Kts.)
A - B	240	-20
B - C	370	-30

Weight at beginning of first sector is 53500 Kg.

Method,
Using the TAS given for cruise (430 Kt.)

Obtain NAM for each leg.   A - B   252
                                 B - C   398

Select table for 0.74 Mach	P. Alt.33000 ft.
Enter with gross weight	= 53500 Kg.
Extract value for cruise distance	= 3796
Subtract first leg NAM	= 252
Obtain new cruise distance value	= 3544
Enter table to find corresponding weight.	= 52100 Kg.
Subtract from start weight to obtain leg fuel used	= 1400 Kg.

Repeat process	
Subtract second leg NAM	398
Obtain new cruise distance value	3146
Enter table to find weight	50000 Kg.
Subtract from start weight to obtain leg fuel used	2100 Kg.

N.B. These tables are based on ISA conditions
If conditions are non-standard use corrections as per instructions given on tables.

Figure 4.5.3.1   Long Range Cruise

All Engines   Maximum Cruise Thrust Limits   A/C Auto

PRESSURE ALTITUDE   27000 Ft.

GROSS WT. KG.	TAS	0	100	200	300	400	500	600	700	800	900
		CRUISE DISTANCE NAUTICAL AIR MILES									
35000	371	0	20	40	61	81	102	122	142	163	183
36000	375	204	224	244	264	284	305	325	345	365	385
37000	379	405	425	445	465	485	505	525	545	565	585
38000	383	605	625	644	664	684	704	723	743	763	782
39000	387	802	822	841	861	880	900	919	939	958	978
40000	391	997	1016	1036	1055	1074	1093	1113	1132	1151	1171
41000	394	1190	1209	1228	1247	1266	1285	1304	1323	1342	1361
42000	398	1381	1399	1418	1437	1456	1475	1494	1513	1531	1550
43000	401	1569	1588	1606	1625	1644	1662	1681	1700	1718	1737
44000	405	1756	1774	1793	1811	1829	1848	1866	1885	1903	1922
45000	408	1940	1958	1977	1995	2013	2031	2050	2068	2086	2104
46000	411	2123	2141	2159	2177	2195	2213	2231	2249	2267	2285
47000	414	2303	2321	2339	2357	2375	2393	2411	2428	2446	2464
48000	417	2482	2500	2517	2535	2553	2570	2588	2606	2624	2641
49000	420	2659	2676	2694	2711	2729	2746	2764	2781	2799	2816
50000	423	2834	2851	2869	2886	2903	2921	2938	2955	2972	2990
51000	426	3007	3024	3041	3059	3076	3093	3110	3127	3144	3161
52000	428	3179	3196	3213	3229	3246	3263	3280	3297	3314	3331
53000	431	3348	3365	3382	3399	3416	3432	3449	3466	3483	3500
54000	433	3516	3533	3550	3566	3583	3600	3616	3633	3650	3666
55000	435	3683	3699	3716	3732	3749	3765	3782	3798	3815	3831
56000	437	3848	3864	3880	3897	3913	3929	3946	3962	3978	3995
57000	438	4011	4027	4043	4059	4075	4092	4108	4124	4140	4156
58000	440	4172	4188	4204	4220	4236	4252	4268	4284	4300	4316
59000	441	4332	4348	4364	4380	4396	4411	4427	4443	4459	4475
60000	442	4491	4506	4522	4538	4553	4569	4585	4600	4616	4632
61000	443	4647	4663	4678	4694	4709	4725	4740	4756	4771	4787
62000	444	4802	4818	4833	4849	4864	4879	4895	4910	4925	4941
63000	444	4956	4971	4986	5002	5017	5032	5047	5063	5078	5093
64000	444	5108	5123	5138	5153	5168	5183	5199	5214	5229	5244
65000	444	5259	5274	5289	5304	5318	5333	5348	5363	5378	5393
66000	444	5408	5423	5437	5452	5467	5482	5497	5511	5526	5541
67000	444	5556	5570	5585	5599	5614	5629	5643	5658	5673	5687

NOTE – OPTIMUM WEIGHT FOR PRESSURE ALTITUDE EXCEEDS STRUCTURAL LIMIT
THRUST LIMITED WEIGHT FOR ISA + 10 AND COLDER EXCEEDS STRUCTURAL LIMIT
THRUST LIMITED WEIGHT FOR ISA + 15   EXCEEDS STRUCTURAL LIMIT
THRUST LIMITED WEIGHT FOR ISA + 20   EXCEEDS STRUCTURAL LIMIT
ADJUSTMENTS FOR OPERATION AT NON-STANDARD TEMPERATURES---
INCREASE FUEL REQUIRED BY   0.5 PERCENT PER 10 DEGREES C ABOVE ISA
DECREASE FUEL REQUIRED BY   0.5 PERCENT PER 10 DEGREES C BELOW ISA
INCREASE TAS BY 1 KNOT PER DEGREE C ABOVE ISA
DECREASE TAS BY 1 KNOT PER DEGREE C BELOW ISA

Figure 4.5.3.1   Long Range Cruise

All Engines   Maximum Cruise Thrust Limits   A/C Auto

PRESSURE ALTITUDE   28000 Ft.

GROSS WT. KG.	TAS	0	100	200	300	400	500	600	700	800	900
		CRUISE DISTANCE NAUTICAL AIR MILES									
35000	376	0	20	41	62	83	104	125	145	166	187
36000	380	208	229	249	270	290	311	332	352	373	393
37000	384	414	434	455	475	495	516	536	557	577	597
38000	388	618	638	658	678	698	718	738	759	779	799
39000	392	819	839	859	879	898	918	938	958	978	998
40000	396	1018	1037	1057	1077	1096	1116	1136	1155	1175	1195
41000	399	1214	1234	1253	1273	1292	1312	1331	1350	1370	1389
42000	403	1409	1428	1447	1466	1486	1505	1524	1543	1563	1582
43000	406	1601	1620	1639	1658	1677	1696	1715	1734	1753	1772
44000	409	1791	1810	1829	1848	1866	1885	1904	1923	1942	1960
45000	413	1979	1998	2016	2035	2054	2072	2091	2109	2128	2147
46000	416	2165	2184	2202	2220	2239	2257	2275	2294	2312	2331
47000	419	2349	2367	2385	2404	2422	2440	2458	2476	2495	2513
48000	422	2531	2549	2567	2585	2603	2621	2639	2657	2675	2693
49000	425	2711	2729	2747	2764	2782	2800	2818	2836	2853	2871
50000	427	2889	2907	2924	2942	2960	2977	2995	3013	3030	3048
51000	429	3065	3083	3100	3118	3135	3153	3170	3188	3205	3222
52000	432	3240	3257	3274	3292	3309	3326	3344	3361	3378	3395
53000	434	3413	3430	3447	3464	3481	3498	3515	3532	3549	3567
54000	436	3584	3601	3617	3634	3651	3668	3685	3702	3719	3736
55000	437	3753	3770	3786	3803	3820	3837	3853	3870	3887	3904
56000	439	3920	3937	3953	3970	3987	4003	4020	4036	4053	4069
57000	440	4086	4102	4119	4135	4152	4168	4184	4201	4217	4234
58000	441	4250	4266	4282	4299	4315	4331	4347	4364	4380	4396
59000	442	4412	4428	4444	4460	4476	4492	4509	4525	4541	4557
60000	442	4573	4589	4605	4620	4636	4652	4668	4684	4700	4716
61000	442	4732	4747	4763	4779	4795	4810	4826	4842	4858	4873
62000	442	4889	4905	4920	4936	4951	4967	4983	4998	5014	5029
63000	443	5045	5060	5076	5091	5106	5122	5137	5153	5168	5184
64000	443	5199	5214	5229	5245	5260	5275	5290	5306	5321	5336
65000	443	5351	5367	5382	5397	5412	5427	5442	5457	5472	5487
66000	443	5502	5517	5532	5547	5562	5577	5592	5607	5622	5637
67000	443	5652	5666	5681	5696	5711	5725	5740	5755	5770	5784

NOTE – OPTIMUM WEIGHT FOR PRESSURE ALTITUDE EXCEEDS STRUCTURAL LIMIT
THRUST LIMITED WEIGHT FOR ISA + 10 AND COLDER EXCEEDS STRUCTURAL LIMIT
THRUST LIMITED WEIGHT FOR ISA + 15   EXCEEDS STRUCTURAL LIMIT
THRUST LIMITED WEIGHT FOR ISA + 20   EXCEEDS STRUCTURAL LIMIT
ADJUSTMENTS FOR OPERATION AT NON-STANDARD TEMPERATURES---
INCREASE FUEL REQUIRED BY   0.6 PERCENT PER 10 DEGREES C ABOVE ISA
DECREASE FUEL REQUIRED BY   0.6 PERCENT PER 10 DEGREES C BELOW ISA
INCREASE TAS BY 1 KNOT PER DEGREE C ABOVE ISA
DECREASE TAS BY 1 KNOT PER DEGREE C BELOW ISA

Figure 4.5.3.1   Long Range Cruise

All Engines   Maximum Cruise Thrust Limits   A/C Auto

PRESSURE ALTITUDE   29000 Ft.

GROSS WT. KG.	TAS	0	100	200	300	400	500	600	700	800	900
		CRUISE DISTANCE NAUTICAL AIR MILES									
35000	381	0	21	42	63	85	106	127	148	170	191
36000	385	212	233	254	275	296	317	338	359	380	401
37000	389	422	443	464	485	506	526	547	568	589	609
38000	393	630	651	671	692	712	733	753	774	794	815
39000	397	835	856	876	896	916	937	957	977	998	1018
40000	400	1038	1058	1078	1098	1118	1138	1158	1178	1198	1218
41000	404	1238	1258	1278	1298	1318	1338	1357	1377	1397	1417
42000	407	1437	1456	1476	1495	1515	1535	1554	1574	1593	1613
43000	411	1632	1652	1671	1691	1710	1729	1749	1768	1787	1807
44000	414	1826	1845	1864	1884	1903	1922	1941	1960	1979	1998
45000	417	2018	2037	2055	2074	2093	2112	2131	2150	2169	2188
46000	420	2207	2226	2244	2263	2282	2301	2319	2338	2357	2375
47000	423	2394	2413	2431	2450	2468	2487	2505	2524	2542	2561
48000	426	2579	2598	2616	2634	2653	2671	2689	2708	2726	2744
49000	428	2763	2781	2799	2817	2835	2853	2871	2890	2908	2926
50000	431	2944	2962	2980	2998	3016	3034	3052	3070	3088	3105
51000	433	3123	3141	3159	3177	3194	3212	3230	3248	3265	3283
52000	435	3301	3319	3336	3354	3371	3389	3406	3424	3441	3459
53000	436	3477	3494	3511	3529	3546	3563	3581	3598	3616	3633
54000	438	3650	3667	3685	3702	3719	3736	3753	3771	3788	3805
55000	439	3822	3839	3856	3873	3890	3907	3924	3941	3958	3975
56000	440	3992	4009	4026	4042	4059	4076	4093	4110	4126	4143
57000	440	4160	4177	4193	4210	4227	4243	4260	4277	4293	4310
58000	440	4326	4343	4359	4376	4392	4409	4425	4442	4458	4475
59000	441	4491	4507	4524	4540	4556	4572	4589	4605	4621	4638
60000	441	4654	4670	4686	4702	4718	4734	4751	4767	4783	4799
61000	441	4815	4831	4847	4863	4879	4895	4911	4927	4942	4958
62000	441	4974	4990	5006	5022	5037	5053	5069	5085	5100	5116
63000	441	5132	5147	5163	5179	5194	5210	5225	5241	5257	5272
64000	441	5288	5303	5319	5334	5349	5365	5380	5396	5411	5426
65000	411	5442	5457	5472	5487	5503	5518	5533	5548	5564	5579
66000	411	5594	5609	5624	5639	5654	5669	5684	5699	5714	5729
67000	411	5745	5759	5774	5789	5804	5819	5834	5849	5863	5878

NOTE – OPTIMUM WEIGHT FOR PRESSURE ALTITUDE EXCEEDS STRUCTURAL LIMIT
THRUST LIMITED WEIGHT FOR ISA + 10 AND COLDER EXCEEDS STRUCTURAL LIMIT
THRUST LIMITED WEIGHT FOR ISA + 15   EXCEEDS STRUCTURAL LIMIT
THRUST LIMITED WEIGHT FOR ISA + 20   EXCEEDS STRUCTURAL LIMIT
ADJUSTMENTS FOR OPERATION AT NON-STANDARD TEMPERATURES---
INCREASE FUEL REQUIRED BY   0.6 PERCENT PER 10 DEGREES C ABOVE ISA
DECREASE FUEL REQUIRED BY   0.6 PERCENT PER 10 DEGREES C BELOW ISA
INCREASE TAS BY 1 KNOT PER DEGREE C ABOVE ISA
DECREASE TAS BY 1 KNOT PER DEGREE C BELOW ISA

Figure 4.5.3.1   Long Range Cruise

All Engines   Maximum Cruise Thrust Limits   A/C Auto

PRESSURE ALTITUDE   30000 Ft.

GROSS WT. KG.	TAS	0	100	200	300	400	500	600	700	800	900
		CRUISE DISTANCE NAUTICAL AIR MILES									
35000	386	0	21	43	65	86	108	130	151	173	195
36000	390	216	238	259	281	302	324	345	366	388	409
37000	394	431	452	473	494	515	536	558	579	600	621
38000	398	642	663	684	705	726	747	768	789	810	831
39000	402	851	872	893	913	934	955	975	996	1017	1037
40000	405	1058	1079	1099	1119	1140	1160	1181	1201	1222	1242
41000	409	1262	1283	1303	1323	1343	1363	1384	1404	1424	1444
42000	412	1464	1484	1504	1524	1544	1564	1584	1604	1624	1644
43000	416	1664	1684	1703	1723	1743	1762	1782	1802	1822	1841
44000	419	1861	1881	1900	1920	1939	1959	1978	1998	2017	2037
45000	422	2056	2075	2095	2114	2133	2152	2172	2191	2210	2230
46000	424	2249	2268	2287	2306	2325	2344	2363	2382	2401	2421
47000	427	2440	2458	2477	2496	2515	2534	2553	2572	2590	2609
48000	429	2628	2647	2666	2684	2703	2721	2740	2759	2777	2796
49000	431	2815	2833	2852	2870	2889	2907	2925	2944	2962	2981
50000	433	2999	3017	3036	3054	3072	3090	3109	3127	3145	3163
51000	435	3182	3200	3218	3236	3254	3272	3290	3308	3326	3344
52000	436	3362	3380	3398	3416	3433	3451	3469	3487	3505	3522
53000	437	3540	3558	3576	3593	3611	3628	3646	3664	3681	3699
54000	438	3717	3734	3751	3769	3786	3804	3821	3839	3856	3873
55000	438	3891	3908	3925	3943	3960	3977	3994	4012	4029	4046
56000	439	4063	4080	4097	4115	4132	4149	4166	4183	4200	4217
57000	439	4234	4251	4268	4285	4301	4318	4335	4352	4369	4386
58000	439	4403	4419	4436	4453	4469	4486	4503	4519	4536	4553
59000	439	4569	4586	4602	4619	4635	4652	4668	4685	4701	4718
60000	439	4734	4750	4767	4783	4799	4816	4832	4848	4865	4881
61000	439	4897	4913	4929	4945	4961	4978	4994	5010	5026	5042
62000	439	5058	5074	5090	5106	5122	5138	5154	5169	5185	5201
63000	439	5217	5233	5249	5264	5280	5296	5311	5327	5343	5359
64000	439	5374	5390	5405	5421	5436	5452	5467	5483	5498	5514
65000	439	5529	5545	5560	5575	5591	5606	5621	5637	5652	5667
66000	439	5683	5698	5713	5728	5743	5758	5773	5788	5804	5819
67000	439	5834	5849	5864	5879	5893	5908	5923	5938	5953	5968

NOTE – OPTIMUM WEIGHT FOR PRESSURE ALTITUDE IS 66600 KG
    THRUST LIMITED WEIGHT FOR ISA + 10 AND COLDER EXCEEDS STRUCTURAL LIMIT
    THRUST LIMITED WEIGHT FOR ISA + 15   EXCEEDS STRUCTURAL LIMIT
    THRUST LIMITED WEIGHT FOR ISA + 20   EXCEEDS STRUCTURAL LIMIT
    ADJUSTMENTS FOR OPERATION AT NON-STANDARD TEMPERATURES---
    INCREASE FUEL REQUIRED BY   0.6 PERCENT PER 10 DEGREES C ABOVE ISA
    DECREASE FUEL REQUIRED BY   0.6 PERCENT PER 10 DEGREES C BELOW ISA
    INCREASE TAS BY 1 KNOT PER DEGREE C ABOVE ISA
    DECREASE TAS BY 1 KNOT PER DEGREE C BELOW ISA

Figure 4.5.3.1   Long Range Cruise

All Engines   Maximum Cruise Thrust Limits   A/C Auto

PRESSURE ALTITUDE   31000 Ft.

GROSS WT. KG.	TAS	0	100	200	300	400	500	600	700	800	900
		CRUISE DISTANCE NAUTICAL AIR MILES									
35000	391	0	22	44	66	88	110	132	154	176	199
36000	395	221	242	264	286	308	330	352	374	395	417
37000	399	439	461	482	504	525	547	569	590 ·	612	633
38000	403	655	676	698	719	740	762	783	804	825	847
39000	406	868	889	910	931	952	973	995	1016	1037	1058
40000	410	1079	1100	1120	1141	1162	1183	1204	1225	1245	1266
41000	414	1287	1308	1328	1349	1369	1390	1410	1431	1452	1472
42000	417	1493	1513	1533	1554	1574	1594	1615	1635	1655	1676
43000	420	1696	1716	1736	1756	1776	1796	1816	1837	1857	1877
44000	423	1897	1917	1936	1956	1976	1996	2016	2036	2056	2076
45000	425	2095	2115	2135	2154	2174	2194	2213	2233	2252	2272
46000	428	2292	2311	2331	2350	2369	2389	2408	2428	2447	2466
47000	430	2486	2505	2524	2543	2563	2582	2601	2620	2639	2658
48000	432	2678	2697	2716	2735	2754	2772	2791	2810	2829	2848
49000	433	2867	2886	2905	2924	2942	2961	2980	2998	3017	3036
50000	435	3055	3073	3092	3110	3129	3147	3166	3184	3203	3221
51000	436	3240	3258	3276	3295	3313	3331	3350	3368	3386	3405
52000	437	3423	3441	3459	3477	3495	3513	3531	3550	3568	3586
53000	437	3604	3622	3640	3657	3675	3693	3711	3729	3747	3765
54000	437	3783	3800	3818	3836	3853	3871	3889	3906	3924	3942
55000	437	3959	3977	3994	4012	4029	4047	4064	4082	4099	4117
56000	437	4134	4152	4169	4186	4203	4221	4238	4255	4272	4290
57000	437	4307	4324	4341	4358	4375	4392	4409	4426	4443	4461
58000	437	4478	4494	4511	4528	4545	4562	4579	4596	4612	4629
59000	437	4646	4663	4679	4696	4713	4729	4746	4763	4779	4796
60000	437	4813	4829	4845	4862	4878	4895	4911	4928	4944	4960
61000	437	4977	4993	5009	5025	5042	5058	5074	5090	5107	5123
62000	437	5139	5155	5171	5187	5203	5219	5235	5251	5267	5283
63000	437	5299	5315	5331	5346	5362	5378	5394	5410	5425	5441 ·
64000	437	5457	5473	5488	5504	5519	5535	5550	5566	5582	5597
65000	437	5613	5628	5643	5659	5674	5690	5705	5720	5736	5751
66000	437	5766	5782	5797	5812	5827	5842	5857	5873	5888	5903
67000	437	5981	5933	5948	5963	5978	5993	6008	6023	6037	6052

NOTE – OPTIMUM WEIGHT FOR PRESSURE ALTITUDE IS 63500 KG
THRUST LIMITED WEIGHT FOR ISA + 10 AND COLDER EXCEEDS STRUCTURAL LIMIT
THRUST LIMITED WEIGHT FOR ISA + 15   EXCEEDS STRUCTURAL LIMIT
THRUST LIMITED WEIGHT FOR ISA + 20   EXCEEDS STRUCTURAL LIMIT
ADJUSTMENTS FOR OPERATION AT NON-STANDARD TEMPERATURES---
INCREASE FUEL REQUIRED BY   0.6 PERCENT PER 10 DEGREES C ABOVE ISA
DECREASE FUEL REQUIRED BY   0.6 PERCENT PER 10 DEGREES C BELOW ISA
INCREASE TAS BY 1 KNOT PER DEGREE C ABOVE ISA
DECREASE TAS BY 1 KNOT PER DEGREE C BELOW ISA

Figure 4.5.3.1   Long Range Cruise

All Engines   Maximum Cruise Thrust Limits   A/C Auto

PRESSURE ALTITUDE   32000 Ft.

GROSS WT. KG.	TAS	0	100	200	300	400	500	600	700	800	900
					CRUISE DISTANCE NAUTICAL AIR MILES						
35000	396	0	22	45	67	90	112	135	157	180	203
36000	400	225	247	270	292	314	337	359	381	403	426
37000	404	448	470	492	514	536	558	580	602	624	646
38000	407	668	690	712	733	755	777	798	820	842	864
39000	411	885	907	928	950	971	993	1014	1036	1057	1079
40000	415	1100	1121	1142	1164	1185	1206	1227	1249	1270	1291
41000	418	1312	1333	1354	1375	1396	1417	1438	1459	1480	1501
42000	421	1522	1542	1563	1584	1604	1625	1646	1667	1687	1708
43000	423	1729	1749	1770	1790	1810	1831	1851	1872	1892	1913
44000	426	1933	1953	1974	1994	2014	2034	2054	2075	2095	2115
45000	428	2135	2155	2175	2195	2215	2235	2255	2275	2295	2315
46000	430	2335	2355	2375	2394	2414	2434	2453	2473	2493	2513
47000	432	2532	2552	2571	2591	2610	2630	2649	2669	2688	2708
48000	433	2727	2746	2766	2785	2804	2823	2843	2862	2881	2900
49000	434	2920	2939	2958	2977	2996	3015	3034	3053	3072	3091
50000	435	3110	3129	3147	3166	3185	3204	3223	3241	3260	3279
51000	435	3298	3316	3335	3353	3372	3391	3409	3428	3446	3465
52000	435	3483	3502	3520	3539	3557	3575	3594	3612	3630	3649
53000	435	3667	3685	3703	3721	3739	3758	3776	3794	3812	3830
54000	435	3848	3866	3884	3902	3920	3938	3956	3973	3991	4009
55000	435	4027	4045	4063	4080	4098	4116	4133	4151	4169	4186
56000	435	4204	4221	4239	4256	4274	4291	4309	4326	4343	4361
57000	435	4378	4396	4413	4430	4447	4464	4482	4499	4516	4533
58000	435	4550	4567	4584	4601	4618	4635	4652	4669	4686	4703
59000	435	4720	4737	4754	4771	4787	4804	4821	4838	4854	4871
60000	435	4888	4904	4921	4937	4954	4970	4987	5003	5020	5036
61000	435	5053	5069	5086	5102	5118	5134	5151	5167	5183	5200
62000	435	5216	5232	5248	5264	5280	5296	5312	5328	5344	5360
63000	435	5377	5392	5408	5424	5440	5456	5471	5487	5503	5519
64000	435	5535	5550	5566	5582	5597	5613	5628	5644	5659	5675
65000	435	5691	5706	5721	5737	5752	5767	5783	5798	5813	5829
66000	435	5844	5859	5874	5889	5904	5920	5935	5950	5965	5980
67000	435	5995	6010	6025	6040	6054	6069	6084	6099	6114	6129

NOTE – OPTIMUM WEIGHT FOR PRESSURE ALTITUDE IS 60700 KG
    THRUST LIMITED WEIGHT FOR ISA + 10 AND COLDER EXCEEDS STRUCTURAL LIMIT
    THRUST LIMITED WEIGHT FOR ISA + 15   EXCEEDS STRUCTURAL LIMIT
    THRUST LIMITED WEIGHT FOR ISA + 20   EXCEEDS STRUCTURAL LIMIT
    ADJUSTMENTS FOR OPERATION AT NON-STANDARD TEMPERATURES---
    INCREASE FUEL REQUIRED BY   0.6 PERCENT PER 10 DEGREES C ABOVE ISA
    DECREASE FUEL REQUIRED BY   0.6 PERCENT PER 10 DEGREES C BELOW ISA
    INCREASE TAS BY 1 KNOT PER DEGREE C ABOVE ISA
    DECREASE TAS BY 1 KNOT PER DEGREE C BELOW ISA

Figure 4.5.3.1 Long Range Cruise

All Engines   Maximum Cruise Thrust Limits   A/C Auto

PRESSURE ALTITUDE   33000 Ft.

GROSS WT. KG.	TAS	0	100	200	300	400	500	600	700	800	900
		CRUISE DISTANCE NAUTICAL AIR MILES									
35000	400	0	23	46	69	92	115	138	161	184	207
36000	405	230	252	275	298	320	343	366	389	411	434
37000	408	457	479	502	524	547	569	591	614	636	659
38000	412	681	703	725	747	770	792	814	836	858	880
39000	415	902	924	946	968	990	1012	1034	1055	1077	1099
40000	419	1121	1143	1164	1186	1207	1229	1251	1272	1294	1315
41000	421	1337	1358	1380	1401	1422	1444	1465	1486	1508	1529
42000	424	1550	1571	1593	1614	1635	1656	1677	1698	1719	1740
43000	426	1761	1782	1803	1823	1844	1865	1886	1907	1928	1948
44000	428	1969	1990	2010	2031	2051	2072	2092	2113	2134	2154
45000	430	2175	2195	2215	2235	2256	2276	2296	2317	2337	2357
46000	432	2377	2397	2417	2437	2458	2478	2498	2518	2538	2558
47000	433	2578	2597	2617	2637	2657	2677	2696	2716	2736	2756
48000	433	2775	2795	2814	2834	2854	2873	2893	2912	2932	2951
49000	433	2971	2990	3009	3029	3048	3067	3087	3106	3125	3144
50000	433	3164	3183	3202	3221	3240	3259	3278	3297	3316	3335
51000	433	3354	3373	3392	3411	3429	3448	3467	3486	3505	3523
52000	433	3542	3561	3579	3598	3617	3635	3654	3672	3691	3709
53000	433	3728	3746	3765	3783	3801	3819	3838	3856	3874	3893
54000	433	3911	3929	3947	3965	3983	4001	4019	4038	4056	4074
55000	433	4092	4110	4127	4145	4163	4181	4199	4216	4234	4252
56000	433	4270	4287	4305	4323	4340	4358	4375	4393	4410	4428
57000	433	4445	4463	4480	4497	4515	4532	4549	4567	4584	4601
58000	433	4619	4636	4653	4670	4687	4704	4721	4738	4755	4772
59000	433	4789	4806	4823	4840	4856	4873	4890	4907	4924	4940
60000	433	4957	4974	4990	5007	5024	5040	5057	5073	5090	5106
61000	433	5123	5139	5155	5172	5188	5204	5221	5237	5253	5270
62000	433	5286	5302	5318	5334	5350	5366	5382	5398	5414	5430
63000	433	5446	5462	5478	5493	5509	5525	5541	5557	5572	5588
64000	433	5604	5619	5635	5650	5666	5681	5697	5712	5728	5743
65000	433	5759	5774	5789	5804	5820	5835	5850	5865	5880	5896
66000	433	5911	5926	5941	5956	5970	5985	6000	6015	6030	6045
67000	433	6060	6075	6089	6104	6118	6133	6148	6162	6177	6191

NOTE – OPTIMUM WEIGHT FOR PRESSURE ALTITUDE IS 58200 KG
        THRUST LIMITED WEIGHT FOR ISA + 10 AND COLDER EXCEEDS STRUCTURAL LIMIT
        THRUST LIMITED WEIGHT FOR ISA + 15   EXCEEDS STRUCTURAL LIMIT
        THRUST LIMITED WEIGHT FOR ISA + 20   IS 66400 KG
        ADJUSTMENTS FOR OPERATION AT NON-STANDARD TEMPERATURES---
        INCREASE FUEL REQUIRED BY   0.6 PERCENT PER 10 DEGREES C ABOVE ISA
        DECREASE FUEL REQUIRED BY   0.6 PERCENT PER 10 DEGREES C BELOW ISA
        INCREASE TAS BY 1 KNOT PER DEGREE C ABOVE ISA
        DECREASE TAS BY 1 KNOT PER DEGREE C BELOW ISA

Figure 4.5.3.1   Long Range Cruise

All Engines   Maximum Cruise Thrust Limits   A/C Auto

PRESSURE ALTITUDE   34000 Ft.

GROSS WT. KG.	TAS	0	100	200	300	400	500	600	700	800	900
		CRUISE DISTANCE NAUTICAL AIR MILES									
35000	405	0	23	46	70	93	117	140	164	187	210
36000	409	234	257	280	303	326	350	373	396	419	442
37000	413	465	488	511	534	557	579	602	625	648	671
38000	416	694	716	739	761	784	806	829	851	874	896
39000	419	919	941	963	986	1008	1030	1053	1075	1097	1119
40000	422	1142	1164	1186	1207	1229	1251	1273	1295	1317	1339
41000	424	1361	1383	1405	1426	1448	1470	1492	1513	1535	1557
42000	427	1578	1600	1621	1643	1664	1685	1707	1728	1750	1771
43000	428	1792	1814	1835	1856	1877	1898	1919	1940	1961	1983
44000	430	2004	2025	2045	2066	2087	2108	2129	2150	2171	2191
45000	431	2212	2233	2253	2274	2295	2315	2336	2356	2377	2398
46000	431	2418	2438	2459	2479	2499	2520	2540	2560	2581	2601
47000	431	2621	2641	2661	2682	2702	2722	2742	2762	2782	2802
48000	431	2822	2842	2862	2881	2901	2921	2941	2960	2980	3000
49000	431	3020	3039	3059	3078	3098	3118	3137	3157	3176	3196
50000	431	3215	3234	3254	3273	3292	3311	3331	3350	3369	3389
51000	431	3408	3427	3446	3465	3484	3503	3522	3541	3560	3579
52000	431	3598	3616	3635	3654	3673	3691	3710	3729	3747	3766
53000	431	3785	3803	3822	3840	3859	3877	3896	3914	3932	3951
54000	431	3969	3987	4006	4024	4042	4060	4078	4096	4115	4133
55000	431	4151	4169	4187	4205	4223	4240	4258	4276	4294	4312
56000	431	4330	4348	4365	4383	4400	4418	4436	4453	4471	4489
57000	431	4506	4524	4541	4558	4576	4593	4610	4628	4645	4662
58000	431	4680	4697	4714	4731	4748	4765	4782	4799	4816	4833
59000	431	4851	4867	4884	4901	4918	4934	4951	4968	4985	5002
60000	431	5018	5035	5051	5068	5084	5101	5117	5134	5150	5167
61000	431	5183	5200	5216	5232	5248	5264	5281	5297	5313	5329
62000	431	5345	5361	5377	5393	5409	5425	5441	5457	5472	5488
63000	431	5504	5520	5535	5551	5566	5582	5598	5613	5629	5644
64000	431	5660	5675	5690	5706	5721	5736	5751	5766	5782	5797
65000	430	5812	5827	5842	5857	5872	5886	5901	5916	5931	5946
66000	430	5961	5975	5990	6004	6019	6033	6048	6062	6077	6091
67000	430	6106	6120	6134	6148	6162	6176	6190	6204	6219	6233

NOTE – OPTIMUM WEIGHT FOR PRESSURE ALTITUDE IS 55500 KG
 THRUST LIMITED WEIGHT FOR ISA + 10 AND COLDER IS 67100 KG
 THRUST LIMITED WEIGHT FOR ISA + 15   IS 65700 KG
 THRUST LIMITED WEIGHT FOR ISA + 20   IS 64000 KG
 ADJUSTMENTS FOR OPERATION AT NON-STANDARD TEMPERATURES---
 INCREASE FUEL REQUIRED BY   0.6 PERCENT PER 10 DEGREES C ABOVE ISA
 DECREASE FUEL REQUIRED BY   0.6 PERCENT PER 10 DEGREES C BELOW ISA
 INCREASE TAS BY 1 KNOT PER DEGREE C ABOVE ISA
 DECREASE TAS BY 1 KNOT PER DEGREE C BELOW ISA

Figure 4.5.3.1 Long Range Cruise

All Engines   Maximum Cruise Thrust Limits   A/C Auto

PRESSURE ALTITUDE   35000 Ft.

GROST WT. KG.	TAS	0	100	200	300	400	500	600	700	800	900
		CRUISE DISTANCE NAUTICAL AIR MILES									
35000	410	0	23	47	71	95	119	143	167	191	214
36000	414	238	262	285	309	333	356	380	403	427	450
37000	417	474	497	521	544	567	590	614	637	660	683
38000	420	707	730	753	776	798	821	844	867	890	913
39000	422	936	959	982	1004	1027	1050	1072	1095	1117	1140
40000	425	1163	1185	1207	1230	1252	1275	1297	1319	1342	1364
41000	426	1386	1408	1430	1452	1474	1496	1519	1541	1563	1585
42000	428	1607	1628	1650	1672	1694	1715	1737	1759	1781	1802
43000	429	1824	1845	1867	1888	1910	1931	1953	1974	1996	2017
44000	429	2039	2060	2081	2102	2123	2144	2165	2187	2208	2229
45000	429	2250	2271	2292	2313	2334	2355	2375	2396	2417	2438
46000	429	2459	2480	2500	2521	2541	2562	2582	2603	2624	2644
47000	429	2665	2685	2705	2726	2746	2766	2787	2807	2827	2848
48000	429	2868	2888	2908	2928	2948	2968	2988	3008	3028	3048
49000	429	3068	3088	3107	3127	3147	3166	3186	3206	3226	3245
50000	429	3265	3284	3304	3323	3343	3362	3381	3401	3420	3440
51000	429	3459	3478	3497	3516	3536	3555	3574	3593	3612	3631
52000	429	3650	3669	3688	3707	3726	3744	3763	3782	3801	3820
53000	429	3838	3857	3875	3894	3913	3931	3950	3968	3987	4005
54000	429	4024	4042	4060	4078	4097	4115	4133	4151	4170	4188
55000	430	4206	4224	4242	4260	4278	4296	4314	4331	4349	4367
56000	430	4385	4403	4420	4438	4456	4473	4491	4509	4526	4544
57000	430	4561	4579	4596	4613	4631	4648	4665	4682	4700	4717
58000	429	4734	4751	4768	4785	4802	4819	4836	4853	4870	4887
59000	429	4904	4921	4937	4954	4971	4987	5004	5021	5037	5054
60000	429	5070	5087	5103	5119	5136	5152	5168	5184	5201	5217
61000	429	5233	5249	5265	5281	5297	5313	5329	5345	5361	5377
62000	429	5393	5408	5424	5439	5455	5470	5486	5501	5517	5532
63000	428	5548	5563	5578	5593	5608	5623	5638	5654	5669	5684
64000	428	5699	5714	5728	5743	5758	5772	5787	5802	5817	5831
65000	427	5846	5860	5874	5889	5903	5917	5932	5946	5960	5974

NOTE – OPTIMUM WEIGHT FOR PRESSURE ALTITUDE IS 53000 KG
　　　　THRUST LIMITED WEIGHT FOR ISA + 10 AND COLDER IS 64500 KG
　　　　THRUST LIMITED WEIGHT FOR ISA + 15　　　　　IS 63100 KG
　　　　THRUST LIMITED WEIGHT FOR ISA + 20　　　　　IS 61600 KG
　　　　ADJUSTMENTS FOR OPERATION AT NON-STANDARD TEMPERATURES---
　　　　INCREASE FUEL REQUIRED BY　　0.6 PERCENT PER 10 DEGREES C ABOVE ISA
　　　　DECREASE FUEL REQUIRED BY　　0.6 PERCENT PER 10 DEGREES C BELOW ISA
　　　　INCREASE TAS BY 1 KNOT PER DEGREE C ABOVE ISA
　　　　DECREASE TAS BY 1 KNOT PER DEGREE C BELOW ISA

Figure 4.5.3.1 Long Range Cruise

All Engines   Maximum Cruise Thrust Limits   A/C Auto

PRESSURE ALTITUDE   36000 Ft.

GROSS WT. KG.	TAS	0	100	200	300	400	500	600	700	800	900
		CRUISE DISTANCE NAUTICAL AIR MILES									
35000	414	0	24	48	73	97	121	146	170	194	219
36000	417	243	267	291	315	339	363	387	411	435	459
37000	420	483	507	531	554	578	602	625	649	673	696
38000	422	720	743	767	790	814	837	860	884	907	930
39000	424	954	977	1000	1023	1046	1069	1092	1115	1138	1161
40000	426	1184	1207	1229	1252	1275	1297	1320	1343	1365	1388
41000	427	1411	1433	1455	1478	1500	1523	1545	1567	1590	1612
42000	427	1634	1657	1679	1701	1723	1745	1767	1789	1811	1833
43000	427	1855	1877	1899	1920	1942	1964	1986	2007	2029	2051
44000	427	2073	2094	2116	2137	2158	2180	2201	2223	2244	2266
45000	427	2287	2308	2329	2350	2372	2393	2414	2435	2456	2477
46000	427	2498	2519	2540	2561	2582	2602	2623	2644	2665	2686
47000	427	2706	2727	2747	2768	2788	2809	2829	2850	2870	2891
48000	427	2911	2931	2951	2972	2992	3012	3032	3052	3072	3093
49000	427	3113	3133	3152	3172	3192	3212	3232	3252	3271	3291
50000	427	3311	3331	3350	3370	3389	3409	3428	3448	3467	3487
51000	427	3506	3525	3545	3564	3583	3602	3621	3641	3660	3679
52000	428	3698	3717	3736	3755	3774	3792	3811	3830	3849	3868
53000	428	3887	3905	3924	3942	3961	3979	3998	4016	4035	4054
54000	428	4072	4090	4108	4127	4145	4163	4181	4199	4217	4236
55000	428	4254	4272	4290	4307	4325	4343	4361	4379	4397	4414
56000	427	4432	4450	4467	4485	4502	4520	4537	4554	4572	4589
57000	427	4607	4624	4641	4658	4675	4692	4709	4727	4744	4761
58000	427	4778	4795	4811	4828	4845	4861	4878	4895	4911	4928
59000	427	4945	4961	4977	4994	5010	5026	5042	5059	5075	5091
60000	426	5107	5123	5139	5155	5171	5187	5202	5218	5234	5250
61000	426	5266	5281	5296	5312	5327	5342	5358	5373	5389	5404
62000	425	5419	5434	5449	5464	5479	5494	5509	5523	5538	5553

NOTE – OPTIMUM WEIGHT FOR PRESSURE ALTITUDE IS 50500 KG
    THRUST LIMITED WEIGHT FOR ISA + 10 AND COLDER IS 61800 KG
    THRUST LIMITED WEIGHT FOR ISA + 15   IS 60500 KG
    THRUST LIMITED WEIGHT FOR ISA + 20   IS 59200 KG
ADJUSTMENTS FOR OPERATION AT NON-STANDARD TEMPERATURES---
    INCREASE FUEL REQUIRED BY   0.6 PERCENT PER 10 DEGREES C ABOVE ISA
    DECREASE FUEL REQUIRED BY   0.6 PERCENT PER 10 DEGREES C BELOW ISA
    INCREASE TAS BY 1 KNOT PER DEGREE C ABOVE ISA
    DECREASE TAS BY 1 KNOT PER DEGREE C BELOW ISA

Figure 4.5.3.1 Long Range Cruise

All Engines   Maximum Cruise Thrust Limits   A/C Auto

PRESSURE ALTITUDE   37000 Ft.

GROST WT. KG.	TAS	0	100	200	300	400	500	600	700	800	900
					CRUISE DISTANCE NAUTICAL AIR MILES						
35000	419	0	24	49	74	99	123	148	173	198	222
36000	422	247	271	296	320	345	369	393	418	442	467
37000	424	491	515	539	563	587	611	635	659	683	707
38000	426	731	755	779	803	826	850	874	897	921	945
39000	427	968	992	1015	1038	1062	1085	1108	1132	1155	1178
40000	427	1202	1225	1248	1271	1294	1317	1340	1363	1386	1409
41000	427	1432	1454	1477	1500	1522	1545	1568	1590	1613	1636
42000	427	1659	1681	1703	1725	1748	1770	1792	1815	1837	1859
43000	427	1882	1904	1926	1948	1970	1992	2014	2036	2058	2080
44000	427	2102	2123	2145	2167	2188	2210	2231	2253	2275	2296
45000	427	2318	2339	2361	2382	2403	2424	2446	2467	2488	2510
46000	427	2531	2552	2573	2594	2615	2635	2656	2677	2698	2719
47000	427	2740	2761	2781	2802	2822	2843	2864	2884	2905	2925
48000	427	2946	2966	2986	3007	3027	3047	3067	3088	3108	3128
49000	427	3148	3168	3188	3208	3228	3248	3268	3288	3307	3327
50000	427	3347	3367	3386	3406	3425	3445	3464	3484	3503	3523
51000	427	3542	3562	3581	3600	3619	3638	3657	3676	3696	3715
52000	427	3734	3753	3771	3790	3809	3828	3846	3865	3884	3903
53000	427	3922	3940	3958	3977	3995	4013	4032	4050	4069	4087
54000	427	4105	4123	4141	4159	4177	4195	4213	4231	4249	4267
55000	427	4285	4303	4320	4338	4355	4373	4390	4408	4425	4443
56000	427	4461	4479	4495	4512	4529	4546	4563	4580	4597	4614
57000	426	4631	4648	4665	4681	4698	4714	4731	4748	4764	4781
58000	426	4797	4813	4830	4845	4862	4878	4894	4910	4926	4942
59000	425	4958	4974	4990	5005	5021	5036	5052	5067	5083	5099

NOTE – OPTIMUM WEIGHT FOR PRESSURE ALTITUDE IS 48000 KG
    THRUST LIMITED WEIGHT FOR ISA + 10 AND COLDER IS 58700 KG
    THRUST LIMITED WEIGHT FOR ISA + 15   IS 57500 KG
    THRUST LIMITED WEIGHT FOR ISA + 20   IS 56300 KG
    ADJUSTMENTS FOR OPERATION AT NON-STANDARD TEMPERATURES---
    INCREASE FUEL REQUIRED BY   0.6 PERCENT PER 10 DEGREES C ABOVE ISA
    DECREASE FUEL REQUIRED BY   0.6 PERCENT PER 10 DEGREES C BELOW ISA
    INCREASE TAS BY 1 KNOT PER DEGREE C ABOVE ISA
    DECREASE TAS BY 1 KNOT PER DEGREE C BELOW ISA

Figure 4.5.3.2　Mach 0.74 Cruise

All Engines　　Maximum Cruise Thrust Limits　　A/C Auto

PRESSURE ALTITUDE　　21000Ft　　　　TAS　453 Kts

GROSS WT. KG	0	100	200	300	400	500	600	700	800	900
	CRUISE DISTANCE NAUTICAL AIR MILES									
35000	0	14	29	44	59	74	89	104	119	134
36000	149	164	179	193	208	223	238	253	268	283
37000	298	313	328	342	357	372	387	402	417	432
38000	447	461	476	491	506	521	536	550	565	580
39000	595	610	625	639	654	669	684	699	714	728
40000	743	758	773	788	802	817	832	847	861	876
41000	891	906	921	935	950	965	980	994	1009	1024
42000	1039	1053	1068	1083	1097	1112	1127	1142	1156	1171
43000	1186	1200	1215	1230	1244	1259	1274	1288	1303	1318
44000	1332	1347	1362	1376	1391	1406	1420	1435	1450	1464
45000	1479	1493	1508	1523	1537	1552	1566	1581	1596	1610
46000	1625	1639	1654	1669	1683	1698	1712	1727	1741	1756
47000	1770	1785	1799	1814	1828	1843	1857	1872	1887	1901
48000	1916	1930	1944	1959	1973	1988	2002	2017	2031	2046
49000	2060	2075	2089	2103	2118	2132	2147	2161	2175	2190
50000	2204	2219	2233	2247	2262	2276	2290	2305	2319	2333
51000	2348	2362	2376	2391	2405	2419	2434	2448	2462	2476
52000	2491	2505	2519	2534	2548	2562	2576	2590	2605	2619
53000	2633	2647	2662	2676	2690	2704	2718	2733	2747	2761
54000	2775	2789	2803	2817	2832	2846	2860	2874	2888	2902
55000	2916	2930	2944	2958	2973	2987	3001	3015	3029	3043
56000	3057	3071	3085	3099	3113	3127	3141	3155	3169	3183
57000	3197	3211	3225	3239	3253	3267	3280	3294	3308	3322
58000	3336	3350	3364	3378	3392	3405	3419	3433	3447	3461
59000	3475	3489	3502	3516	3530	3544	3558	3571	3585	3599
60000	3613	3626	3640	3654	3668	3681	3695	3709	3722	3736
61000	3750	3764	3777	3791	3804	3818	3832	3845	3859	3873
62000	3886	3900	3913	3927	3941	3954	3968	3981	3995	4008
63000	4022	4036	4049	4063	4076	4090	4103	4117	4130	4144
64000	4157	4170	4184	4197	4211	4224	4238	4251	4264	4278
65000	4291	4305	4318	4331	4345	4358	4371	4385	4398	4411
66000	4425	4438	4451	4465	4478	4491	4504	4518	4531	4544
67000	4558	4571	4584	4597	4610	4623	4637	4650	4663	4676

NOTE – OPTIMUM WEIGHT FOR PRESSURE ALTITUDE EXCEEDS STRUCTURAL LIMIT
　　　THRUST LIMITED WEIGHT FOR ISA + 10 AND COLDER EXCEEDS STRUCTURAL LIMIT
　　　THRUST LIMITED WEIGHT FOR ISA + 15　　　　EXCEEDS STRUCTURAL LIMIT
　　　THRUST LIMITED WEIGHT FOR ISA + 20　　　　EXCEEDS STRUCTURAL LIMIT
　　ADJUSTMENTS FOR OPERATION AT NON-STANDARD TEMPERATURES---
　　INCREASE FUEL REQUIRED BY　　0.6 PERCENT PER 10 DEGREES C ABOVE ISA
　　DECREASE FUEL REQUIRED BY　　0.6 PERCENT PER 10 DEGREES C BELOW ISA
　　INCREASE TAS BY 1 KNOT PER DEGREE C ABOVE ISA
　　DECREASE TAS BY 1 KNOT PER DEGREE C BELOW ISA

Figure 4.5.3.2   Mach 0.74 Cruise

All Engines   Maximum Cruise Thrust Limits   A/C Auto

PRESSURE ALTITUDE   22000Ft   **TAS   451 Kts**

GROSS WT. KG	0	100	200	300	400	500	600	700	800	900
	CRUISE DISTANCE NAUTICAL AIR MILES									
35000	0	15	30	46	61	77	92	108	123	139
36000	154	170	185	201	216	232	247	263	278	293
37000	309	324	340	355	371	386	401	417	432	448
38000	463	479	494	509	525	540	555	571	586	602
39000	617	632	648	663	678	694	709	725	740	755
40000	771	786	801	817	832	847	862	878	893	908
41000	924	939	954	970	985	1000	1015	1031	1046	1061
42000	1077	1092	1107	1122	1137	1153	1168	1183	1198	1214
43000	1229	1244	1259	1274	1290	1305	1320	1335	1350	1366
44000	1381	1396	1411	1426	1441	1457	1472	1487	1502	1517
45000	1532	1547	1562	1578	1593	1608	1623	1638	1653	1668
46000	1683	1698	1713	1728	1743	1758	1773	1789	1804	1819
47000	1834	1849	1864	1879	1894	1909	1924	1939	1954	1969
48000	1984	1998	2013	2028	2043	2058	2073	2088	2103	2118
49000	2133	2148	2163	2177	2192	2207	2222	2237	2252	2267
50000	2282	2296	2311	2326	2341	2356	2370	2385	2400	2415
51000	2430	2444	2459	2474	2489	2503	2518	2533	2548	2562
52000	2577	2592	2606	2621	2636	2650	2665	2680	2695	2709
53000	2724	2738	2753	2768	2782	2797	2812	2826	2841	2855
54000	2870	2885	2899	2914	2928	2943	2957	2972	2986	3001
55000	3015	3030	3044	3059	3073	3088	3102	3117	3131	3146
56000	3160	3174	3189	3203	3218	3232	3246	3261	3275	3290
57000	3304	3318	3333	3347	3361	3376	3390	3404	3418	3433
58000	3447	3461	3476	3490	3504	3518	3533	3547	3561	3575
59000	3589	3604	3618	3632	3646	3660	3674	3689	3703	3717
60000	3731	3745	3759	3773	3787	3801	3816	3830	3844	3858
61000	3872	3886	3900	3914	3928	3942	3956	3970	3984	3998
62000	4012	4026	4040	4054	4068	4081	4095	4109	4123	4137
63000	4151	4165	4179	4193	4206	4220	4234	4248	4262	4276
64000	4289	4303	4317	4331	4344	4358	4372	4386	4399	4413
65000	4427	4441	4454	4468	4482	4495	4509	4523	4536	4550
66000	4564	4577	4591	4604	4618	4632	4645	4659	4672	4686
67000	4699	4713	4726	4740	4753	4767	4780	4794	4807	4821

NOTE – OPTIMUM WEIGHT FOR PRESSURE ALTITUDE EXCEEDS STRUCTURAL LIMIT
THRUST LIMITED WEIGHT FOR ISA + 10 AND COLDER EXCEEDS STRUCTURAL LIMIT
THRUST LIMITED WEIGHT FOR ISA + 15   EXCEEDS STRUCTURAL LIMIT
THRUST LIMITED WEIGHT FOR ISA + 20   EXCEEDS STRUCTURAL LIMIT
ADJUSTMENTS FOR OPERATION AT NON-STANDARD TEMPERATURES---
INCREASE FUEL REQUIRED BY   0.6 PERCENT PER 10 DEGREES C ABOVE ISA
DECREASE FUEL REQUIRED BY   0.6 PERCENT PER 10 DEGREES C BELOW ISA
INCREASE TAS BY 1 KNOT PER DEGREE C ABOVE ISA
DECREASE TAS BY 1 KNOT PER DEGREE C BELOW ISA

Figure 4.5.3.2   Mach 0.74 Cruise

All Engines   Maximum Cruise Thrust Limits   A/C Auto

PRESSURE ALTITUDE   23000Ft   **TAS   449 Kts**

GROSS WT. KG	0	100	200	300	400	500	600	700	800	900
	CRUISE DISTANCE NAUTICAL AIR MILES									
35000	0	16	32	48	64	80	96	112	128	144
36000	160	176	192	208	224	240	256	272	288	304
37000	320	336	352	368	384	400	416	432	448	464
38000	480	496	512	528	544	560	576	592	608	624
39000	640	656	671	687	703	719	735	751	767	783
40000	799	815	830	846	862	878	894	910	926	941
41000	957	973	989	1005	1021	1036	1052	1068	1084	1100
42000	1115	1131	1147	1163	1179	1194	1210	1226	1242	1257
43000	1273	1289	1305	1320	1336	1352	1367	1383	1399	1415
44000	1430	1446	1462	1477	1493	1509	1524	1540	1555	1571
45000	1587	1602	1618	1634	1649	1665	1680	1696	1712	1727
46000	1743	1758	1774	1789	1805	1820	1836	1851	1867	1883
47000	1898	1914	1929	1945	1960	1975	1991	2006	2022	2037
48000	2053	2068	2084	2099	2114	2130	2145	2161	2176	2191
49000	2207	2222	2237	2253	2268	2283	2299	2314	2329	2345
50000	2360	2375	2391	2406	2421	2436	2452	2467	2482	2497
51000	2513	2528	2543	2558	2574	2589	2604	2619	2634	2649
52000	2665	2680	2695	2710	2725	2740	2755	2770	2786	2801
53000	2816	2831	2846	2861	2876	2891	2906	2921	2936	2951
54000	2966	2981	2996	3011	3026	3041	3056	3071	3086	3101
55000	3116	3130	3145	3160	3175	3190	3205	3220	3234	3249
56000	3264	3279	3294	3309	3323	3338	3353	3368	3382	3397
57000	3412	3427	3441	3456	3471	3486	3500	3515	3530	3544
58000	3559	3574	3588	3603	3617	3632	3647	3661	3676	3690
59000	3705	3720	3734	3749	3763	3778	3792	3807	3821	3836
60000	3850	3865	3879	3894	3908	3923	3937	3951	3966	3980
61000	3995	4009	4023	4038	4052	4066	4081	4095	4109	4124
62000	4138	4152	4167	4181	4195	4209	4224	4238	4252	4266
63000	4281	4295	4309	4323	4337	4351	4366	4380	4394	4408
64000	4422	4436	4450	4464	4479	4493	4507	4521	4535	4549
65000	4563	4577	4591	4605	4619	4633	4647	4661	4675	4689
66000	4703	4716	4730	4744	4758	4772	4786	4800	4814	4827
67000	4841	4855	4869	4883	4896	4910	4924	4938	4952	4965

NOTE – OPTIMUM WEIGHT FOR PRESSURE ALTITUDE EXCEEDS STRUCTURAL LIMIT
THRUST LIMITED WEIGHT FOR ISA + 10 AND COLDER EXCEEDS STRUCTURAL LIMIT
THRUST LIMITED WEIGHT FOR ISA + 15   EXCEEDS STRUCTURAL LIMIT
THRUST LIMITED WEIGHT FOR ISA + 20   EXCEEDS STRUCTURAL LIMIT
ADJUSTMENTS FOR OPERATION AT NON-STANDARD TEMPERATURES---
INCREASE FUEL REQUIRED BY   0.6 PERCENT PER 10 DEGREES C ABOVE ISA
DECREASE FUEL REQUIRED BY   0.6 PERCENT PER 10 DEGREES C BELOW ISA
INCREASE TAS BY 1 KNOT PER DEGREE C ABOVE ISA
DECREASE TAS BY 1 KNOT PER DEGREE C BELOW ISA

Figure 4.5.3.2   Mach 0.74 Cruise

All Engines   Maximum Cruise Thrust Limits   A/C Auto

PRESSURE ALTITUDE   24000Ft   **TAS   447 Kts**

GROSS WT. KG	0	100	200	300	400	500	600	700	800	900
	CRUISE DISTANCE NAUTICAL AIR MILES									
35000	0	16	33	49	66	83	99	116	133	149
36000	166	183	199	216	232	249	266	282	299	315
37000	332	349	365	382	398	415	431	448	465	481
38000	498	514	531	547	564	580	597	613	630	646
39000	663	679	696	712	729	745	762	778	795	811
40000	828	844	860	877	893	910	926	942	959	975
41000	992	1008	1024	1041	1057	1074	1090	1106	1123	1139
42000	1155	1172	1188	1204	1221	1237	1253	1269	1286	1302
43000	1318	1335	1351	1367	1383	1399	1416	1432	1448	1464
44000	1481	1497	1513	1529	1545	1561	1578	1594	1610	1626
45000	1642	1658	1675	1691	1707	1723	1739	1755	1771	1787
46000	1803	1819	1835	1851	1867	1883	1899	1915	1932	1948
47000	1964	1980	1995	2011	2027	2043	2059	2075	2091	2107
48000	2123	2139	2155	2171	2187	2202	2218	2234	2250	2266
49000	2282	2298	2313	2329	2345	2361	2377	2392	2408	2424
50000	2440	2455	2471	2487	2503	2518	2534	2550	2565	2581
51000	2597	2612	2628	2644	2659	2675	2691	2706	2722	2737
52000	2753	2769	2784	2800	2815	2831	2846	2862	2877	2893
53000	2908	2924	2939	2955	2970	2986	3001	3017	3032	3047
54000	3063	3078	3094	3109	3124	3140	3155	3170	3186	3201
55000	3216	3232	3247	3262	3277	3293	3308	3323	3338	3354
56000	3369	3384	3399	3414	3430	3445	3460	3475	3490	3506
57000	3521	3536	3551	3566	3581	3596	3611	3626	3641	3656
58000	3671	3686	3701	3716	3731	3746	3761	3776	3791	3806
59000	3821	3836	3851	3866	3881	3896	3910	3925	3940	3995
60000	3970	3985	4000	4014	4029	4044	4059	4073	4088	4103
61000	4118	4132	4147	4162	4176	4191	4206	4220	4235	4250
62000	4264	4279	4294	4308	4323	4337	4352	4366	4381	4396
63000	4410	4425	4439	4454	4468	4483	4497	4511	4526	4540
64000	4555	4569	4584	4598	4612	4627	4641	4655	4670	4684
65000	4698	4713	4727	4741	4756	4770	4784	4798	4813	4827
66000	4841	4855	4869	4884	4898	4912	4926	4940	4954	4968
67000	4983	4997	5011	5025	5039	5053	5067	5081	5095	5109

NOTE – OPTIMUM WEIGHT FOR PRESSURE ALTITUDE EXCEEDS STRUCTURAL LIMIT
THRUST LIMITED WEIGHT FOR ISA + 10 AND COLDER EXCEEDS STRUCTURAL LIMIT
THRUST LIMITED WEIGHT FOR ISA + 15   EXCEEDS STRUCTURAL LIMIT
THRUST LIMITED WEIGHT FOR ISA + 20   EXCEEDS STRUCTURAL LIMIT
ADJUSTMENTS FOR OPERATION AT NON-STANDARD TEMPERATURES---
INCREASE FUEL REQUIRED BY   0.6 PERCENT PER 10 DEGREES C ABOVE ISA
DECREASE FUEL REQUIRED BY   0.6 PERCENT PER 10 DEGREES C BELOW ISA
INCREASE TAS BY 1 KNOT PER DEGREE C ABOVE ISA
DECREASE TAS BY 1 KNOT PER DEGREE C BELOW ISA

Figure 4.5.3.2   Mach 0.74 Cruise

All Engines   Maximum Cruise Thrust Limits   A/C Auto

PRESSURE ALTITUDE   25000Ft   TAS   445 Kts

GROSS WT. KG	0	100	200	300	400	500	600	700	800	900
					CRUISE DISTANCE NAUTICAL AIR MILES					
35000	0	17	34	51	69	86	103	120	138	155
36000	172	189	206	224	241	258	275	293	310	327
37000	344	361	378	396	413	430	447	464	481	499
38000	516	533	550	567	584	601	618	635	652	670
39000	687	704	721	738	755	772	789	806	823	840
40000	857	874	891	908	925	942	959	976	993	1010
41000	1027	1044	1061	1078	1094	1111	1128	1145	1162	1179
42000	1196	1213	1230	1246	1263	1280	1297	1314	1331	1347
43000	1364	1381	1398	1414	1431	1448	1465	1482	1498	1515
44000	1532	1548	1565	1582	1599	1615	1632	1649	1665	1682
45000	1699	1715	1732	1748	1765	1782	1798	1815	1831	1848
46000	1865	1881	1898	1914	1931	1947	1964	1980	1997	2013
47000	2030	2046	2063	2079	2096	2112	2128	2145	2161	2178
48000	2194	2210	2227	2243	2259	2276	2292	2308	2325	2341
49000	2358	2374	2390	2406	2422	2439	2455	2471	2487	2504
50000	2520	2536	2552	2568	2585	2601	2617	2633	2649	2665
51000	2681	2698	2714	2730	2746	2762	2778	2794	2810	2826
52000	2842	2858	2874	2890	2906	2922	2938	2954	2970	2986
53000	3002	3017	3033	3049	3065	3081	3097	3113	3128	3144
54000	3160	3176	3192	3207	3223	3239	3255	3270	3286	3302
55000	3318	3333	3349	3365	3380	3396	3411	3427	3443	3458
56000	3474	3490	3505	3521	3536	3552	3567	3583	3598	3614
57000	3630	3645	3660	3676	3691	3707	3722	3738	3753	3768
58000	3784	3799	3815	3830	3845	3861	3876	3891	3906	3922
59000	3937	3952	3968	3983	3998	4013	4028	4044	4059	4074
60000	4089	4104	4119	4135	4150	4165	4180	4195	4210	4225
61000	4240	4255	4270	4285	4300	4315	4330	4345	4360	4375
62000	4390	4405	4420	4435	4450	4465	4479	4494	4509	4524
63000	4539	4554	4568	4583	4598	4613	4628	4642	4657	4672
64000	4687	4701	4716	4731	4745	4760	4774	4789	4804	4818
65000	4833	4848	4862	4877	4891	4906	4920	4935	4949	4964
66000	4978	4993	5007	5022	5036	5050	5065	5079	5094	5108
67000	5122	5137	5151	5165	5180	5194	5208	5222	5237	5251

NOTE – OPTIMUM WEIGHT FOR PRESSURE ALTITUDE EXCEEDS STRUCTURAL LIMIT
THRUST LIMITED WEIGHT FOR ISA + 10 AND COLDER EXCEEDS STRUCTURAL LIMIT
THRUST LIMITED WEIGHT FOR ISA + 15   EXCEEDS STRUCTURAL LIMIT
THRUST LIMITED WEIGHT FOR ISA + 20   EXCEEDS STRUCTURAL LIMIT
ADJUSTMENTS FOR OPERATION AT NON-STANDARD TEMPERATURES---
INCREASE FUEL REQUIRED BY   0.6 PERCENT PER 10 DEGREES C ABOVE ISA
DECREASE FUEL REQUIRED BY   0.6 PERCENT PER 10 DEGREES C BELOW ISA
INCREASE TAS BY 1 KNOT PER DEGREE C ABOVE ISA
DECREASE TAS BY 1 KNOT PER DEGREE C BELOW ISA

Figure 4.5.3.2   Mach 0.74 Cruise

All Engines   Maximum Cruise Thrust Limits   A/C Auto

PRESSURE ALTITUDE   26000Ft   **TAS   443 Kts**

GROSS WT. KG	0	100	200	300	400	500	600	700	800	900
	CRUISE DISTANCE NAUTICAL AIR MILES									
35000	0	17	35	53	71	89	107	125	143	160
36000	178	196	214	232	250	267	285	303	321	339
37000	356	374	392	410	427	445	463	481	498	516
38000	534	552	569	587	605	622	640	658	675	693
39000	711	728	746	764	781	799	817	834	852	869
40000	887	905	922	940	957	975	992	1010	1027	1045
41000	1062	1080	1097	1115	1132	1150	1167	1185	1202	1219
42000	1237	1254	1272	1289	1306	1324	1341	1359	1376	1393
43000	1411	1428	1445	1463	1480	1497	1514	1532	1549	1566
44000	1584	1601	1618	1635	1652	1669	1687	1704	1721	1738
45000	1755	1773	1790	1807	1824	1841	1858	1875	1892	1909
46000	1926	1943	1960	1977	1994	2011	2028	2045	2062	2080
47000	2097	2113	2130	2147	2164	2181	2198	2215	2232	2249
48000	2266	2282	2299	2316	2333	2350	2366	2383	2400	2417
49000	2433	2450	2467	2484	2500	2517	2534	2550	2567	2584
50000	2600	2617	2634	2650	2667	2683	2700	2716	2733	2750
51000	2766	2783	2799	2816	2832	2849	2865	2882	2898	2915
52000	2931	2947	2964	2980	2996	3013	3029	3046	3062	3078
53000	3095	3111	3127	3143	3160	3176	3192	3208	3225	3241
54000	3257	3273	3289	3306	3322	3338	3354	3370	3386	3402
55000	3419	3435	3451	3467	3483	3499	3515	3531	3547	3563
56000	3579	3595	3611	3626	3642	3658	3674	3690	3706	3722
57000	3738	3754	3769	3785	3801	3817	3832	3848	3864	3880
58000	3896	3911	3927	3943	3958	3974	3990	4005	4021	4036
59000	4052	4068	4083	4099	4114	4130	4145	4161	4176	4192
60000	4207	4223	4238	4254	4269	4285	4300	4315	4331	4346
61000	4362	4377	4392	4407	4423	4438	4453	4469	4484	4499
62000	4514	4530	4545	4560	4575	4590	4605	4621	4636	4651
63000	4666	4681	4696	4711	4726	4741	4756	4771	4786	4801
64000	4816	4831	4846	4861	4876	4891	4906	4921	4936	4950
65000	4965	4980	4995	5010	5024	5039	5054	5069	5083	5098
66000	5113	5128	5142	5157	5172	5186	5201	5215	5230	5245
67000	5259	5274	5288	5303	5317	5332	5346	5361	5375	5390

NOTE – OPTIMUM WEIGHT FOR PRESSURE ALTITUDE EXCEEDS STRUCTURAL LIMIT
   THRUST LIMITED WEIGHT FOR ISA + 10 AND COLDER EXCEEDS STRUCTURAL LIMIT
   THRUST LIMITED WEIGHT FOR ISA + 15   EXCEEDS STRUCTURAL LIMIT
   THRUST LIMITED WEIGHT FOR ISA + 20   EXCEEDS STRUCTURAL LIMIT
   ADJUSTMENTS FOR OPERATION AT NON-STANDARD TEMPERATURES---
   INCREASE FUEL REQUIRED BY   0.6 PERCENT PER 10 DEGREES C ABOVE ISA
   DECREASE FUEL REQUIRED BY   0.6 PERCENT PER 10 DEGREES C BELOW ISA
   INCREASE TAS BY 1 KNOT PER DEGREE C ABOVE ISA
   DECREASE TAS BY 1 KNOT PER DEGREE C BELOW ISA

Figure 4.5.3.2   Mach 0.74 Cruise

All Engines   Maximum Cruise Thrust Limits   A/C Auto

PRESSURE ALTITUDE   27000Ft   TAS   442 Kts

GROSS WT. KG	0	100	200	300	400	500	600	700	800	900
	CRUISE DISTANCE NAUTICAL AIR MILES									
35000	0	18	37	55	74	92	111	129	148	166
36000	185	203	221	240	258	277	295	314	332	351
37000	369	387	406	424	442	461	479	497	516	534
38000	553	571	589	607	626	644	662	680	699	717
39000	735	754	772	790	808	826	844	863	881	899
40000	917	935	953	972	990	1008	1026	1044	1062	1080
41000	1098	1116	1134	1152	1170	1188	1206	1224	1242	1260
42000	1278	1296	1314	1332	1350	1368	1386	1404	1422	1440
43000	1457	1475	1493	1511	1529	1546	1564	1582	1600	1618
44000	1636	1653	1671	1689	1706	1724	1742	1759	1777	1795
45000	1813	1830	1848	1865	1883	1900	1918	1936	1953	1971
46000	1988	2006	2023	2041	2058	2076	2093	2111	2128	2146
47000	2163	2181	2198	2215	2233	2250	2267	2285	2302	2320
48000	2337	2354	2371	2389	2406	2423	2440	2458	2475	2492
49000	2509	2527	2544	2561	2578	2595	2612	2629	2646	2664
50000	2681	2698	2715	2732	2749	2766	2783	2800	2817	2834
51000	2851	2868	2885	2902	2918	2935	2952	2969	2986	3003
52000	3020	3036	3053	3070	3087	3104	3120	3137	3154	3171
53000	3187	3204	3221	3237	3254	3271	3287	3304	3320	3337
54000	3354	3370	3387	3403	3420	3436	3453	3469	3486	3502
55000	3519	3535	3552	3568	3584	3601	3617	3633	3650	3666
56000	3683	3699	3715	3731	3747	3764	3780	3796	3812	3829
57000	3845	3861	3877	3893	3909	3925	3942	3958	3974	3990
58000	4006	4022	4038	4054	4070	4086	4102	4118	4134	4150
59000	4166	4182	4197	4213	4229	4245	4261	4277	4292	4308
60000	4324	4340	4356	4371	4387	4403	4418	4434	4450	4465
61000	4481	4497	4512	4528	4543	4559	4574	4590	4606	4621
62000	4637	4652	4668	4683	4698	4714	4729	4745	4760	4775
63000	4791	4806	4821	4837	4852	4867	4882	4898	4913	4928
64000	4943	4959	4974	4989	5004	5019	5034	5049	5064	5080
65000	5095	5110	5125	5140	5155	5170	5185	5200	5214	5229
66000	5244	5259	5274	5289	5304	5319	5333	5348	5363	5378
67000	5393	5407	5422	5437	5451	5466	5481	5495	5510	5525

NOTE – OPTIMUM WEIGHT FOR PRESSURE ALTITUDE EXCEEDS STRUCTURAL LIMIT
THRUST LIMITED WEIGHT FOR ISA + 10 AND COLDER EXCEEDS STRUCTURAL LIMIT
THRUST LIMITED WEIGHT FOR ISA + 15   EXCEEDS STRUCTURAL LIMIT
THRUST LIMITED WEIGHT FOR ISA + 20   EXCEEDS STRUCTURAL LIMIT
ADJUSTMENTS FOR OPERATION AT NON-STANDARD TEMPERATURES---
INCREASE FUEL REQUIRED BY   0.6 PERCENT PER 10 DEGREES C ABOVE ISA
DECREASE FUEL REQUIRED BY   0.6 PERCENT PER 10 DEGREES C BELOW ISA
INCREASE TAS BY 1 KNOT PER DEGREE C ABOVE ISA
DECREASE TAS BY 1 KNOT PER DEGREE C BELOW ISA

Figure 4.5.3.2   Mach 0.74 Cruise

All Engines   Maximum Cruise Thrust Limits   A/C Auto

PRESSURE ALTITUDE   28000Ft   **TAS   440 Kts**

GROSS WT. KG	0	100	200	300	400	500	600	700	800	900
	CRUISE DISTANCE NAUTICAL AIR MILES									
35000	0	19	38	57	76	95	114	134	153	172
36000	191	210	229	248	267	286	305	324	343	362
37000	382	401	419	438	457	476	495	514	533	552
38000	571	590	609	628	647	666	684	703	722	741
39000	760	779	797	816	835	854	873	891	910	929
40000	948	966	985	1004	1022	1041	1060	1078	1097	1116
41000	1134	1153	1171	1190	1209	1227	1246	1264	1283	1301
42000	1320	1338	1357	1375	1394	1412	1431	1449	1467	1486
43000	1504	1523	1541	1559	1578	1596	1614	1633	1651	1669
44000	1688	1706	1724	1742	1760	1779	1797	1815	1833	1851
45000	1870	1888	1906	1924	1942	1960	1978	1996	2014	2032
46000	2050	2068	2086	2104	2122	2140	2158	2176	2194	2212
47000	2230	2248	2265	2283	2301	2319	2337	2355	2372	2390
48000	2408	2426	2443	2461	2479	2497	2514	2532	2550	2567
49000	2585	2603	2620	2638	2655	2673	2690	2708	2726	2743
50000	2761	2778	2795	2813	2830	2848	2865	2883	2900	2917
51000	2935	2952	2969	2987	3004	3021	3039	3056	3073	3090
52000	3108	3125	3142	3159	3176	3193	3211	3228	3245	3262
53000	3279	3296	3313	3330	3347	3364	3381	3398	3415	3432
54000	3449	3466	3483	3500	3517	3533	3550	3567	3584	3601
55000	3618	3634	3651	3668	3685	3701	3718	3735	3751	3768
56000	3785	3801	3818	3835	3851	3868	3884	3901	3917	3934
57000	3951	3967	3983	4000	4016	4033	4049	4065	4082	4098
58000	4115	4131	4147	4164	4180	4196	4212	4229	4245	4261
59000	4277	4293	4310	4326	4324	4358	4374	4390	4406	4422
60000	4438	4454	4470	4486	4502	4518	4534	4550	4566	4582
61000	4598	4614	4630	4645	4661	4677	4693	4709	4724	4740
62000	4756	4772	4787	4803	4818	4834	4850	4865	4881	4897
63000	4912	4928	4943	4959	4974	4990	5005	5021	5036	5052
64000	5067	5082	5098	5113	5128	5144	5159	5174	5190	5205
65000	5220	5235	5250	5266	5281	5296	5311	5326	5341	5357
66000	5372	5387	5402	5417	5432	5447	5462	5477	5492	5507
67000	5522	5536	5551	5566	5581	5596	5611	5625	5640	5655

NOTE – OPTIMUM WEIGHT FOR PRESSURE ALTITUDE EXCEEDS STRUCTURAL LIMIT
THRUST LIMITED WEIGHT FOR ISA + 10 AND COLDER EXCEEDS STRUCTURAL LIMIT
THRUST LIMITED WEIGHT FOR ISA + 15   EXCEEDS STRUCTURAL LIMIT
THRUST LIMITED WEIGHT FOR ISA + 20   EXCEEDS STRUCTURAL LIMIT
ADJUSTMENTS FOR OPERATION AT NON-STANDARD TEMPERATURES---
INCREASE FUEL REQUIRED BY   0.6 PERCENT PER 10 DEGREES C ABOVE ISA
DECREASE FUEL REQUIRED BY   0.6 PERCENT PER 10 DEGREES C BELOW ISA
INCREASE TAS BY 1 KNOT PER DEGREE C ABOVE ISA
DECREASE TAS BY 1 KNOT PER DEGREE C BELOW ISA

Figure 4.5.3.2   Mach 0.74 Cruise

All Engines   Maximum Cruise Thrust Limits   A/C Auto

PRESSURE ALTITUDE   29000Ft   TAS   438 Kts

GROSS WT. KG	0	100	200	300	400	500	600	700	800	900
	CRUISE DISTANCE NAUTICAL AIR MILES									
35000	0	19	39	59	79	98	118	138	158	178
36000	197	217	237	256	276	296	315	335	355	375
37000	394	414	433	453	473	492	512	531	551	570
38000	590	609	629	648	668	687	707	726	746	765
39000	785	804	823	843	862	881	901	920	939	959
40000	978	997	1017	1036	1055	1074	1093	1113	1132	1151
41000	1170	1189	1209	1228	1247	1266	1285	1304	1323	1342
42000	1361	1380	1399	1418	1437	1456	1475	1494	1513	1532
43000	1551	1570	1589	1608	1626	1645	1664	1683	1702	1721
44000	1739	1758	1777	1795	1814	1833	1852	1870	1889	1908
45000	1926	1945	1963	1982	2001	2019	2038	2056	2075	2093
46000	2112	2130	2149	2167	2186	2204	2222	2241	2259	2278
47000	2296	2314	2333	2351	2369	2387	2406	2424	2442	2461
48000	2479	2497	2515	2533	2551	2569	2588	2606	2624	2642
49000	2660	2678	2696	2714	2732	2750	2768	2786	2804	2822
50000	2840	2858	2875	2893	2911	2929	2947	2964	2982	3000
51000	3018	3036	3053	3071	3089	3106	3124	3142	3159	3177
52000	3195	3212	3230	3247	3265	3282	3300	3317	3335	3352
53000	3370	3387	3404	3422	3439	3456	3474	3491	3508	3526
54000	3543	3560	3578	3595	3612	3629	3646	3664	3681	3698
55000	3715	3732	3749	3766	3783	3800	3817	3834	3851	3868
56000	3885	3902	3919	3936	3953	3970	3987	4003	4020	4037
57000	4054	4071	4087	4104	4121	4137	4154	4171	4187	4204
58000	4221	4237	4254	4270	4287	4303	4320	4337	4353	4370
59000	4386	4402	4419	4435	4451	4468	4484	4501	4517	4533
60000	4550	4566	4582	4598	4614	4630	4647	4663	4679	4695
61000	4711	4727	4743	4759	4775	4791	4807	4823	4839	4855
62000	4871	4887	4903	4919	4935	4950	4966	4982	4998	5014
63000	5030	5045	5061	5077	5092	5108	5123	5139	5155	5170
64000	5186	5202	5217	5233	5248	5263	5279	5294	5310	5325
65000	5341	5356	5371	5387	5402	5417	5433	5448	5463	5479
66000	5494	5509	5524	5539	5554	5569	5585	5600	5615	5630
67000	5645	5660	5675	5690	5705	5720	5735	5750	5765	5780

NOTE – OPTIMUM WEIGHT FOR PRESSURE ALTITUDE EXCEEDS STRUCTURAL LIMIT
         THRUST LIMITED WEIGHT FOR ISA + 10 AND COLDER EXCEEDS STRUCTURAL LIMIT
         THRUST LIMITED WEIGHT FOR ISA + 15   EXCEEDS STRUCTURAL LIMIT
         THRUST LIMITED WEIGHT FOR ISA + 20   EXCEEDS STRUCTURAL LIMIT
         ADJUSTMENTS FOR OPERATION AT NON-STANDARD TEMPERATURES---
         INCREASE FUEL REQUIRED BY   0.6 PERCENT PER 10 DEGREES C ABOVE ISA
         DECREASE FUEL REQUIRED BY   0.6 PERCENT PER 10 DEGREES C BELOW ISA
         INCREASE TAS BY 1 KNOT PER DEGREE C ABOVE ISA
         DECREASE TAS BY 1 KNOT PER DEGREE C BELOW ISA

Figure 4.5.3.2 Mach 0.74 Cruise

All Engines   Maximum Cruise Thrust Limits   A/C Auto

PRESSURE ALTITUDE   30000Ft   **TAS   436 Kts**

GROSS WT. KG	0	100	200	300	400	500	600	700	800	900
	CRUISE DISTANCE NAUTICAL AIR MILES									
35000	0	20	40	61	81	102	122	143	163	183
36000	204	224	244	265	285	305	326	346	366	387
37000	407	427	447	468	488	508	528	548	568	589
38000	609	629	649	669	689	709	729	749	769	789
39000	809	829	849	869	889	909	929	949	969	989
40000	1009	1028	1048	1068	1088	1108	1127	1147	1167	1187
41000	1206	1226	1246	1265	1285	1305	1324	1344	1363	1383
42000	1403	1422	1442	1461	1481	1500	1520	1539	1559	1578
43000	1598	1617	1636	1656	1675	1694	1714	1733	1752	1772
44000	1791	1810	1829	1849	1868	1887	1906	1925	1944	1964
45000	1983	2002	2021	2040	2059	2078	2097	2116	2135	2154
46000	2173	2192	2211	2230	2249	2267	2286	2305	2324	2343
47000	2362	2380	2399	2418	2437	2455	2474	2493	2511	2530
48000	2549	2567	2586	2604	2623	2641	2660	2679	2697	2716
49000	2734	2753	2771	2789	2808	2826	2844	2863	2881	2899
50000	2918	2936	2954	2972	2991	3009	3027	3045	3063	3082
51000	3100	3118	3136	3154	3172	3190	3208	3226	3244	3262
52000	3280	3298	3316	3334	3352	3369	3387	3405	3423	3441
53000	3459	3476	3494	3512	3529	3547	3565	3582	3600	3618
54000	3635	3653	3670	3688	3705	3723	3740	3758	3775	3793
55000	3810	3828	3845	3862	3879	3897	3914	3931	3949	3966
56000	3983	4000	4018	4035	4052	4069	4086	4103	4120	4137
57000	4155	4171	4188	4205	4222	4239	4256	4273	4290	4307
58000	4324	4341	4357	4374	4391	4408	4424	4441	4458	4475
59000	4491	4508	4524	4541	4558	4574	4591	4607	4624	4640
60000	4657	4673	4690	4706	4722	4739	4755	4771	4788	4804
61000	4820	4837	4853	4869	4885	4901	4918	4934	4950	4966
62000	4982	4998	5014	5030	5046	5062	5078	5094	5110	5126
63000	5142	5158	5174	5189	5205	5221	5237	5253	5268	5284
64000	5300	5316	5331	5347	5362	5378	5393	5409	5425	5440
65000	5456	5471	5487	5502	5517	5533	5548	5564	5579	5594
66000	5610	5625	5640	5655	5671	5686	5701	5716	5731	5747
67000	5762	5777	5792	5807	5822	5837	5852	5867	5882	5897

NOTE – OPTIMUM WEIGHT FOR PRESSURE ALTITUDE IS 66600 KG
THRUST LIMITED WEIGHT FOR ISA + 10 AND COLDER EXCEEDS STRUCTURAL LIMIT
THRUST LIMITED WEIGHT FOR ISA + 15   EXCEEDS STRUCTURAL LIMIT
THRUST LIMITED WEIGHT FOR ISA + 20   EXCEEDS STRUCTURAL LIMIT
ADJUSTMENTS FOR OPERATION AT NON-STANDARD TEMPERATURES---
INCREASE FUEL REQUIRED BY   0.6 PERCENT PER 10 DEGREES C ABOVE ISA
DECREASE FUEL REQUIRED BY   0.6 PERCENT PER 10 DEGREES C BELOW ISA
INCREASE TAS BY 1 KNOT PER DEGREE C ABOVE ISA
DECREASE TAS BY 1 KNOT PER DEGREE C BELOW ISA

Figure 4.5.3.2   Mach 0.74 Cruise

All Engines   Maximum Cruise Thrust Limits   A/C Auto

PRESSURE ALTITUDE   31000Ft   TAS   434 Kts

GROSS WT. KG	0	100	200	300	400	500	600	700	800	900
	CRUISE DISTANCE NAUTICAL AIR MILES									
35000	0	21	42	63	84	105	126	147	168	189
36000	210	231	252	273	294	315	336	357	378	399
37000	420	440	461	482	503	524	544	565	586	607
38000	628	648	669	689	710	731	751	772	793	813
39000	834	854	875	895	916	936	957	977	998	1018
40000	1039	1059	1079	1100	1120	1140	1161	1181	1201	1222
41000	1242	1262	1282	1303	1323	1343	1363	1383	1403	1423
42000	1444	1464	1484	1504	1524	1544	1564	1584	1604	1624
43000	1644	1663	1683	1703	1723	1743	1763	1782	1802	1822
44000	1842	1862	1881	1901	1921	1940	1960	1979	1999	2019
45000	2038	2058	2077	2097	2116	2136	2155	2175	2194	2214
46000	2233	2252	2272	2291	2310	2330	2349	2368	2388	2407
47000	2426	2445	2464	2483	2503	2522	2541	2560	2579	2598
48000	2617	2636	2655	2674	2693	2712	2731	2750	2769	2788
49000	2807	2825	2844	2863	2882	2900	2919	2938	2956	2975
50000	2994	3013	3031	3050	3068	3087	3105	3124	3142	3161
51000	3179	3198	3216	3235	3253	3271	3290	3308	3326	3345
52000	3363	3381	3399	3417	3436	3454	3472	3490	3508	3526
53000	3545	3563	3580	3598	3616	3634	3652	3670	3688	3706
54000	3724	3742	3760	3777	3795	3813	3831	3848	3866	3884
55000	3902	3919	3937	3954	3972	3989	4007	4024	4042	4060
56000	4077	4094	4112	4129	4146	4164	4181	4198	4216	4233
57000	4251	4268	4285	4302	4319	4336	4353	4370	4388	4405
58000	4422	4439	4456	4473	4490	4507	4523	4540	4557	4574
59000	4591	4608	4625	4641	4658	4675	4691	4708	4725	4742
60000	4758	4775	4791	4808	4824	4841	4857	4874	4890	4907
61000	4923	4940	4956	4972	4989	5005	5021	5038	5054	5070
62000	5086	5103	5119	5135	5151	5167	5183	5199	5215	5231
63000	5247	5263	5279	5295	5311	5327	5343	5358	5374	5390
64000	5406	5422	5437	5453	5469	5484	5500	5516	5531	5547
65000	5563	5578	5594	5609	5624	5640	5655	5671	5686	5702
66000	5717	5732	5748	5763	5778	5793	5809	5824	5839	5854
67000	5870	5885	5900	5915	5930	5945	5960	5975	5990	6005

NOTE – OPTIMUM WEIGHT FOR PRESSURE ALTITUDE IS 63500 KG
    THRUST LIMITED WEIGHT FOR ISA + 10 AND COLDER EXCEEDS STRUCTURAL LIMIT
    THRUST LIMITED WEIGHT FOR ISA + 15   EXCEEDS STRUCTURAL LIMIT
    THRUST LIMITED WEIGHT FOR ISA + 20   EXCEEDS STRUCTURAL LIMIT
    ADJUSTMENTS FOR OPERATION AT NON-STANDARD TEMPERATURES---
    INCREASE FUEL REQUIRED BY   0.6 PERCENT PER 10 DEGREES C ABOVE ISA
    DECREASE FUEL REQUIRED BY   0.6 PERCENT PER 10 DEGREES C BELOW ISA
    INCREASE TAS BY 1 KNOT PER DEGREE C ABOVE ISA
    DECREASE TAS BY 1 KNOT PER DEGREE C BELOW ISA

Figure 4.5.3.2   Mach 0.74 Cruise

All Engines   Maximum Cruise Thrust Limits   A/C Auto

PRESSURE ALTITUDE   32000Ft   **TAS   432 Kts**

GROSS WT. KG	0	100	200	300	400	500	600	700	800	900
	CRUISE DISTANCE NAUTICAL AIR MILES									
35000	0	21	43	65	86	108	130	152	173	195
36000	217	238	260	281	303	324	346	368	389	411
37000	432	454	475	496	518	539	561	582	603	625
38000	646	667	689	710	731	752	773	795	816	837
39000	858	879	900	921	942	963	984	1005	1027	1048
40000	1069	1089	1110	1131	1152	1173	1194	1215	1236	1256
41000	1277	1298	1319	1339	1360	1381	1401	1422	1443	1463
42000	1484	1505	1525	1546	1566	1586	1607	1627	1648	1668
43000	1689	1709	1730	1750	1770	1790	1811	1831	1851	1872
44000	1892	1912	1932	1952	1972	1992	2012	2033	2053	2073
45000	2093	2113	2133	2153	2172	2192	2212	2232	2252	2272
46000	2292	2312	2331	2351	2371	2390	2410	2430	2450	2469
47000	2489	2508	2528	2547	2567	2586	2606	2625	2645	2664
48000	2684	2703	2722	2742	2761	2780	2800	2819	2838	2857
49000	2877	2896	2915	2934	2953	2972	2991	3010	3029	3048
50000	3068	3086	3105	3124	3143	3162	3181	3200	3218	3237
51000	3256	3275	3293	3312	3331	3349	3368	3387	3405	3424
52000	3442	3461	3479	3498	3516	3535	3553	3571	3590	3608
53000	3627	3645	3663	3681	3699	3718	3736	3754	3772	3790
54000	3809	3827	3845	3863	3881	3899	3916	3934	3952	3970
55000	3988	4006	4024	4042	4059	4077	4095	4113	4130	4148
56000	4166	4183	4201	4218	4236	4254	4271	4289	4306	4324
57000	4341	4358	4376	4393	4410	4428	4445	4462	4480	4497
58000	4514	4531	4548	4565	4582	4599	4617	4634	4651	4668
59000	4685	4702	4718	4735	4752	4769	4786	4803	4819	4836
60000	4853	4870	4886	4903	4920	4936	4953	4969	4986	5003
61000	5019	5036	5052	5068	5085	5101	5117	5134	5150	5167
62000	5183	5199	5215	5231	5248	5264	5280	5296	5312	5328
63000	5344	5360	5376	5392	5408	5424	5440	5456	5472	5488
64000	5504	5519	5535	5551	5566	5582	5598	5613	5629	5645
65000	5660	5676	5691	5707	5722	5738	5753	5769	5784	5799
66000	5815	5830	5845	5861	5876	5891	5906	5921	5937	5952
67000	5967	5982	5997	6012	6027	6042	6057	6072	6087	6101

NOTE – OPTIMUM WEIGHT FOR PRESSURE ALTITUDE IS 60700 KG
THRUST LIMITED WEIGHT FOR ISA + 10 AND COLDER EXCEEDS STRUCTURAL LIMIT
THRUST LIMITED WEIGHT FOR ISA + 15   EXCEEDS STRUCTURAL LIMIT
THRUST LIMITED WEIGHT FOR ISA + 20   EXCEEDS STRUCTURAL LIMIT
ADJUSTMENTS FOR OPERATION AT NON-STANDARD TEMPERATURES---
INCREASE FUEL REQUIRED BY   0.6 PERCENT PER 10 DEGREES C ABOVE ISA
DECREASE FUEL REQUIRED BY   0.6 PERCENT PER 10 DEGREES C BELOW ISA
INCREASE TAS BY 1 KNOT PER DEGREE C ABOVE ISA
DECREASE TAS BY 1 KNOT PER DEGREE C BELOW ISA

Figure 4.5.3.2 Mach 0.74 Cruise

All Engines   Maximum Cruise Thrust Limits   A/C Auto

PRESSURE ALTITUDE   33000Ft   TAS 430 Kts

GROST WT. KG	0	100	200	300	400	500	600	700	800	900
	CRUISE DISTANCE NAUTICAL AIR MILES									
35000	0	22	44	67	89	111	134	156	178	201
36000	223	245	267	289	312	334	356	378	400	422
37000	444	466	488	510	532	554	576	598	620	642
38000	664	686	708	730	751	773	795	817	838	860
39000	882	904	925	947	968	990	1011	1033	1055	1076
40000	1098	1119	1141	1162	1183	1205	1226	1247	1269	1290
41000	1312	1333	1354	1375	1396	1417	1439	1460	1481	1502
42000	1523	1544	1565	1586	1607	1628	1649	1670	1691	1712
43000	1733	1753	1774	1795	1816	1836	1857	1878	1899	1919
44000	1940	1961	1981	2002	2022	2043	2063	2084	2104	2125
45000	2145	2166	2186	2206	2227	2247	2267	2287	2308	2328
46000	2348	2368	2388	2409	2429	2449	2469	2489	2509	2529
47000	2549	2569	2589	2609	2628	2648	2668	2688	2708	2728
48000	2747	2767	2787	2806	2826	2845	2865	2885	2904	2924
49000	2943	2963	2982	3002	3021	3040	3060	3079	3098	3118
50000	3137	3156	3175	3194	3214	3233	3252	3271	3290	3309
51000	3328	3347	3366	3385	3404	3423	3442	3461	3479	3498
52000	3517	3536	3554	3573	3592	3610	3629	3648	3666	3685
53000	3704	3722	3740	3759	3777	3796	3814	3832	3851	3869
54000	3888	3906	3924	3942	3960	3978	3996	4015	4033	4051
55000	4069	4087	4105	4123	4141	4159	4176	4194	4212	4230
56000	4248	4266	4283	4301	4319	4336	4354	4372	4389	4407
57000	4425	4442	4459	4477	4494	4512	4529	4546	4564	4581
58000	4599	4616	4633	4650	4667	4684	4701	4719	4736	4753
59000	4770	4787	4804	4821	4838	4855	4871	4888	4905	4922
60000	4939	4956	4972	4989	5006	5022	5039	5056	5072	5089
61000	5106	5122	5138	5155	5171	5188	5204	5220	5237	5253
62000	5270	5286	5302	5318	5334	5350	5366	5383	5399	5415
63000	5431	5447	5463	5479	5494	5510	5526	5542	5558	5574
64000	5590	5605	5621	5636	5652	5668	5683	5699	5714	5730
65000	5746	5761	5776	5791	5807	5822	5837	5853	5868	5883
66000	5899	5914	5929	5944	5959	5974	5989	6004	6019	6034
67000	6049	6063	6078	6093	6107	6122	6137	6151	6166	6181

NOTE – OPTIMUM WEIGHT FOR PRESSURE ALTITUDE IS 58200 KG
    THRUST LIMITED WEIGHT FOR ISA + 10 AND COLDER EXCEEDS STRUCTURAL LIMIT
    THRUST LIMITED WEIGHT FOR ISA + 15   EXCEEDS STRUCTURAL LIMIT
    THRUST LIMITED WEIGHT FOR ISA + 20   IS 66400 KG
    ADJUSTMENTS FOR OPERATION AT NON-STANDARD TEMPERATURES---
    INCREASE FUEL REQUIRED BY   0.6 PERCENT PER 10 DEGREES C ABOVE ISA
    DECREASE FUEL REQUIRED BY   0.6 PERCENT PER 10 DEGREES C BELOW ISA
    INCREASE TAS BY 1 KNOT PER DEGREE C ABOVE ISA
    DECREASE TAS BY 1 KNOT PER DEGREE C BELOW ISA

Figure 4.5.3.2   Mach 0.74 Cruise

All Engines   Maximum Cruise Thrust Limits   A/C Auto

PRESSURE ALTITUDE   34000Ft   **TAS   428 Kts**

GROSS WT. KG	0	100	200	300	400	500	600	700	800	900
	CRUISE DISTANCE NAUTICAL AIR MILES									
35000	0	22	45	68	91	114	137	160	183	206
36000	229	252	275	297	320	343	365	388	411	434
37000	456	479	502	524	547	569	592	614	637	659
38000	682	704	726	749	771	793	816	838	860	883
39000	905	927	949	971	993	1015	1037	1060	1082	1104
40000	1126	1148	1170	1191	1213	1235	1257	1279	1301	1323
41000	1344	1366	1388	1409	1431	1453	1474	1496	1517	1539
42000	1561	1582	1603	1625	1646	1668	1689	1710	1732	1753
43000	1775	1796	1817	1838	1859	1880	1902	1923	1944	1965
44000	1986	2007	2028	2049	2070	2091	2112	2132	2153	2174
45000	2195	2216	2236	2257	2278	2298	2319	2340	2360	2381
46000	2402	2422	2442	2463	2483	2504	2524	2544	2565	2585
47000	2606	2626	2646	2666	2686	2706	2726	2747	2767	2787
48000	2807	2827	2847	2867	2886	2906	2926	2946	2966	2986
49000	3006	3025	3045	3064	3084	3104	3123	3143	3163	3182
50000	3202	3221	3240	3260	3279	3298	3318	3337	3356	3376
51000	3395	3414	3433	3452	3471	3491	3510	3529	3548	3567
52000	3586	3605	3624	3642	3661	3680	3699	3718	3736	3755
53000	3774	3792	3811	3830	3848	3867	3885	3904	3922	3941
54000	3959	3978	3996	4014	4032	4051	4069	4087	4105	4124
55000	4142	4160	4178	4196	4214	4232	4250	4268	4286	4304
56000	4322	4339	4357	4375	4393	4410	4428	4446	4464	4481
57000	4499	4516	4534	4551	4569	4586	4604	4621	4639	4656
58000	4673	4691	4708	4725	4742	4759	4777	4794	4811	4828
59000	4845	4862	4879	4896	4913	4930	4947	4963	4980	4997
60000	5014	5031	5047	5064	5081	5097	5114	5130	5147	5164
61000	5180	5196	5213	5229	5245	5262	5278	5294	5311	5327
62000	5343	5359	5375	5391	5407	5423	5439	5455	5471	5487
63000	5503	5519	5534	5550	5566	5581	5597	5613	5628	5644
64000	5660	5675	5690	5705	5721	5736	5751	5767	5782	5797
65000	5812	5827	5842	5857	5872	5887	5902	5917	5932	5947
66000	5962	5976	5991	6005	6020	6034	6049	6063	6078	6093
67000	6107	6121	6135	6150	6164	6178	6192	6206	6220	6234

NOTE – OPTIMUM WEIGHT FOR PRESSURE ALTITUDE IS 55500 KG
   THRUST LIMITED WEIGHT FOR ISA + 10 AND COLDER IS 67100 KG
   THRUST LIMITED WEIGHT FOR ISA + 15   IS 65700 KG
   THRUST LIMITED WEIGHT FOR ISA + 20   IS 64000 KG
   ADJUSTMENTS FOR OPERATION AT NON-STANDARD TEMPERATURES---
   INCREASE FUEL REQUIRED BY   0.6 PERCENT PER 10 DEGREES C ABOVE ISA
   DECREASE FUEL REQUIRED BY   0.6 PERCENT PER 10 DEGREES C BELOW ISA
   INCREASE TAS BY 1 KNOT PER DEGREE C ABOVE ISA
   DECREASE TAS BY 1 KNOT PER DEGREE C BELOW ISA

Figure 4.5.3.2   Mach 0.74 Cruise

All Engines   Maximum Cruise Thrust Limits   A/C Auto

PRESSURE ALTITUDE   35000Ft   **TAS   426 Kts**

GROSS WT. KG	0	100	200	300	400	500	600	700	800	900
	CRUISE DISTANCE NAUTICAL AIR MILES									
35000	0	23	47	70	94	117	141	164	188	212
36000	235	258	282	305	328	352	375	398	422	445
37000	468	491	514	537	561	584	607	630	653	676
38000	699	722	745	768	790	813	836	859	882	904
39000	927	950	972	995	1018	1040	1063	1085	1108	1131
40000	1153	1175	1198	1220	1242	1265	1287	1309	1332	1354
41000	1376	1398	1420	1443	1465	1487	1509	1531	1553	1575
42000	1597	1619	1641	1662	1684	1706	1728	1749	1771	1793
43000	1815	1836	1858	1879	1901	1922	1944	1965	1987	2009
44000	2030	2051	2073	2094	2115	2136	2157	2179	2200	2221
45000	2242	2263	2284	2305	2326	2347	2368	2389	2410	2431
46000	2452	2473	2493	2514	2535	2555	2576	2597	2617	2638
47000	2659	2679	2699	2720	2740	2761	2781	2801	2822	2842
48000	2862	2883	2903	2923	2943	2963	2983	3003	3023	3043
49000	3063	3083	3103	3123	3143	3162	3182	3202	3222	3242
50000	3261	3281	3300	3320	3339	3359	3378	3398	3417	3437
51000	3456	3476	3495	3514	3533	3552	3572	3591	3610	3629
52000	3648	3667	3686	3705	3724	3743	3762	3781	3800	3819
53000	3838	3856	3875	3893	3912	3931	3949	3968	3987	4005
54000	4024	4042	4060	4079	4097	4115	4134	4152	4170	4189
55000	4207	4225	4243	4261	4279	4297	4315	4333	4351	4369
56000	4387	4405	4423	4441	4458	4476	4494	4511	4529	4547
57000	4565	4582	4599	4617	4634	4652	4669	4686	4704	4721
58000	4739	4756	4773	4790	4807	4824	4841	4858	4875	4892
59000	4910	4926	4943	4960	4976	4993	5010	5027	5043	5060
60000	5077	5093	5110	5126	5142	5159	5175	5192	5208	5224
61000	5241	5257	5273	5289	5305	5321	5337	5353	5369	5385
62000	5401	5416	5432	5447	5463	5479	5494	5510	5525	5541
63000	5556	5572	5587	5602	5617	5632	5647	5663	5678	5693
64000	5708	5723	5738	5752	5767	5782	5796	5811	5826	5841
65000	5855	5870	5884	5898	5912	5927	5941	5955	5969	5984

NOTE – OPTIMUM WEIGHT FOR PRESSURE ALTITUDE IS 53000 KG
THRUST LIMITED WEIGHT FOR ISA + 10 AND COLDER IS 64500 KG
THRUST LIMITED WEIGHT FOR ISA + 15   IS 63100 KG
THRUST LIMITED WEIGHT FOR ISA + 20   IS 61600 KG
ADJUSTMENTS FOR OPERATION AT NON-STANDARD TEMPERATURES---
INCREASE FUEL REQUIRED BY   0.6 PERCENT PER 10 DEGREES C ABOVE ISA
DECREASE FUEL REQUIRED BY   0.6 PERCENT PER 10 DEGREES C BELOW ISA
INCREASE TAS BY 1 KNOT PER DEGREE C ABOVE ISA
DECREASE TAS BY 1 KNOT PER DEGREE C BELOW ISA

Figure 4.5.3.2 Mach 0.74 Cruise

All Engines   Maximum Cruise Thrust Limits   A/C Auto

PRESSURE ALTITUDE   36000Ft   TAS   425 Kts

GROSS WT. KG	0	100	200	300	400	500	600	700	800	900
	CRUISE DISTANCE NAUTICAL AIR MILES									
35000	0	24	48	72	96	120	144	168	193	217
36000	241	265	289	312	336	360	384	408	432	456
37000	480	503	527	550	574	598	621	645	668	692
38000	716	739	762	786	809	832	855	879	902	925
39000	949	927	995	1018	1041	1064	1087	1110	1133	1156
40000	1179	1202	1225	1247	1270	1293	1316	1338	1361	1384
41000	1407	1429	1451	1474	1496	1519	1541	1564	1586	1609
42000	1631	1653	1675	1697	1720	1742	1764	1786	1808	1830
43000	1852	1874	1896	1918	1940	1962	1984	2005	2027	2049
44000	2071	2092	2114	2135	2157	2178	2200	2222	2243	2265
45000	2286	2307	2329	2350	2371	2392	2413	2435	2456	2477
46000	2498	2519	2540	2561	2582	2603	2624	2644	2665	2686
47000	2707	2728	2748	2769	2789	2810	2831	2851	2872	2892
48000	2913	2933	2953	2974	2994	3014	3034	3055	3075	3095
49000	3115	3135	3155	3175	3195	3215	3235	3255	3275	3295
50000	3315	3334	3354	3374	3393	3413	3432	3452	3472	3491
51000	3511	3530	3549	3569	3588	3607	3627	3646	3665	3684
52000	3704	3723	3742	3761	3780	3799	3818	3837	3856	3874
53000	3893	3912	3931	3949	3968	3987	4005	4024	4043	4061
54000	4080	4098	4116	4135	4153	4171	4190	4208	4226	4245
55000	4263	4281	4299	4317	4335	4353	4371	4389	4406	4424
56000	4442	4460	4478	4495	4513	4530	4548	4565	4583	4601
57000	4618	4635	4653	4670	4687	4704	4721	4738	4756	4773
58000	4790	4807	4824	4840	4857	4874	4891	4907	4924	4941
59000	4958	4974	4990	5007	5023	5039	5056	5072	5088	5105
60000	5121	5137	5153	5169	5185	5200	5216	5232	5248	5264
61000	5280	5295	5310	5326	5341	5357	5372	5387	5403	5418
62000	5434	5448	5463	5478	5493	5508	5523	5538	5553	5568

NOTE – OPTIMUM WEIGHT FOR PRESSURE ALTITUDE IS 50500 KG
THRUST LIMITED WEIGHT FOR ISA + 10 AND COLDER IS 61800 KG
THRUST LIMITED WEIGHT FOR ISA + 15   IS 60500 KG
THRUST LIMITED WEIGHT FOR ISA + 20   IS 59200 KG
ADJUSTMENTS FOR OPERATION AT NON-STANDARD TEMPERATURES---
INCREASE FUEL REQUIRED BY   0.6 PERCENT PER 10 DEGREES C ABOVE ISA
DECREASE FUEL REQUIRED BY   0.6 PERCENT PER 10 DEGREES C BELOW ISA
INCREASE TAS BY 1 KNOT PER DEGREE C ABOVE ISA
DECREASE TAS BY 1 KNOT PER DEGREE C BELOW ISA

Figure 4.5.3.2   Mach 0.74 Cruise

All Engines   Maximum Cruise Thrust Limits   A/C Auto

PRESSURE ALTITUDE   37000Ft   **TAS   424 Kts**

GROSS WT. KG	0	100	200	300	400	500	600	700	800	900
				CRUISE DISTANCE NAUTICAL AIR MILES						
35000	0	24	49	73	98	123	147	172	197	221
36000	246	270	295	319	344	368	392	417	441	465
37000	490	514	538	562	586	610	634	658	682	706
38000	730	754	778	801	825	849	873	896	920	944
39000	968	991	1014	1038	1061	1085	1108	1131	1155	1178
40000	1202	1225	1248	1271	1294	1317	1340	1363	1386	1409
41000	1433	1455	1478	1501	1524	1546	1569	1592	1615	1637
42000	1660	1682	1705	1727	1750	1772	1795	1817	1839	1862
43000	1884	1906	1928	1950	1973	1995	2017	2039	2061	2083
44000	2105	2127	2148	2170	2192	2214	2235	2257	2279	2301
45000	2322	2344	2365	2386	2408	2429	2451	2472	2493	2515
46000	2536	2557	2578	2599	2620	2641	2662	2683	2704	2725
47000	2746	2767	2788	2808	2829	2850	2870	2891	2912	2932
48000	2953	2973	2994	3014	3035	3055	3075	3096	3116	3136
49000	3157	3177	3197	3217	3237	3257	3277	3297	3317	3336
50000	3356	3376	3396	3415	3435	3455	3474	3494	3514	3533
51000	3553	3572	3591	3611	3630	3649	3668	3688	3707	3726
52000	3746	3764	3783	3802	3821	3840	3859	3878	3897	3916
53000	3934	3953	3971	3990	4008	4027	4045	4064	4082	4101
54000	4119	4138	4156	4174	4192	4210	4228	4246	4264	4282
55000	4300	4318	4335	4353	4371	4388	4406	4424	4441	4459
56000	4477	4494	4511	4528	4545	4562	4579	4597	4614	4631
57000	4648	4665	4681	4698	4715	4731	4748	4765	4781	4798
58000	4815	4831	4847	4863	4879	4895	4911	4927	4944	4960
59000	4976	4991	5007	5023	5038	5054	5069	5085	5101	5116

NOTE -- OPTIMUM WEIGHT FOR PRESSURE ALTITUDE IS 48000 KG
THRUST LIMITED WEIGHT FOR ISA + 10 AND COLDER IS 58700 KG
THRUST LIMITED WEIGHT FOR ISA + 15   IS 57500 KG
THRUST LIMITED WEIGHT FOR ISA + 20   IS 56300 KG
ADJUSTMENTS FOR OPERATION AT NON-STANDARD TEMPERATURES---
INCREASE FUEL REQUIRED BY   0.6 PERCENT PER 10 DEGREES C ABOVE ISA
DECREASE FUEL REQUIRED BY   0.6 PERCENT PER 10 DEGREES C BELOW ISA
INCREASE TAS BY 1 KNOT PER DEGREE C ABOVE ISA
DECREASE TAS BY 1 KNOT PER DEGREE C BELOW ISA

Figure 4.5.3.3   Mach 0.78 Cruise

All Engines   Maximum Cruise Thrust Limits   A/C Auto

PRESSURE ALTITUDE   29000Ft   **TAS   462 Kts**

GROSS WT. KG	0	100	200	300	400	500	600	700	800	900
	CRUISE DISTANCE NAUTICAL AIR MILES									
35000	0	18	37	55	74	93	111	130	149	167
36000	186	205	223	242	260	279	298	316	335	353
37000	372	390	409	427	446	464	483	501	520	538
38000	557	575	594	612	631	649	667	686	704	723
39000	741	759	778	796	814	833	851	869	888	906
40000	924	943	961	979	997	1015	1034	1052	1070	1088
41000	1106	1125	1143	1161	1179	1197	1215	1233	1251	1269
42000	1288	1306	1324	1342	1360	1378	1396	1414	1432	1450
43000	1468	1485	1503	1521	1539	1557	1575	1593	1611	1628
44000	1646	1664	1682	1700	1717	1735	1753	1771	1788	1806
45000	1824	1842	1859	1877	1894	1912	1930	1947	1965	1982
46000	2000	2018	2035	2053	2070	2087	2105	2122	2140	2157
47000	2175	2192	2210	2227	2244	2262	2279	2296	2314	2331
48000	2348	2365	2383	2400	2417	2434	2451	2469	2486	2503
49000	2520	2537	2554	2571	2588	2605	2622	2639	2656	2673
50000	2690	2707	2724	2741	2758	2775	2792	2808	2825	2842
51000	2859	2876	2892	2909	2926	2942	2959	2976	2993	3009
52000	3026	3042	3059	3075	3092	3108	3125	3141	3158	3175
53000	3191	3207	3224	3240	3256	3273	3289	3305	3322	3338
54000	3354	3371	3387	3403	3419	3435	3451	3467	3484	3500
55000	3516	3532	3548	3564	3580	3596	3612	3628	3644	3660
56000	3676	3691	3707	3723	3739	3755	3770	3786	3802	3818
57000	3834	3849	3865	3880	3896	3912	3927	3943	3958	3974
58000	3989	4005	4020	4036	4051	4066	4082	4097	4113	4128
59000	4143	4159	4174	4189	4204	4220	4235	4250	4265	4280
60000	4296	4311	4326	4341	4356	4371	4386	4401	4416	4431
61000	4446	4460	4475	4490	4505	4520	4535	4549	4564	4579
62000	4594	4608	4623	4638	4652	4667	4682	4696	4711	4726
63000	4740	4755	4769	4783	4798	4812	4827	4841	4856	4870
64000	4885	4899	4913	4927	4942	4956	4970	4984	4999	5013
65000	5027	5041	5055	5069	5083	5098	5112	5126	5140	5154
66000	5168	5182	5196	5210	5223	5237	5251	5265	5279	5293
67000	5307	5320	5334	5348	5361	5375	5389	5402	5416	5430

NOTE – OPTIMUM WEIGHT FOR PRESSURE ALTITUDE IS 67000 KG
THRUST LIMITED WEIGHT FOR ISA + 10 AND COLDER EXCEEDS STRUCTURAL LIMIT
THRUST LIMITED WEIGHT FOR ISA + 15   EXCEEDS STRUCTURAL LIMIT
THRUST LIMITED WEIGHT FOR ISA + 20   EXCEEDS STRUCTURAL LIMIT
ADJUSTMENTS FOR OPERATION AT NON-STANDARD TEMPERATURES---
INCREASE FUEL REEQUIRED BY   .6 PERCENT PER 10 DEGREES C ABOVE ISA
DECREASE FUEL REQUIRED BY   .6 PERCENT PER 10 DEGREES C BELOW ISA
INCREASE TAS BY 1 KNOT PER DEGREE C ABOVE ISA
DECREASE TAS BY 1 KNOT PER DEGREE C BELOW ISA

Figure 4.5.3.3   Mach 0.78 Cruise

All Engines   Maximum Cruise Thrust Limits   A/C Auto

PRESSURE ALTITUDE   30000Ft   **TAS   460 Kts**

GROSS WT. KG	0	100	200	300	400	500	600	700	800	900
				CRUISE DISTANCE NAUTICAL AIR MILES						
35000	0	19	38	57	77	96	115	135	154	173
36000	193	212	231	250	269	289	308	327	346	366
37000	385	404	423	442	461	480	499	519	538	557
38000	576	595	614	633	652	671	690	709	728	747
39000	766	785	804	823	842	860	879	898	917	936
40000	955	974	993	1011	1030	1049	1068	1086	1105	1124
41000	1143	1161	1180	1199	1217	1236	1255	1273	1292	1311
42000	1329	1348	1366	1385	1403	1422	1440	1459	1477	1496
43000	1514	1533	1551	1569	1588	1606	1624	1643	1661	1679
44000	1698	1716	1734	1752	1771	1789	1807	1825	1844	1862
45000	1880	1898	1916	1934	1952	1970	1988	2006	2024	2042
46000	2061	2078	2096	2114	2132	2150	2168	2186	2204	2222
47000	2239	2257	2275	2293	2310	2328	2346	2363	2381	2399
48000	2417	2434	2452	2469	2487	2504	2522	2539	2557	2574
49000	2592	2609	2627	2644	2661	2679	2696	2713	2731	2748
50000	2765	2783	2800	2817	2834	2851	2868	2886	2903	2920
51000	2937	2954	2971	2988	3005	3022	3039	3056	3073	3090
52000	3107	3123	3140	3157	3174	3191	3207	3224	3241	3258
53000	3274	3291	3308	3324	3341	3357	3374	3390	3407	3423
54000	3440	3456	3473	3489	3506	3522	3538	3555	3571	3587
55000	3604	3620	3636	3652	3668	3684	3701	3717	3733	3749
56000	3765	3781	3797	3813	3829	3845	3861	3877	3893	3909
57000	3925	3940	3956	3972	3988	4003	4019	4035	4050	4066
58000	4082	4097	4113	4128	4144	4160	4175	4191	4206	4222
59000	4237	4252	4268	4283	4298	4314	4329	4344	4360	4375
60000	4390	4405	4421	4436	4451	4466	4481	4496	4511	4526
61000	4541	4556	4571	4586	4601	4616	4631	4646	4661	4676
62000	4691	4705	4720	4735	4749	4764	4779	4794	4808	4823
63000	4838	4852	4867	4881	4896	4910	4925	4939	4954	4968
64000	4983	4997	5011	5025	5040	5054	5068	5083	5097	5111
65000	5125	5140	5154	5168	5182	5196	5210	5224	5238	5252
66000	5266	5280	5294	5308	5322	5335	5349	5363	5377	5391
67000	5405	5418	5432	5446	5459	5473	5487	5500	5514	5528

NOTE - OPTIMUM WEIGHT FOR PRESSURE ALTITUDE IS 64200 KG

THRUST LIMITED WEIGHT FOR ISA + 10 AND COLDER EXCEEDS STRUCTURAL LIMIT
THRUST LIMITED WEIGHT FOR ISA + 15   EXCEEDS STRUCTURAL LIMIT
THRUST LIMITED WEIGHT FOR ISA + 20   EXCEEDS STRUCTURAL LIMIT
ADJUSTMENTS FOR OPERATION AT NON-STANDARD TEMPERATURES---
INCREASE FUEL REQUIRED BY   .6 PERCENT PER 10 DEGREES C ABOVE ISA
DECREASE FUEL REQUIRED BY   .6 PERCENT PER 10 DEGREES C BELOW ISA
INCREASE TAS BY 1 KNOT PER DEGREE C ABOVE ISA
DECREASE TAS BY 1 KNOT PER DEGREE C BELOW ISA.

Figure 4.5.3.3   Mach 0.78 Cruise

All Engines   Maximum Cruise Thrust Limits   A/C Auto

PRESSURE ALTITUDE   31000Ft   TAS   458 Kts

GROSS WT. KG	0	100	200	300	400	500	600	700	800	900
	CRUISE DISTANCE NAUTICAL AIR MILES									
35000	0	19	39	59	79	99	119	139	159	179
36000	199	219	239	259	278	298	318	338	358	378
37000	398	417	437	457	476	496	516	536	555	575
38000	595	614	634	654	673	693	712	732	752	771
39000	791	810	830	849	869	888	908	927	946	966
40000	985	1005	1024	1043	1063	1082	1101	1121	1140	1159
41000	1178	1198	1217	1236	1255	1274	1293	1313	1332	1351
42000	1370	1389	1408	1427	1446	1465	1484	1503	1522	1541
43000	1560	1579	1598	1616	1635	1654	1673	1692	1711	1729
44000	1748	1767	1785	1804	1823	1841	1860	1879	1897	1916
45000	1935	1953	1971	1990	2008	2027	2045	2064	2082	2101
46000	2119	2137	2156	2174	2192	2210	2229	2247	2265	2283
47000	2302	2320	2338	2356	2374	2392	2410	2428	2446	2464
48000	2482	2500	2518	2536	2554	2571	2589	2607	2625	2643
49000	2661	2678	2696	2713	2731	2749	2766	2784	2802	2819
50000	2837	2854	2872	2889	2907	2924	2941	2959	2976	2994
51000	3011	3028	3046	3063	3080	3097	3114	3`131	3149	3166
52000	3183	3200	3217	3234	3251	3268	3285	3302	3319	3336
53000	3353	3370	3386	3403	3420	3437	3453	3470	3487	3504
54000	3520	3537	3553	3570	3586	3603	3619	3636	3652	3669
55000	3685	3702	3718	3734	3751	3767	3783	3799	3816	3832
56000	3848	3864	3880	3896	3913	3929	3945	3961	3977	3993
57000	4009	4025	4041	4056	4072	4088	4104	4120	4136	4152
58000	4167	4183	4199	4214	4230	4245	4261	4277	4292	4308
59000	4324	4339	4354	4370	4385	4401	4416	4431	4447	4462
60000	4478	4493	4508	4523	4538	4553	4569	4584	4599	4614
61000	4629	4644	4659	4674	4689	4704	4719	4734	4749	4764
62000	4779	4793	4808	4823	4838	4852	4867	4882	4896	4911
63000	4926	4940	4955	4969	4984	4998	5013	5027	5042	5056
64000	5071	5085	5099	5113	5128	5142	5156	5170	5184	5199
65000	5213	5227	5241	5255	5269	5283	5297	5311	5325	5339
66000	5353	5367	5381	5394	5408	5422	5436	5449	5463	5477
67000	5491	5504	5518	5531	5545	5558	5572	5585	5599	5612

NOTE - OPTIMUM WEIGHT FOR PRESSURE ALTITUDE IS 61300 KG
      THRUST LIMITED WEIGHT FOR ISA + 10 AND COLDER EXCEEDS STRUCTURAL LIMIT
      THRUST LIMITED WEIGHT FOR ISA + 15   EXCEEDS STRUCTURAL LIMIT
      THRUST LIMITED WEIGHT FOR ISA + 20   IS 63500KG
    ADJUSTMENTS FOR OPERATION AT NON-STANDARD TEMPERATURES---
      INCREASE FUEL REQUIRED BY   .6 PERCENT PER 10 DEGREES C ABOVE ISA
      DECREASE FUEL REQUIRED BY   .6 PERCENT PER 10 DEGREES C BELOW ISA
      INCREASE TAS BY 1 KNOT PER DEGREE C ABOVE ISA
      DECREASE TAS BY 1 KNOT PER DEGREE C BELOW ISA.

Figure 4.5.3.3   Mach 0.78 Cruise

All Engines   Maximum Cruise Thrust Limits   A/C Auto

PRESSURE ALTITUDE   33000 ft   TAS   454 Kts

GROSS WT. KG	0	100	200	300	400	500	600	700	800	900
	CRUISE DISTANCE NAUTICAL AIR MILES									
35000	0	21	42	63	84	106	127	148	169	191
36000	212	233	254	275	296	317	338	359	381	402
37000	423	444	465	485	506	527	548	569	590	611
38000	632	652	673	694	715	735	756	777	797	818
39000	839	859	880	900	921	941	962	982	1003	1023
40000	1044	1064	1084	1105	1125	1145	1166	1186	1206	1227
41000	1247	1267	1287	1307	1327	1347	1367	1387	1407	1428
42000	1448	1467	1487	1507	1527	1547	1567	1587	1606	1626
43000	1646	1666	1685	1705	1724	1744	1764	1783	1803	1822
44000	1842	1861	1881	1900	1920	1939	1958	1978	1997	2016
45000	2036	2055	2074	2093	2112	2131	2150	2169	2189	2208
46000	2227	2246	2264	2283	2302	2321	2340	2359	2378	2396
47000	2415	2434	2452	2471	2490	2508	2527	2545	2564	2582
48000	2601	2619	2638	2656	2674	2693	2711	2729	2748	2766
49000	2784	2802	2820	2838	2856	2874	2893	2911	2929	2947
50000	2965	2982	3000	3018	3036	3054	3071	3089	3107	3125
51000	3142	3160	3177	3195	3212	3230	3248	3265	3283	3300
52000	3318	3335	3352	3369	3387	3404	3421	3438	3456	3473
53000	3490	3507	3524	3541	3558	3575	3592	3609	3626	3643
54000	3660	3677	3693	3710	3727	3744	3760	3777	3794	3810
55000	3827	3844	3860	3877	3893	3909	3926	3942	3959	3975
56000	3992	4008	4024	4040	4056	4073	4089	4105	4121	4137
57000	4153	4169	4185	4201	4217	4233	4249	4265	4281	4296
58000	4312	4328	4344	4359	4375	4390	4406	4422	4437	4453
59000	4469	4484	4499	4515	4530	4545	4561	4576	4591	4607
60000	4622	4637	4652	4667	4682	4697	4712	4727	4742	4757
61000	4772	4787	4802	4817	4832	4846	4861	4876	4891	4905
62000	4920	4935	4949	4963	4978	4992	5007	5021	5036	5050
63000	5065	5079	5093	5107	5121	5136	5150	5164	5178	5192
64000	5206	5220	5234	5248	5262	5275	5289	5303	5317	5331

NOTE - OPTIMUM WEIGHT FOR PRESSURE ALTITUDE IS 56000 KG
THRUST LIMITED WEIGHT FOR ISA + 10 AND COLDER IS 63700 KG
THRUST LIMITED WEIGHT FOR ISA + 15   IS 61600 KG
THRUST LIMITED WEIGHT FOR ISA + 20   IS 59500 KG
ADJUSTMENTS FOR OPERATION AT NON-STANDARD TEMPERATURES---
INCREASE FUEL REQUIRED BY   .6 PERCENT PER 10 DEGREES C ABOVE ISA
DECREASE FUEL REQUIRED BY   .6 PERCENT PER 10 DEGREES C BELOW ISA
INCREASE TAS BY 1 KNOT PER DEGREE C ABOVE ISA
DECREASE TAS BY 1 KNOT PER DEGREE C BELOW ISA.

Figure 4.5.3.3　Mach 0.78 Cruise

All Engines　　Maximum Cruise Thrust Limits　　A/C Auto

PRESSURE ALTITUDE　　35000 ft　　　　TAS　449 Kts

GROSS WT. KG	0	100	200	300	400	500	600	700	800	900
	CRUISE DISTANCE NAUTICAL AIR MILES									
35000	0	22	44	67	89	112	134	156	179	201
36000	224	246	268	290	313	335	357	379	401	423
37000	446	468	490	511	533	555	577	599	621	643
38000	665	687	708	730	752	773	795	817	838	860
39000	882	903	924	946	967	988	1010	1031	1053	1074
40000	1095	1116	1137	1159	1180	1201	1222	1243	1264	1285
41000	1306	1327	1348	1368	1389	1410	1431	1452	1472	1493
42000	1514	1534	1555	1575	1596	1616	1637	1657	1678	1698
43000	1719	1739	1759	1779	1799	1820	1840	1860	1880	1900
44000	1920	1940	1960	1980	2000	2020	2040	2059	2079	2099
45000	2119	2138	2158	2178	2197	2217	2236	2256	2275	2295
46000	2314	2333	2353	2372	2391	2410	2430	2449	2468	2487
47000	2506	2525	2544	2563	2582	2601	2620	2639	2658	2677
48000	2695	2714	2733	2751	2770	2788	2807	2826	2844	2863
49000	2881	2900	2918	2936	2954	2973	2991	3009	3028	3046
50000	3064	3082	3100	3118	3136	3154	3172	3190	3208	3226
51000	3244	3261	3279	3297	3314	3332	3349	3367	3385	3402
52000	3420	3437	3454	3472	3489	3506	3524	3541	3558	3576
53000	3593	3610	3627	3644	3661	3678	3695	3712	3729	3745
54000	3762	3779	3796	3812	3829	3846	3862	3879	3895	3912
55000	3929	3945	3961	3978	3994	4010	4026	4043	4059	4075
56000	4092	4107	4123	4139	4155	4171	4187	4203	4219	4235
57000	4251	4266	4282	4297	4313	4329	4344	4360	4375	4391
58000	4406	4422	4437	4452	4467	4482	4498	4513	4528	4543
59000	4558	4573	4588	4603	4618	4632	4647	4662	4677	4692

NOTE - OPTIMUM WEIGHT FOR PRESSURE ALTITUDE IS 51100 KG
　　　　THRUST LIMITED WEIGHT FOR ISA + 10 AND COLDER IS 58800 KG
　　　　THRUST LIMITED WEIGHT FOR ISA + 15　　　　IS 57200 KG
　　　　THRUST LIMITED WEIGHT FOR ISA + 20　　　　IS 55500 KG
　　　ADJUSTMENTS FOR OPERATION AT NON-STANDARD TEMPERATURES---
　　　INCREASE FUEL REQUIRED BY　　.6 PERCENT PER 10 DEGREES C ABOVE ISA
　　　DECREASE FUEL REQUIRED BY　　.6 PERCENT PER 10 DEGREES C BELOW ISA
　　　INCREASE TAS BY 1 KNOT PER DEGREE C ABOVE ISA
　　　DECREASE TAS BY 1 KNOT PER DEGREE C BELOW ISA.

Figure 4.5.3.3   Mach 0.78 Cruise

All Engines   Maximum Cruise Thrust Limits   A/C Auto

PRESSURE ALTITUDE   37000 ft   **TAS   447 Kts**

GROSS WT. KG	0	100	200	300	400	500	600	700	800	900
	CRUISE DISTANCE NAUTICAL AIR MILES									
35000	0	23	46	70	93	116	140	163	187	210
36000	233	256	279	303	326	349	372	395	418	441
37000	464	487	509	532	555	577	600	623	645	668
38000	691	713	736	758	780	803	825	847	870	892
39000	914	936	958	980	1002	1024	1046	1068	1090	1112
40000	1134	1156	1177	1199	1220	1242	1263	1285	1307	1328
41000	1350	1371	1392	1413	1435	1456	1477	1498	1519	1541
42000	1562	1583	1604	1624	1645	1666	1687	1708	1729	1750
43000	1770	1791	1811	1832	1852	1873	1893	1914	1934	1955
44000	1975	1995	2015	2035	2056	2076	2096	2116	2136	2156
45000	2176	2196	2216	2235	2255	2275	2295	2314	2334	2354
46000	2373	2393	2412	2431	2451	2470	2489	2509	2528	2547
47000	2567	2586	2605	2624	2642	2661	2680	2699	2718	2737
48000	2756	2775	2793	2812	2830	2849	2867	2886	2905	2923
49000	2942	2960	2978	2996	3014	3032	3050	3069	3087	3105
50000	3123	3141	3159	3176	3194	3212	3229	3247	3265	3283
51000	3300	3318	3335	3352	3370	3387	3404	3422	3439	3456
52000	3474	3490	3507	3524	3541	3558	3575	3592	3608	3625
53000	3642	3659	3675	3691	3708	3724	3741	3757	3774	3790
54000	3806	3822	3838	3854	3870	3886	3902	3918	3934	3950

NOTE - OPTIMUM WEIGHT FOR PRESSURE ALTITUDE IS 46500 KG
    THRUST LIMITED WEIGHT FOR ISA + 10 AND COLDER IS 53700 KG
    THRUST LIMITED WEIGHT FOR ISA + 15   IS 52300 KG
    THRUST LIMITED WEIGHT FOR ISA + 20   IS 50900 KG
    ADJUSTMENTS FOR OPERATION AT NON-STANDARD TEMPERATURES---
    INCREASE FUEL REQUIRED BY   .6 PERCENT PER 10 DEGREES C ABOVE ISA
    DECREASE FUEL REQUIRED BY   .6 PERCENT PER 10 DEGREES C BELOW ISA
    INCREASE TAS BY 1 KNOT PER DEGREE C ABOVE ISA
    DECREASE TAS BY 1 KNOT PER DEGREE C BELOW ISA.

Figure 4.5.3.4   LOW LEVEL CRUISE   300 KIAS

All Engines   Maximum Cruise Thrust Limits   A/C Auto

PRESSURE ALTITUDE   14000Ft   **TAS   366 Kts**

GROSS WT KG	0	100	200	300	400	500	600	700	800	900
				CRUISE DISTANCE NAUTICAL AIR MILES						
35000	0	14	28	43	57	71	86	100	114	129
36000	143	158	172	186	200	215	229	243	258	272
37000	286	301	315	329	344	358	372	386	401	415
38000	429	444	458	472	486	500	515	529	543	557
39000	572	586	600	614	628	643	657	671	685	699
40000	713	728	742	756	770	784	798	812	826	841
41000	855	869	883	897	911	925	939	953	967	981
42000	995	1009	1023	1037	1051	1065	1079	1093	1107	1121
43000	1135	1149	1163	1177	1191	1205	1219	1233	1247	1261
44000	1275	1289	1303	1317	1330	1344	1358	1372	1386	1400
45000	1414	1427	1441	1455	1469	1483	1496	1510	1524	1538
46000	1552	1565	1579	1593	1607	1620	1634	1648	1662	1675
47000	1689	1703	1716	1730	1744	1757	1771	1785	1798	1812
48000	1826	1839	1853	1866	1880	1894	1907	1921	1934	1948
49000	1962	1975	1989	2002	2016	2029	2043	2056	2070	2083
50000	2097	2110	2124	2137	2151	2164	2177	2191	2204	2218
51000	2231	2245	2258	2271	2285	2298	2311	2325	2338	2352
52000	2365	2378	2391	2405	2418	2431	2445	2458	2471	2484
53000	2498	2511	2524	2537	2551	2564	2577	2590	2603	2617
54000	2630	2643	2656	2669	2682	2695	2709	2722	2735	2748
55000	2761	2774	2787	2800	2813	2826	2839	2852	2865	2879
56000	2892	2905	2917	2930	2943	2956	2969	2982	2995	3008
57000	3021	3034	3047	3060	3073	3086	3098	3111	3124	3137
58000	3150	3163	3176	3188	3201	3214	3227	3240	3252	3265
59000	3278	3291	3303	3316	3329	3341	3354	3367	3380	3392
60000	3405	3418	3430	3443	3455	3468	3481	3493	3506	3519
61000	3531	3544	3556	3569	3581	3594	3606	3619	3631	3644
62000	3656	3669	3681	3694	3706	3719	3731	3744	3756	3768
63000	3781	3793	3806	3818	3830	3843	3855	3867	3880	3892
64000	3904	3917	3929	3941	3953	3966	3978	3990	4002	4015
65000	4027	4039	4051	4063	4076	4088	4100	4112	4124	4136
66000	4149	4161	4173	4185	4197	4209	4221	4233	4245	4257
67000	4269	4281	4293	4305	4317	4329	4341	4353	4365	4377

NOTE – OPTIMUM WEIGHT FOR PRESSURE ALTITUDE EXCEEDS STRUCTURAL LIMIT
   THRUST LIMITED WEIGHT FOR ISA + 10 AND COLDER EXCEEDS STRUCTURAL LIMIT
   THRUST LIMITED WEIGHT FOR ISA + 15   EXCEEDS STRUCTURAL LIMIT
   THRUST LIMITED WEIGHT FOR ISA + 20   EXCEEDS STRUCTURAL LIMIT
   ADJUSTMENTS FOR OPERATION AT NON-STANDARD TEMPERATURES---
   INCREASE FUEL REQUIRED BY   0.5 PERCENT PER 10 DEGREES C ABOVE ISA
   DECREASE FUEL REQUIRED BY   0.5 PERCENT PER 10 DEGREES C BELOW ISA
   INCREASE TAS BY 1 KNOT PER DEGREE C ABOVE ISA
   DECREASE TAS BY 1 KNOT PER DEGREE C BELOW ISA

Figure 4.5.3.4   LOW LEVEL CRUISE   300 KIAS

All Engines   Maximum Cruise Thrust Limits   A/C Auto

PRESSURE ALTITUDE   **15000Ft**   **TAS   371 Kts**

GROSS WT KG	0	100	200	300	400	500	600	700	800	900
	CRUISE DISTANCE NAUTICAL AIR MILES									
35000	0	14	29	43	58	73	87	102	117	131
36000	146	161	175	190	205	219	234	248	263	278
37000	292	307	321	336	350	365	380	394	409	423
38000	438	452	467	481	496	510	525	540	554	569
39000	583	598	612	626	641	655	670	684	699	713
40000	728	742	757	771	785	800	814	829	843	857
41000	872	886	900	915	929	943	958	972	987	1001
42000	1015	1029	1044	1058	1072	1087	1101	1115	1129	1144
43000	1158	1172	1186	1201	1215	1229	1243	1257	1272	1286
44000	1300	1314	1328	1343	1357	1371	1385	1399	1413	1427
45000	1442	1456	1470	1484	1498	1512	1526	1540	1554	1568
46000	1582	1596	1610	1624	1638	1652	1666	1680	1694	1708
47000	1722	1736	1750	1764	1778	1792	1806	1820	1834	1848
48000	1862	1875	1889	1903	1917	1931	1945	1959	1972	1986
49000	2000	2014	2028	2041	2055	2069	2083	2097	2110	2124
50000	2138	2152	2165	2179	2193	2206	2220	2234	2247	2261
51000	2275	2288	2302	2316	2329	2343	2357	2370	2384	2397
52000	2411	2425	2438	2452	2465	2479	2492	2506	2519	2533
53000	2546	2560	2573	2587	2600	2614	2627	2641	2654	2668
54000	2681	2694	2708	2721	2734	2748	2761	2775	2788	2801
55000	2815	2828	2841	2855	2868	2881	2894	2908	2921	2934
56000	2948	2961	2974	2987	3000	3014	3027	3040	3053	3066
57000	3080	3093	3106	3119	3132	3145	3158	3171	3185	3198
58000	3211	3224	3237	3250	3263	3276	3289	3302	3315	3328
59000	3341	3354	3367	3380	3393	3406	3419	3432	3445	3458
60000	3470	3483	3496	3509	3522	3535	3548	3560	3573	3586
61000	3599	3612	3624	3637	3650	3663	3675	3688	3701	3714
62000	3727	3739	3752	3765	3777	3790	3803	3815	3828	3841
63000	3853	3866	3878	3891	3903	3916	3929	3941	3954	3966
64000	3979	3991	4004	4016	4029	4041	4054	4066	4079	4091
65000	4104	4116	4129	4141	4153	4166	4178	4190	4203	4215
66000	4228	4240	4252	4264	4277	4289	4301	4314	4326	4338
67000	4351	4363	4375	4387	4399	4411	4424	4436	4448	4460

NOTE – OPTIMUM WEIGHT FOR PRESSURE ALTITUDE EXCEEDS STRUCTURAL LIMIT
　　　THRUST LIMITED WEIGHT FOR ISA + 10 AND COLDER EXCEEDS STRUCTURAL LIMIT
　　　THRUST LIMITED WEIGHT FOR ISA + 15　　　　EXCEEDS STRUCTURAL LIMIT
　　　THRUST LIMITED WEIGHT FOR ISA + 20　　　　EXCEEDS STRUCTURAL LIMIT
　　　ADJUSTMENTS FOR OPERATION AT NON-STANDARD TEMPERATURES---
　　　INCREASE FUEL REQUIRED BY　　0.5 PERCENT PER 10 DEGREES C ABOVE ISA
　　　DECREASE FUEL REQUIRED BY　　0.5 PERCENT PER 10 DEGREES C BELOW ISA
　　　INCREASE TAS BY 1 KNOT PER DEGREE C ABOVE ISA
　　　DECREASE TAS BY 1 KNOT PER DEGREE C BELOW ISA

Figure 4.5.3.4   LOW LEVEL CRUISE   300 KIAS

All Engines   Maximum Cruise Thrust Limits   A/C Auto

PRESSURE ALTITUDE   16000Ft   **TAS   377 Kts**

GROSS WT KG	0	100	200	300	400	500	600	700	800	900
				CRUISE DISTANCE NAUTICAL AIR MILES						
35000	0	14	29	44	59	74	89	104	119	134
36000	149	164	179	194	209	224	239	254	269	284
37000	299	313	328	343	358	373	388	403	418	433
38000	447	462	477	492	507	522	536	551	566	581
39000	596	611	625	640	655	670	684	699	714	729
40000	744	758	773	788	802	817	832	847	861	876
41000	891	905	920	935	949	964	979	993	1008	1023
42000	1037	1052	1066	1081	1096	1110	1125	1139	1154	1168
43000	1183	1198	1212	1227	1241	1256	1270	1285	1299	1314
44000	1328	1343	1357	1371	1386	1400	1415	1429	1444	1458
45000	1473	1487	1501	1516	1530	1544	1559	1573	1588	1602
46000	1616	1631	1645	1659	1673	1688	1702	1716	1731	1745
47000	1759	1773	1788	1802	1816	1830	1845	1859	1873	1887
48000	1901	1916	1930	1944	1958	1972	1986	2000	2015	2029
49000	2043	2057	2071	2085	2099	2113	2127	2141	2155	2169
50000	2183	2197	2211	2225	2239	2253	2267	2281	2295	2309
51000	2323	2337	2351	2365	2379	2393	2407	2421	2435	2448
52000	2462	2476	2490	2504	2518	2531	2545	2559	2573	2587
53000	2600	2614	2628	2642	2655	2669	2683	2697	2710	2724
54000	2738	2751	2765	2779	2792	2806	2820	2833	2847	2861
55000	2874	2888	2901	2915	2928	2942	2956	2969	2983	2996
56000	3010	3023	3037	3050	3064	3077	3091	3104	3118	3131
57000	3145	3158	3171	3185	3198	3211	3225	3238	3252	3265
58000	3278	3292	3305	3318	3331	3345	3358	3371	3385	3398
59000	3411	3424	3438	3451	3464	3477	3490	3504	3517	3530
60000	3543	3556	3569	3582	3596	3609	3622	3635	3648	3661
61000	3674	3687	3700	3713	3726	3739	3752	3765	3778	3791
62000	3804	3817	3830	3843	3856	3869	3882	3895	3908	3920
63000	3933	3946	3959	3972	3985	3997	4010	4023	4036	4049
64000	4062	4074	4087	4100	4112	4125	4138	4151	4163	4176
65000	4189	4201	4214	4227	4239	4252	4265	4277	4290	4302
66000	4315	4328	4340	4353	4365	4378	4390	4403	4415	4428
67000	4440	4453	4465	4478	4490	4503	4515	4527	4540	4552

NOTE – OPTIMUM WEIGHT FOR PRESSURE ALTITUDE EXCEEDS STRUCTURAL LIMIT
THRUST LIMITED WEIGHT FOR ISA + 10 AND COLDER EXCEEDS STRUCTURAL LIMIT
THRUST LIMITED WEIGHT FOR ISA + 15   EXCEEDS STRUCTURAL LIMIT
THRUST LIMITED WEIGHT FOR ISA + 20   EXCEEDS STRUCTURAL LIMIT
ADJUSTMENTS FOR OPERATION AT NON-STANDARD TEMPERATURES---
INCREASE FUEL REQUIRED BY   0.5 PERCENT PER 10 DEGREES C ABOVE ISA
DECREASE FUEL REQUIRED BY   0.5 PERCENT PER 10 DEGREES C BELOW ISA
INCREASE TAS BY 1 KNOT PER DEGREE C ABOVE ISA
DECREASE TAS BY 1 KNOT PER DEGREE C BELOW ISA

Figure 4.5.3.4   LOW LEVEL CRUISE   300 KIAS

All Engines   Maximum Cruise Thrust Limits   A/C Auto

PRESSURE ALTITUDE   **17000Ft**   **TAS   382 Kts**

GROSS WT KG	0	100	200	300	400	500	600	700	800	900
	CRUISE DISTANCE NAUTICAL AIR MILES									
35000	0	15	30	45	61	76	91	107	122	137
36000	152	168	183	198	213	229	244	259	274	290
37000	305	320	335	351	366	381	396	411	427	442
38000	457	472	487	502	518	533	548	563	578	593
39000	608	624	639	654	669	684	699	714	729	744
40000	759	774	789	804	819	834	849	864	879	894
41000	910	924	939	954	969	984	999	1014	1029	1044
42000	1059	1074	1089	1104	1119	1134	1148	1163	1178	1193
43000	1208	1223	1238	1252	1267	1282	1297	1312	1326	1341
44000	1356	1371	1386	1400	1415	1430	1445	1459	1474	1489
45000	1504	1518	1533	1547	1562	1577	1591	1606	1621	1635
46000	1650	1665	1679	1694	1709	1723	1738	1752	1767	1781
47000	1796	1811	1825	1840	1854	1869	1883	1898	1912	1927
48000	1941	1956	1970	1984	1999	2013	2028	2042	2057	2071
49000	2085	2100	2114	2128	2143	2157	2172	2186	2200	2215
50000	2229	2243	2257	2272	2286	2300	2315	2329	2343	2357
51000	2372	2386	2400	2414	2428	2442	2457	2471	2485	2499
52000	2513	2527	2542	2556	2570	2584	2598	2612	2626	2640
53000	2654	2668	2682	2696	2710	2724	2738	2752	2766	2780
54000	2794	2808	2822	2836	2850	2864	2878	2892	2906	2920
55000	2933	2947	2961	2975	2989	3003	3016	3030	3044	3058
56000	3072	3085	3099	3113	3127	3140	3154	3168	3182	3195
57000	3209	3223	3236	3250	3264	3277	3291	3304	3318	3332
58000	3345	3359	3372	3386	3400	3413	3427	3440	3454	3467
59000	3481	3494	3508	3521	3535	3548	3561	3575	3588	3602
60000	3615	3629	3642	3655	3669	3682	3695	3709	3722	3735
61000	3749	3762	3775	3789	3802	3815	3828	3842	3855	3868
62000	3881	3894	3908	3921	3934	3947	3960	3973	3987	4000
63000	4013	4026	4039	4052	4065	4078	4091	4104	4117	4130
64000	4143	4156	4169	4182	4195	4208	4221	4234	4247	4260
65000	4273	4286	4299	4312	4325	4337	4350	4363	4376	4389
66000	4402	4415	4427	4440	4453	4466	4478	4491	4504	4517
67000	4529	4542	4555	4567	4580	4593	4605	4618	4631	4643

NOTE – OPTIMUM WEIGHT FOR PRESSURE ALTITUDE EXCEEDS STRUCTURAL LIMIT
THRUST LIMITED WEIGHT FOR ISA + 10 AND COLDER EXCEEDS STRUCTURAL LIMIT
THRUST LIMITED WEIGHT FOR ISA + 15   EXCEEDS STRUCTURAL LIMIT
THRUST LIMITED WEIGHT FOR ISA + 20   EXCEEDS STRUCTURAL LIMIT
ADJUSTMENTS FOR OPERATION AT NON-STANDARD TEMPERATURES---
INCREASE FUEL REQUIRED BY   0.5 PERCENT PER 10 DEGREES C ABOVE ISA
DECREASE FUEL REQUIRED BY   0.5 PERCENT PER 10 DEGREES C BELOW ISA
INCREASE TAS BY 1 KNOT PER DEGREE C ABOVE ISA
DECREASE TAS BY 1 KNOT PER DEGREE C BELOW ISA

Figure 4.5.3.4   LOW LEVEL CRUISE   300 KIAS

All Engines   Maximum Cruise Thrust Limits   A/C Auto

PRESSURE ALTITUDE   18000Ft   **TAS   388 Kts**

GROSS WT KG	0	100	200	300	400	500	600	700	800	900
	CRUISE DISTANCE NAUTICAL AIR MILES									
35000	0	15	31	46	62	78	93	109	124	140
36000	156	171	187	202	218	234	249	265	280	296
37000	311	327	342	358	373	389	404	420	435	451
38000	466	482	497	513	528	544	559	575	590	606
39000	621	636	652	667	683	698	713	729	744	760
40000	775	790	806	821	836	852	867	882	898	913
41000	928	944	959	974	989	1005	1020	1035	1050	1066
42000	1081	1096	1111	1127	1142	1157	1172	1187	1202	1218
43000	1233	1248	1263	1278	1293	1308	1324	1339	1354	1369
44000	1384	1399	1414	1429	1444	1459	1474	1489	1504	1519
45000	1534	1549	1564	1579	1594	1609	1624	1639	1654	1669
46000	1684	1699	1714	1729	1743	1758	1773	1788	1803	1818
47000	1833	1848	1862	1877	1892	1907	1922	1936	1951	1966
48000	1981	1995	2010	2025	2040	2054	2069	2084	2098	2113
49000	2128	2143	2157	2172	2186	2201	2216	2230	2245	2260
50000	2274	2289	2303	2318	2332	2347	2362	2376	2391	2405
51000	2420	2434	2449	2463	2478	2492	2506	2521	2535	2550
52000	2564	2579	2593	2607	2622	2636	2650	2665	2679	2694
53000	2708	2722	2737	2751	2765	2779	2794	2808	2822	2836
54000	2851	2865	2879	2893	2907	2922	2936	2950	2964	2978
55000	2993	3007	3021	3035	3049	3063	3077	3091	3105	3119
56000	3133	3147	3161	3175	3189	3203	3217	3231	3245	3259
57000	3273	3287	3301	3315	3329	3343	3357	3371	3385	3398
58000	3412	3426	3440	3454	3468	3481	3495	3509	3523	3537
59000	3550	3564	3578	3591	3605	3619	3633	3646	3660	3674
60000	3687	3701	3715	3728	3742	3755	3769	3783	3796	3810
61000	3823	3837	3850	3864	3877	3891	3904	3918	3931	3945
62000	3958	3972	3985	3999	4012	4025	4039	4052	4066	4079
63000	4092	4106	4119	4132	4146	4159	4172	4186	4199	4212
64000	4225	4239	4252	4265	4278	4291	4305	4318	4331	4344
65000	4357	4371	4384	4397	4410	4423	4436	4449	4462	4475
66000	4488	4501	4514	4527	4540	4553	4566	4579	4592	4605
67000	4618	4631	4644	4657	4670	4683	4696	4708	4721	4734

NOTE – OPTIMUM WEIGHT FOR PRESSURE ALTITUDE EXCEEDS STRUCTURAL LIMIT
THRUST LIMITED WEIGHT FOR ISA + 10 AND COLDER EXCEEDS STRUCTURAL LIMIT
THRUST LIMITED WEIGHT FOR ISA + 15   EXCEEDS STRUCTURAL LIMIT
THRUST LIMITED WEIGHT FOR ISA + 20   EXCEEDS STRUCTURAL LIMIT
ADJUSTMENTS FOR OPERATION AT NON-STANDARD TEMPERATURES---
INCREASE FUEL REQUIRED BY   0.5 PERCENT PER 10 DEGREES C ABOVE ISA
DECREASE FUEL REQUIRED BY   0.5 PERCENT PER 10 DEGREES C BELOW ISA
INCREASE TAS BY 1 KNOT PER DEGREE C ABOVE ISA
DECREASE TAS BY 1 KNOT PER DEGREE C BELOW ISA

Figure 4.5.3.4   LOW LEVEL CRUISE   300 KIAS

All Engines   Maximum Cruise Thrust Limits   A/C Auto

PRESSURE ALTITUDE   19000Ft   TAS   394 Kts

GROSS WT KG	0	100	200	300	400	500	600	700	800	900
	CRUISE DISTANCE NAUTICAL AIR MILES									
35000	0	15	31	47	63	79	95	111	127	143
36000	159	175	191	207	222	238	254	270	286	302
37000	318	334	349	365	381	397	413	429	444	460
38000	476	492	508	523	539	555	571	586	602	618
39000	634	649	665	681	697	712	728	744	759	775
40000	791	806	822	838	853	869	885	900	916	931
41000	947	963	978	994	1009	1025	1041	1056	1072	1087
42000	1103	1118	1134	1149	1165	1180	1196	1211	1227	1242
43000	1258	1273	1288	1304	1319	1335	1350	1366	1381	1396
44000	1412	1427	1442	1458	1473	1488	1504	1519	1534	1550
45000	1565	1580	1596	1611	1626	1641	1657	1672	1687	1702
46000	1718	1733	1748	1763	1778	1793	1809	1824	1839	1854
47000	1869	1884	1900	1915	1930	1945	1960	1975	1990	2005
48000	2020	2035	2050	2065	2080	2095	2110	2125	2140	2155
49000	2170	2185	2200	2215	2230	2245	2260	2275	2290	2305
50000	2319	2334	2349	2364	2379	2394	2408	2423	2438	2453
51000	2468	2482	2497	2512	2527	2541	2556	2571	2586	2600
52000	2615	2630	2644	2659	2674	2688	2703	2718	2732	2747
53000	2762	2776	2791	2805	2820	2834	2849	2863	2878	2892
54000	2907	2921	2936	2950	2965	2979	2994	3008	3023	3037
55000	3051	3066	3080	3095	3109	3123	3138	3152	3166	3181
56000	3195	3209	3224	3238	3252	3266	3281	3295	3309	3323
57000	3338	3352	3366	3380	3394	3408	3423	3437	3451	3465
58000	3479	3493	3507	3521	3535	3550	3564	3578	3592	3606
59000	3620	3634	3648	3662	3676	3690	3704	3718	3732	3746
60000	3759	3773	3787	3801	3815	3829	3843	3856	3870	3884
61000	3898	3912	3926	3939	3953	3967	3980	3994	4008	4022
62000	4035	4049	4063	4076	4090	4104	4117	4131	4145	4158
63000	4172	4185	4199	4212	4226	4240	4253	4267	4280	4294
64000	4307	4321	4334	4348	4361	4374	4388	4401	4415	4428
65000	4442	4455	4468	4482	4495	4508	4522	4535	4548	4562
66000	4575	4588	4601	4615	4628	4641	4654	4667	4681	4694
67000	4707	4720	4733	4746	4760	4773	4786	4799	4812	4825

NOTE – OPTIMUM WEIGHT FOR PRESSURE ALTITUDE EXCEEDS STRUCTURAL LIMIT
THRUST LIMITED WEIGHT FOR ISA + 10 AND COLDER EXCEEDS STRUCTURAL LIMIT
THRUST LIMITED WEIGHT FOR ISA + 15   EXCEEDS STRUCTURAL LIMIT
THRUST LIMITED WEIGHT FOR ISA + 20   EXCEEDS STRUCTURAL LIMIT
ADJUSTMENTS FOR OPERATION AT NON-STANDARD TEMPERATURES---
INCREASE FUEL REQUIRED BY   0.5 PERCENT PER 10 DEGREES C ABOVE ISA
DECREASE FUEL REQUIRED BY   0.5 PERCENT PER 10 DEGREES C BELOW ISA
INCREASE TAS BY 1 KNOT PER DEGREE C ABOVE ISA
DECREASE TAS BY 1 KNOT PER DEGREE C BELOW ISA

Figure 4.5.3.4   LOW LEVEL CRUISE   300 KIAS

All Engines   Maximum Cruise Thrust Limits   A/C Auto

PRESSURE ALTITUDE   20000Ft   **TAS   400 Kts**

GROSS WT KG	0	100	200	300	400	500	600	700	800	900
	CRUISE DISTANCE NAUTICAL AIR MILES									
35000	0	16	32	48	65	81	97	113	130	146
36000	162	178	194	211	227	243	259	275	292	308
37000	324	340	356	372	389	405	421	437	453	469
38000	485	502	518	534	550	566	582	598	614	630
39000	646	662	678	694	710	726	742	758	774	790
40000	806	822	838	854	870	886	902	918	934	950
41000	966	982	998	1013	1029	1045	1061	1077	1093	1109
42000	1124	1140	1156	1172	1188	1203	1219	1235	1251	1267
43000	1282	1298	1314	1329	1345	1361	1377	1392	1408	1424
44000	1439	1455	1471	1486	1502	1518	1533	1549	1564	1580
45000	1596	1611	1627	1642	1658	1673	1689	1705	1720	1736
46000	1751	1767	1782	1798	1813	1829	1844	1859	1875	1890
47000	1906	1921	1937	1952	1967	1983	1998	2013	2029	2044
48000	2060	2075	2090	2105	2121	2136	2151	2167	2182	2197
49000	2213	2228	2243	2258	2273	2289	2304	2319	2334	2349
50000	2365	2380	2395	2410	2425	2440	2455	2470	2485	2500
51000	2516	2531	2546	2561	2576	2591	2606	2621	2636	2651
52000	2666	2681	2696	2710	2725	2740	2755	2770	2785	2800
53000	2815	2830	2844	2859	2874	2889	2904	2918	2933	2948
54000	2963	2978	2992	3007	3022	3036	3051	3066	3081	3095
55000	3110	3125	3139	3154	3169	3183	3198	3212	3227	3242
56000	3256	3271	3285	3300	3314	3329	3343	3358	3372	3387
57000	3401	3416	3430	3445	3459	3474	3488	3502	3517	3531
58000	3546	3560	3574	3589	3603	3617	3632	3646	3660	3674
59000	3689	3703	3717	3731	3746	3760	3774	3788	3803	3817
60000	3831	3845	3859	3873	3887	3901	3916	3930	3944	3958
61000	3972	3986	4000	4014	4028	4042	4056	4070	4084	4098
62000	4112	4126	4140	4153	4167	4181	4195	4209	4223	4237
63000	4251	4264	4278	4292	4306	4320	4333	4347	4361	4375
64000	4389	4402	4416	4430	4443	4457	4471	4484	4498	4512
65000	4525	4539	4552	4566	4580	4593	4607	4620	4634	4647
66000	4661	4674	4688	4701	4715	4728	4742	4755	4769	4782
67000	4795	4809	4822	4835	4849	4862	4875	4889	4902	4915

NOTE – OPTIMUM WEIGHT FOR PRESSURE ALTITUDE EXCEEDS STRUCTURAL LIMIT
THRUST LIMITED WEIGHT FOR ISA + 10 AND COLDER EXCEEDS STRUCTURAL LIMIT
THRUST LIMITED WEIGHT FOR ISA + 15   EXCEEDS STRUCTURAL LIMIT
THRUST LIMITED WEIGHT FOR ISA + 20   EXCEEDS STRUCTURAL LIMIT
ADJUSTMENTS FOR OPERATION AT NON-STANDARD TEMPERATURES---
INCREASE FUEL REQUIRED BY   0.5 PERCENT PER 10 DEGREES C ABOVE ISA
DECREASE FUEL REQUIRED BY   0.5 PERCENT PER 10 DEGREES C BELOW ISA
INCREASE TAS BY 1 KNOT PER DEGREE C ABOVE ISA
DECREASE TAS BY 1 KNOT PER DEGREE C BELOW ISA

Figure 4.5.3.4   LOW LEVEL CRUISE   300 KIAS

All Engines   Maximum Cruise Thrust Limits   A/C Auto

PRESSURE ALTITUDE   21000Ft   **TAS   406 Kts**

GROSS WT KG	0	100	200	300	400	500	600	700	800	900
	CRUISE DISTANCE NAUTICAL AIR MILES									
35000	0	16	33	49	66	82	99	115	132	148
36000	165	181	198	214	231	247	264	280	297	313
37000	330	346	363	379	396	412	429	445	461	478
38000	494	511	527	543	560	576	592	609	625	642
39000	658	674	691	707	723	739	756	772	788	805
40000	821	837	853	870	886	902	918	935	951	967
41000	983	999	1016	1032	1048	1064	1080	1096	1112	1129
42000	1145	1161	1177	1193	1209	1225	1241	1257	1273	1289
43000	1305	1321	1337	1353	1369	1385	1401	1417	1433	1449
44000	1465	1481	1497	1513	1529	1545	1561	1577	1593	1609
45000	1624	1640	1656	1672	1688	1704	1719	1735	1751	1767
46000	1783	1798	1814	1830	1846	1861	1877	1893	1909	1924
47000	1940	1956	1971	1987	2003	2018	2034	2050	2065	2081
48000	2096	2112	2128	2143	2159	2174	2190	2205	2221	2236
49000	2252	2267	2283	2298	2314	2329	2345	2360	2376	2391
50000	2407	2422	2437	2453	2468	2483	2499	2514	2530	2545
51000	2560	2576	2591	2606	2621	2637	2652	2667	2682	2698
52000	2713	2728	2743	2758	2774	2789	2804	2819	2834	2849
53000	2865	2880	2895	2910	2925	2940	2955	2970	2985	3000
54000	3015	3030	3045	3060	3075	3090	3105	3120	3135	3150
55000	3165	3180	3195	3209	3224	3239	3254	3269	3284	3299
56000	3313	3328	3343	3358	3373	3387	3402	3417	3432	3446
57000	3461	3476	3490	3505	3520	3534	3549	3564	3578	3593
58000	3608	3622	3637	3651	3666	3681	3695	3710	3724	3739
59000	3753	3768	3782	3797	3811	3826	3840	3854	3869	3883
60000	3898	3912	3926	3941	3955	3969	3984	3998	4012	4027
61000	4041	4055	4070	4084	4098	4112	4126	4141	4155	4169
62000	4183	4197	4212	4226	4240	4254	4268	4282	4296	4310
63000	4324	4338	4352	4366	4381	4395	4409	4423	4437	4451
64000	4465	4478	4492	4506	4520	4534	4548	4562	4576	4590
65000	4604	4617	4631	4645	4659	4672	4686	4700	4714	4728
66000	4741	4755	4769	4782	4796	4810	4823	4837	4851	4864
67000	4878	4892	4905	4919	4932	4946	4959	4973	4987	5000

NOTE – OPTIMUM WEIGHT FOR PRESSURE ALTITUDE EXCEEDS STRUCTURAL LIMIT
THRUST LIMITED WEIGHT FOR ISA + 10 AND COLDER EXCEEDS STRUCTURAL LIMIT
THRUST LIMITED WEIGHT FOR ISA + 15   EXCEEDS STRUCTURAL LIMIT
THRUST LIMITED WEIGHT FOR ISA + 20   EXCEEDS STRUCTURAL LIMIT
ADJUSTMENTS FOR OPERATION AT NON-STANDARD TEMPERATURES---
INCREASE FUEL REQUIRED BY   0.5 PERCENT PER 10 DEGREES C ABOVE ISA
DECREASE FUEL REQUIRED BY   0.5 PERCENT PER 10 DEGREES C BELOW ISA
INCREASE TAS BY 1 KNOT PER DEGREE C ABOVE ISA
DECREASE TAS BY 1 KNOT PER DEGREE C BELOW ISA

5.4 Descent

These tables (Fig. 4.5.4) provide tabulations of time, fuel and distance for "flight idle" thrust at 0.74 mach/250 KIAS (economy) and 0.70 Mach/280 KIAS (turbulence penetration)

Allowances are made for a straight in approach with gear down.

Figure 4.5.4 Descent

.74M/250 KIAS

PRES. ALT. FT.	TIME MIN.	FUEL KG.	DISTANCE NAM				
			LANDING WEIGHT KG.				
			35000	45000	55000	65000	75000
37000	23	295	98	109	114	114	110
35000	22	290	94	105	110	110	106
33000	21	285	89	99	103	104	101
31000	20	280	83	93	97	98	95
29000	19	275	78	87	91	91	89
27000	19	270	73	81	85	85	83
25000	18	260	68	75	79	79	77
23000	16	255	63	69	72	73	71
21000	15	245	58	64	66	67	66
19000	14	235	53	58	60	61	60
17000	13	225	48	52	54	55	54
15000	12	215	43	46	48	49	48
10000	9	185	30	32	33	34	33
5000	6	140	18	18	18	18	18
3700	5	130	14	14	14	14	14

.70M/280/250 KIAS

PRES. ALT. FT.	TIME MIN.	FUEL KG.	DISTANCE NAM				
			LANDING WEIGHT KG.				
			35000	45000	55000	65000	75000
37000	21	280	88	100	107	110	109
35000	20	275	84	96	102	105	105
33000	20	275	80	91	98	101	101
31000	19	270	76	86	93	96	96
29000	18	265	72	82	88	91	92
27000	17	260	69	78	84	87	87
25000	17	255	64	73	78	80	81
23000	16	250	60	67	72	74	74
21000	15	240	55	62	66	68	68
19000	14	230	51	57	60	62	62
17000	13	225	46	52	55	56	56
15000	12	215	42	46	49	50	50
10000	9	185	30	32	33	34	33
5000	6	140	18	18	18	18	18
3700	5	130	14	14	14	14	14

BASED ON IDLE THRUST.
ALLOWANCES FOR A STRAIGHT-IN APPROACH ARE INCLUDED.

6.   NON-NORMAL OPERATION

Simplified Flight Planning
Gear Down   220 KIAS (Fig.4.6.1)

This graph is similar in use to those in paragraph 3.1, giving fuel and time required for a flight with 'gear down'.
Climb and descent are included.

Figure 4.6.1   Non Normal Operation - 'Gear Down' Ferry Flight

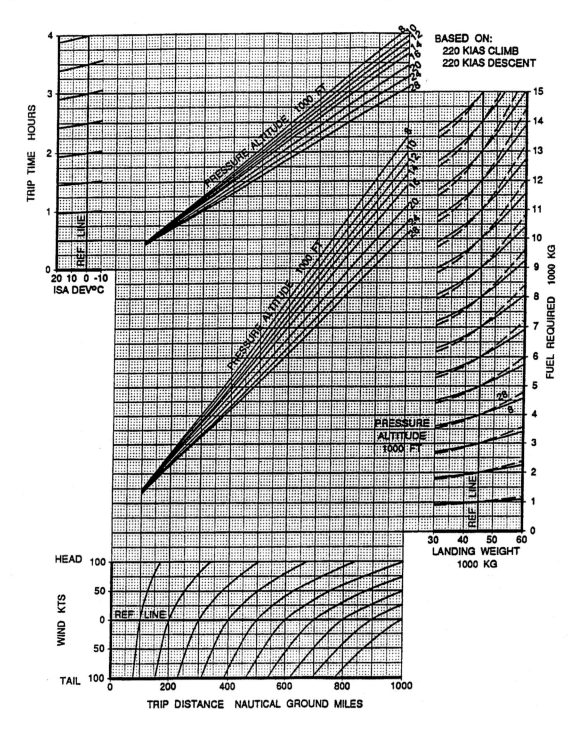

7.   EXTENDED RANGE OPERATIONS (EROPS)

This paragraph provides planning information necessary for the conduct of EROPS.

7.1   Critical Fuel Reserve   (Fig. 4.7.1)

These graphs permit the determination of the minimum fuel reserve at the critical point.
If this fuel reserve exceeds the predicted (planned) fuel remaining at that point, the fuel load must be adjusted accordingly.

7.2   Area of Operation - Diversion Distance   (Fig. 4.7.2)

The area of operation is defined as the region within which the operator is authorised to conduct extended range operation. The distance to the diversion airport from any point along the route must be covered within the approved time using the single engine cruise speed and assuming still air and ISA conditions. The maximum diversion distance used to establish the area of operation may be obtained from this chart.
Enter the chart for the appropriate speed with the weight at the point of diversion. Select the appropriate time and read off the maximum diversion distance.

7.3   In Flight Diversion (LRC)   (Fig. 4.7.3)

Figure 4.7.3 is a simplified flight planning method of determining the fuel required and time for the flight from a point of diversion to a selected alternate.
The graph is similar in layout and use to those in paragraph 3

Figure 4.7.1a   Critical Fuel Reserve

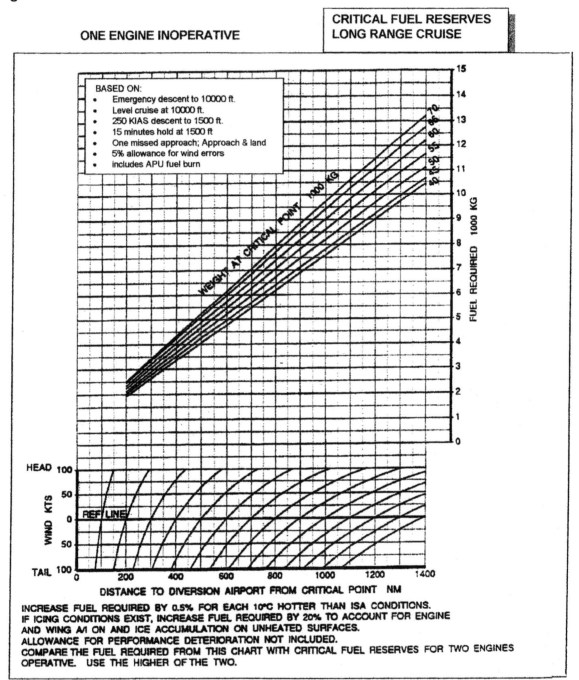

Figure 4.7.1 b   Critical Fuel Reserve

CRITICAL FUEL RESERVES
LONG RANGE CRUISE

ALL ENGINES OPERATIVE

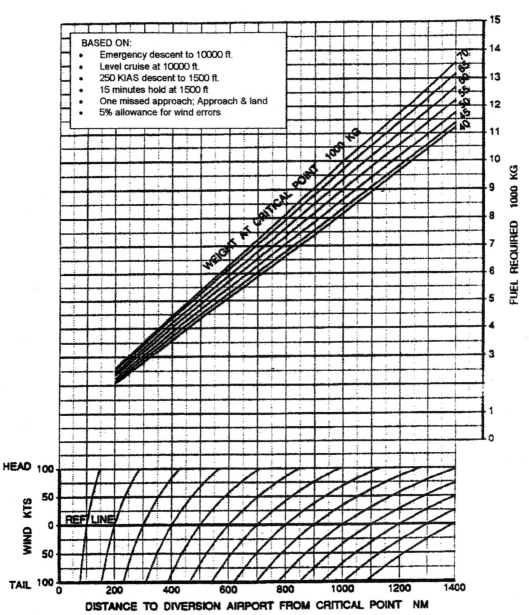

BASED ON:
- Emergency descent to 10000 ft.
- Level cruise at 10000 ft.
- 250 KIAS descent to 1500 ft.
- 15 minutes hold at 1500 ft
- One missed approach; Approach & land
- 5% allowance for wind errors

DISTANCE TO DIVERSION AIRPORT FROM CRITICAL POINT NM

INCREASE FUEL REQUIRED BY 0.5% FOR EACH 10°C HOTTER THAN ISA CONDITIONS.
IF ICING CONDITIONS EXIST, INCREASE FUEL REQUIRED BY 18% TO ACCOUNT FOR ENGINE AND WING
A/I ON AND ICE ACCUMULATION ON UNHEATED SURFACES.
ALLOWANCE FOR PERFORMANCE DETERIORATION NOT INCLUDED.
COMPARE THE FUEL REQUIRED FROM THIS CHART WITH CRITICAL FUEL RESERVES FOR ONE ENGINE
INOPERATIVE. USE THE HIGHER OF THE TWO.

Figure 4.7.2   Area of Operation - Diversion Distance
ONE ENGINE INOPERATIVE

Speed M/KIAS	Div. Wt 1000Kg	TIME MINUTES														
		60	70	80	90	100	110	120	130	140	150	160	170	180	190	200
.70/280	35	406	472	539	605	672	738	805	871	938	1004	1071	1137	1204	1271	1337
	40	402	467	533	598	663	729	794	860	925	990	1056	1121	1187	1252	1318
	45	397	462	526	590	654	718	782	846	910	975	1039	1103	1167	1231	1295
	50	392	454	517	580	642	705	768	830	893	956	1018	1081	1144	1207	1269
	55	385	446	507	568	630	691	752	813	875	936	997	1058	119	1181	1242
	60	377	437	497	557	616	676	736	796	855	915	975	1035	1094	1154	1214
	65	369	427	486	544	602	660	718	776	835	893	951	1009	1067	1125	1183
	70	363	419	476	532	589	645	702	758	815	871	928	985	1041	1098	1154
.74/290	35	412	478	545	612	678	745	811	878	945	1011	1078	1145	1211	1278	1345
	40	409	474	540	606	672	737	803	869	935	1000	1066	1132	1198	1263	1329
	45	404	469	533	598	663	727	792	856	921	986	1050	1115	1180	1244	1309
	50	400	463	526	590	653	717	780	844	907	970	1034	1097	1161	1224	1288
	55	393	455	517	579	641	704	766	828	890	952	1014	1077	1139	1201	1263
	60	386	447	508	568	629	690	751	812	872	933	994	1055	1116	1176	1237
	65	378	437	497	556	615	675	734	793	853	912	971	1031	1090	1149	1209
	70	372	430	488	546	603	661	719	777	835	893	950	1008	1066	1124	1182
.74/310	35	415	482	548	615	681	748	814	881	948	1014	1081	1147	1214	1280	1347
	40	413	479	545	611	677	743	810	876	942	1008	1074	1140	1206	1272	1338
	45	410	476	541	607	672	737	803	868	933	999	1064	1130	1195	1260	1326
	50	407	472	536	601	665	730	794	859	923	988	1052	1116	1181	1245	1310
	55	402	466	529	592	656	719	783	846	9098	973	1036	1100	1163	1226	1290
	60	397	459	521	583	646	708	770	833	95	957	1019	1082	1144	1206	1269
	65	391	452	513	574	635	696	757	818	879	940	1002	1063	1124	1185	1246
	70	385	445	505	565	625	685	744	804	864	924	984	1044	1103	1163	1223
.74/330	35	416	482	548	614	680	746	811	877	943	1009	1075	1141	1207	1273	1339
	40	415	481	547	613	678	744	810	875	941	1007	1072	1138	1204	1270	1335
	45	414	480	545	610	676	741	806	871	937	1002	1067	1133	1198	1263	1328
	50	412	477	542	607	671	736	801	865	930	995	1059	1124	1189	1254	1318
	55	408	472	536	600	664	728	792	856	920	984	1048	1112	1176	1240	1304
	60	404	467	530	593	656	719	783	846	909	972	1035	1098	1161	1224	1287
	65	399	461	523	586	648	710	772	834	896	958	1020	1082	1144	1207	1269
	70	395	457	518	579	640	701	762	823	884	945	1006	1067	1128	1190	1251
LRC	35	368	428	488	548	608	668	728	787	847	906	965	1024	1083	1141	1200
	40	372	433	493	554	614	674	735	794	854	914	973	1032	1092	1151	1209
	45	376	437	497	558	619	679	739	799	859	919	979	1038	1097	1157	1216
	50	379	440	501	561	622	682	742	803	862	922	982	1041	1101	1160	1219
	55	380	441	502	562	623	683	743	803	863	922	982	1041	1100	1159	1218
	60	381	442	503	563	624	684	744	804	863	923	982	1041	1100	1159	1218
	65	381	442	502	563	623	683	742	802	861	921	980	1038	1097	1156	1214
	70	383	444	504	564	623	683	742	802	860	919	978	1036	1094	1152	1210

ISA
BASED ON DRIFTDOWN STARTING AT OR NEAR OPTIMUM ALTITUDE

Figure 4.7.3   In Flight Diversion (LRC) ONE ENGINE INOPERATIVE

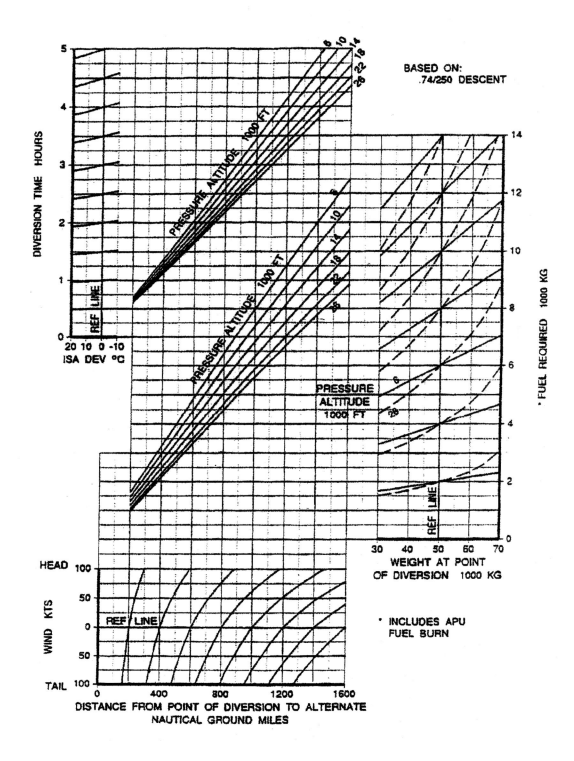

8.   FUEL TANKERING

Because of fuel cost differentials between those at departure airport and those at destination airport, economic benefit. can sometimes be gained from carrying excess fuel (i.e. a fuel load greater than that required for the flight). The following graphs provide a ready means of determining whether such an action is beneficial.

Fig. 4.8.1   Fuel Tankering (LRC)

This graph shows the surplus fuel burn required for carriage of additional fuel. In the example shown, a trip has a distance of 1600 NAM and is to be conducted at FL 330. If excess fuel is carried, 13.2% of that excess will be consumed as a "fuel penalty".

Fig. 4.8.2   Fuel Price Differential

Using the % value for the surplus fuel burn (as obtained from Fig. 4.8.1) and fuel price at the departure airport, the break-even price at the destination airport can be determined.

Figure 4.8.1   Fuel Tankering (LRC and O.74M)

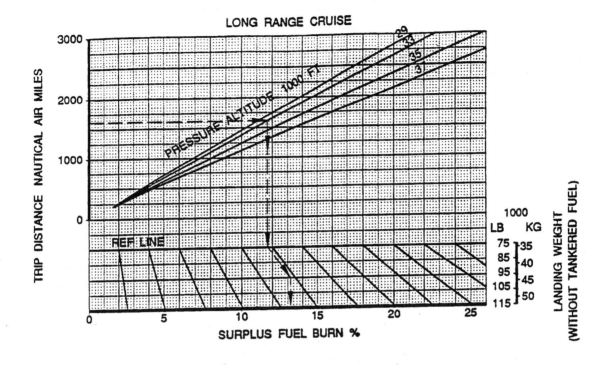

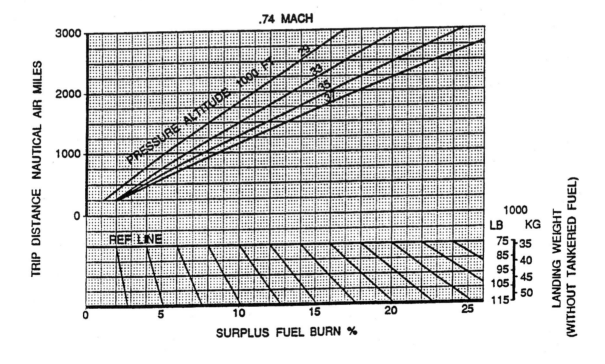

Figure 4.8.2   Fuel Price Differential

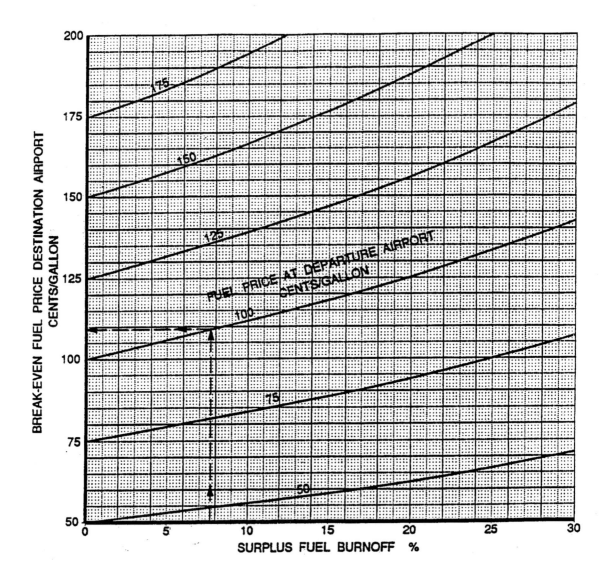

CIVIL AVIATION
AUTHORITY

CAP 698

CIVIL AVIATION AUTHORITY
JAR FCL EXAMINATIONS
PERFORMANCE MANUAL

CIVIL AVIATION AUTHORITY, LONDON

CAP 698

CIVIL AVIATION AUTHORITY
JAR FCL EXAMINATIONS
PERFORMANCE MANUAL

CIVIL AVIATION AUTHORITY, LONDON, AUGUST 1999

ISBN 0 86039 771 8

Printed and distributed by
Westward Digital Limited, 37 Windsor Street, Cheltenham, England

Contents

INTENTIONALLY BLANK

PERFORMANCE AEROPLANES

SECTION 1-GENERAL NOTES

INTRODUCTION

IMPORTANT NOTICE

These data sheets are intended for the use of candidates for the European Professional Pilot's Licence Examinations.

The data contained within these sheets is for **examination purposes only**. The data must not be used for any other purpose and, specifically, **are not to be used for the purpose of planning activities associated with the operation of any aircraft in use now or in the future**.

AIRCRAFT DESCRIPTION

The aircraft used in these data sheets are of generic types related to the classes of aircraft on which the appropriate examinations are based.

Candidates must select the correct class of aircraft for the question being attempted. To assist in this, the data for each class is presented on different coloured paper.

Generic Aircraft

Single engine piston	not certified under JAR 25 (Light Aeroplanes) Performance Class B	**SEP1**
Multi engine piston	not certified under JAR 25 (Light Aeroplanes) Performance Class B	**MEP1**
Medium range jet transport	certified under JAR 25 Performance Class A	**MRJT**

The same set of generic aircraft will be utilised in the following subjects:
- 031 - Mass and Balance - Aeroplanes
- 032 - Performance - Aeroplanes
- 033 - Flight Planning and Monitoring - Aeroplanes

LAYOUT OF DATA SHEETS

Each set of data sheets will consist of an introduction that will contain some pertinent information relating to the aircraft and the subject being examined. This data will include (but not be limited to) a list of abbreviations and some conversion factors.

This will be followed by a selection of graphs and/or tables that will provide coverage suitable for the syllabus to be examined. A worked example will accompany each graph/table and will demonstrate typical usage.

Data sheets for each type will appear on different colour paper as follows:-

- SEP1 green paper
- MEP1 blue paper
- MRJT white paper

DEFINITIONS

Definitions given in italics are not given in ICAO or JAA documentation but are in common use.

Take-Off Mass (TOM)	is the mass of an aeroplane including everything and everyone contained within it at the start of the take-off run.
Maximum Structural Take-Off Mass	the maximum permissible total aeroplane mass at the start of the take-off run.
Maximum Structural Landing Mass	the maximum permissible total aeroplane mass on landing (under normal circumstances).

Outside Air Temperature (OAT or SAT) - the free air static (ambient) temperature.

Total Air Temperature (TAT) - static air temperature plus adiabatic compression (ram) rise as indicated on the Total Air Temperature indicator.

Altitude
The altitude shown on the charts is pressure altitude. This is the height in the International Standard Atmosphere at which the prevailing pressure occurs. It may be obtained by setting the sub-scale of a pressure altimeter to 1013 hPa (29.92 inches or 760 mm. of mercury).

International Standard Atmosphere (ISA)
A structure of **assumed** conditions relating to the change of pressure, temperature and density with height in the atmosphere.

Elevation
The vertical distance of an object above mean sea level. This may be given in metres or feet.

Height
The vertical distance between the lowest part of the aeroplane and the relevant datum.

Gross Height
The true height attained at any point in the take-off flight path using gross climb performance. Gross height is used for calculating pressure altitudes for purposes of obstacle clearance and the height at which wing flap retraction is initiated.

Net Height
The true height attained at any point in the take-off flight path using net climb performance. Net height is used to determine the net flight path that must clear all obstacles by at least 35 feet to comply with the operating Regulations.

Climb Gradient
The ratio, in the same units, and expressed as a percentage as obtained from the FORMULA:-

$$\frac{\text{Change in Height}}{\text{Horizontal Distance Travelled}} \times 100 \%$$

IAS
The indicated airspeed is the reading obtained on a pilot-static airspeed indicator having no instrument calibration error. If the calibration error of the particular instrument is not known, then the actual reading may be taken to be equal to IAS because the tolerances permitted on the instrument are small.

TAS
The 'true airspeed' is the speed of the aeroplane relative to the undisturbed air.

Gross Performance
The average performance that a fleet of aeroplanes should achieve if satisfactorily maintained and flown in accordance with the techniques described in the manual.

Net Performance
Net performance is the gross performance diminished to allow for various contingencies that cannot be accounted for operationally e.g., variations in piloting technique, temporary below average performance, etc. It is improbable that the net performance will not be achieved in operation, provided the aeroplane is flown in accordance with the recommended techniques.

N.B.
Within these data sheets the term 'weight' should be considered to have the same meaning as 'mass'.

CONVERSIONS

All conversions are taken from ICAO Annex

Mass conversions

Pounds (LB) to Kilograms (KG)   LB x 0.45359237 KG

Kilograms (KG) to Pounds (LB)   KG x 2.20462262 LB

Volumes (Liquid)

Imperial Gallons to Litres (L)   Imp. Gall x 4.546092

US Gallons to Litres (L)   US Gall x 3.785412

Lengths

Feet (ft) to Metres (m)   Feet x 0.3048

Distances

Nautical mile (NM) to metres (m)   NM x 1852.0

PERFORMANCE - AEROPLANES

SECTION II - DATA FOR CLASS B SINGLE ENGINE AEROPLANE

CONTENT

1.   GENERAL CONSIDERATIONS

Performance Classification and Limitations
The specimen aeroplane is a low wing monoplane with retractable undercarriage. It is powered by a single reciprocating engine and a constant speed propeller.

The aeroplane, which is not certified under JAR/FAR 25, is a landplane classified in Performance Class B.

1.1   General Requirements
An operator shall not operate a single-engine aeroplane:
(1)   At night
(2)   In instrument meteorological conditions except under special visual flight rules
(3)   Unless surfaces are available which permit a safe forced landing to be executed.
(4)   Above a cloud layer that extends below the relevant minimum safe altitude.

1.2   Aeroplane Limitations
Structural Limitations
Maximum Take-off Mass   3650 lb.
Maximum Landing Mass   3650 lb.
Maximum Runway Cross Wind   17 kts.

2.   TAKE-OFF REQUIREMENTS

2.1   Requirements
There are two requirements for take-off with which compliance is necessary. They are the minimum field-length and climb gradient requirements.

Field-Length Requirements
a.   When no stopway or clearway is available the take-off distance must not exceed 1.25 x TORA.
b.   When a stopway and/or clearway is available the take-off distance must not exceed:
(1)   TORA
(2)   1.3 x ASDA
(3)   1.15 x TODA
c.   If the runway surface is other than dry and paved the following factors must be used when determining the take-off distance in a. or b. above:

SURFACE TYPE	CONDITION	FACTOR
Grass (on firm soil)	Dry	X 1.2
Up to 20 cm. Long	Wet	X 1.3
Paved	Wet	X 1.0

d.   Take-off distance should be increased by 5% for each 1% upslope. No factorisation is permitted for downslope.

NOTE: the same surface and slope correction factors should be used when calculating TOR of ASD.

2.2 Use of Take-Off Graphs
There are two take-off distance graphs. One with flaps up (Fig. 2.1) and the other with flaps approach (Fig. 2.2). These graphs are used in exactly the same manner.

Distance Calculation
To determine the take-off distance:
a. Select the graph appropriate to the flap setting.
b. Enter the left carpet at the OAT. Move vertically up to the aerodrome pressure altitude.
c. From this point, travel horizontally right to the weight reference line. Parallel the grid lines to the take-off weight input.
d. Continue horizontally right to the wind component reference line. Parallel the grid lines to the wind component input.
e. Proceed horizontally right to the right vertical axis to read the take-off distance.
f. Factorise for surface and slope.

Example: Flaps Up

Aerodrome Pressure Altitude	5653 ft.
Ambient Temperature	+15°C
Take-Off Weight	3650 lb.
Wind Component	10 Kt. Head
Runway Slope	1.5 % Uphill
Runway Surface	Grass
Runway Condition	Wet

Calculate: Take-off Distance

Solution:

Graphical Distance	3475 ft.
Surface Factor	x 1.3
Slope Factor	x 1.075
Take-off Distance	4856 ft.

Weight Calculation

To calculate the field-length limited take-off weight it is necessary to apply the requirements of JAR-OPS. Only the take-off distance graph is used but the right vertical axis is entered with shortest available de-factored distance. The factors to be considered are those of slope, surface, condition and regulation.

1. Enter the left carpet at the ambient temperature. Move vertically to the aerodrome pressure altitude.
2. From this point, travel horizontally right to the weight reference line. Mark this position with a pencil.
3. Enter the right vertical axis at the de-factored distance.
4. Now travel horizontally left to the appropriate wind component input. Parallel the grid lines to the wind component reference line.
5. From this point, draw a horizontal line left through the weight grid.
6. From the position marked in 2, above, parallel the grid lines to intersect the horizontal line from 5, above.
7. At the intersection, drop vertically to the carpet to read the field-length limited TOW.

Example: Flaps Approach

Aerodrome Pressure Altitude	5653 ft.
Ambient Temperature	+15°C
Wind Component	10 Kt. Head
Runway Slope	2% Uphill
Runway Surface	Grass
Runway Condition	Dry

TORA 4250 ft.; ASDA 4470 ft.; TODA 4600 ft.

Calculate the Field-Length Limited TOW.

	TORA	ASDA	TODA
Given Distances	4250 ft.	4470 ft.	4600 ft.
Slope Factor	1.1	1.1	1.1
Surface/Condition Factor	1.2	1.2	1.2
Regulation Factor	1.0	1.3	1.15
De-Factored Distance	3220 ft.	2605 ft.	3030 ft.

Field Length Limited TOW   3520 lb. Using 2605 ft.

Figure 2.1   **Take-Off Distance**   **Flaps Up**

TAKE-OFF DISTANCE – FLAPS UP

ASSOCIATED CONDITIONS:

POWER	TAKE-OFF POWER SET BEFORE BRAKE RELEASE
MIXTURE	FULL RICH
FLAPS	UP
LANDING GEAR	RETRACT AFTER POSITIVE CLIMB ESTABLISHED
COWL FLAPS	OPEN
RUNWAY	PAVED, LEVEL, DRY SURFACE

EXAMPLE:

OAT	15°C
PRESSURE ALTITUDE	5653 FT
TAKE-OFF WEIGHT	3650 LBS
HEAD WIND COMPONENT	10 KTS
GROUND ROLL	1900 FT
TOTAL DISTANCE OVER 50-FT OBSTACLE	3475 FT
TAKE-OFF SPEED AT	
ROTATION	73 KTS
50-FT	84 KTS

WEIGHT ~ POUNDS	TAKE-OFF SPEED	
	ROTATION KNOTS	50 FT KNOTS
3650	73	84
3600	72	83
3400	71	82
3200	70	80
3000	68	78
2800	65	75

Figure 2.2   **Take-Off Distance**   **Flaps Approach**

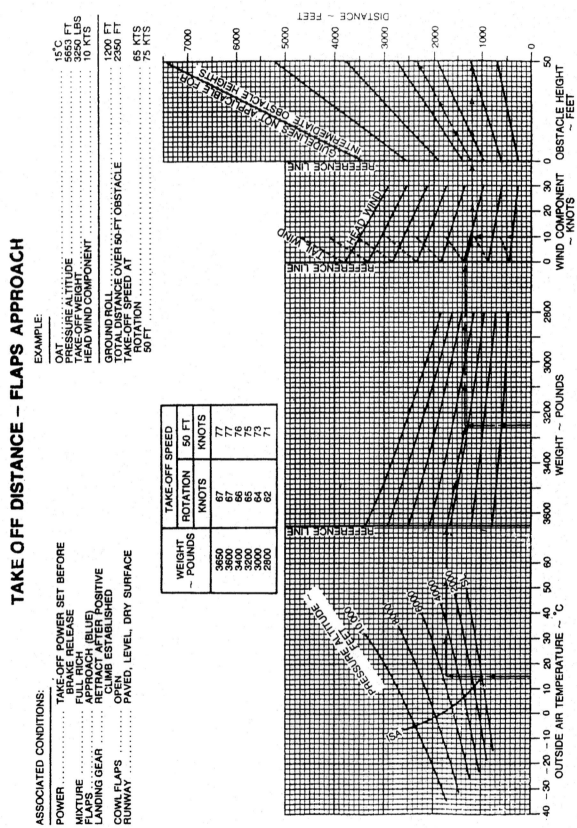

3.   CLIMB

3.1   Requirements
        There are no obstacle clearance limits but the minimum acceptable climb
        gradient is 4 %.

3.2   Use of Climb Graph

        Climb Gradient and Rate of Climb.

        To determine the climb gradient and rate of climb:
        1.   Enter the left carpet of the graph at the ambient temperature.
                Move vertically up to the pressure altitude.
        2.   From this point, travel horizontally right to the weight reference line.
                Parallel the grid lines to the appropriate weight input.
        3.   Now continue horizontally right to the first vertical axis to read the rate
                of climb. Continue horizontally to the second vertical axis to read the
                climb gradient.

Example:
        Pressure Altitude   11,550 ft.
        Ambient Temperature   -5°C
        Weight   3600 lb.

Solution:
        Graphical ROC   510 fpm
        Climb Gradient   3.8 %

        WAT Limit

        To determine the Weight -Altitude -Temperature limitation on take-off weight:
        1.   Enter the left carpet of the graph at the ambient temperature.
                Move vertically up to the pressure altitude.
        2.   From this point, travel horizontally right to the weight reference line
                and mark with a pencil.
        3.   Enter the right vertical axis at 4 %. Draw a horizontal line left through
                the weight.
        4.   From the pencil mark in 2, above, parallel the grid lines to intersect the
                horizontal line drawn in 3 above. Drop a vertical to the carpet to read
                the WAT limited TOW.

Example:
        Aerodrome Pressure Altitude   11000 ft.
        Ambient Temperature   +25°C

Solution: 3360 lb.

Distance to Reach given height.

To calculate the ground distance travelled in order to attain a given height above reference zero:

1. Convert the IAS 100 Kt to a TAS.
2. Apply the wind component to the TAS to obtain the ground speed.
3. Determine the climb gradient from the graph.
4. Calculate the still air distance using the formula:

$$\text{Still air Distance} = \frac{\text{Height Difference}}{\text{Gradient}} \times 100$$

5. Calculate ground distance using the formula:

$$\text{Ground Distance} = \text{Still Air Distance} \times \frac{\text{Groundspeed}}{\text{TAS}}$$

Example:

Aerodrome Pressure Altitude	4000 ft.
Ambient Temperature	+ 30° C
Wind Component	30 Kt. Tail
Take-Off Weight	3200 lb.

Calculate the ground distance to reach 950 ft. above reference zero from the end of TODR.

Solution:

100 Kt. IAS = 112 Kt. TAS
Groundspeed = 142 Kt.
Graph   Gradient = 9.2 %

$$\text{Still Air Distance} = \frac{900}{9.2} \times 100 = 9782 \text{ ft.}$$

$$\text{Ground Distance} = 9782 \times \frac{142}{112} = 12402 \text{ ft.} = 2.04 \text{ NM.}$$

Figure 2.3   **Climb**

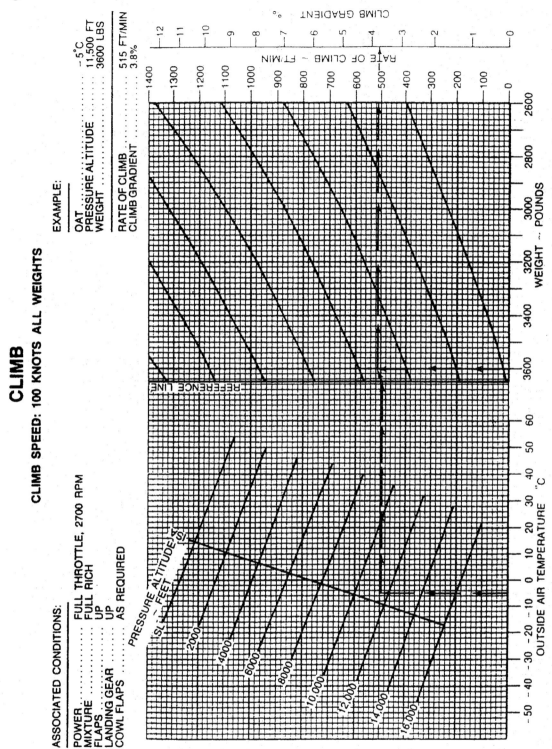

4.   CRUISE

    4.1   Requirements
          The aeroplane may not be assumed to be flying above the altitude at which a
          rate of climb of 300 ft/minute is attained.

          The net gradient of descent, in the event of engine failure, is the gross
          gradient + 0.5 %.

5.   LANDING

   5.1   Requirements

      Field Length Requirements
      a.   The landing distance, from a screen height of 50 ft., must not exceed 70 % of the landing distance available, i.e. a factor of 1.43.
      b.   If the landing surface is grass up to 20cm. long on firm soil, the landing distance should be multiplied by a factor of 1.15.
      c.   If the METAR or TAF or combination of both indicate that the runway may be wet at the estimated time of arrival, the landing distance should be multiplied by a factor of 1.15.
      d.   The landing distance should be increased by 5% for each 1% downslope. No allowance is permitted for upslope.
      e.   The despatch rules for scheduled (planned) landing calculations are in JAR - OPS 1.550 (c).

   5.2   Use of the Landing Field-Length Graph

      Distance Calculations
      1.   Enter left carpet at the ambient temperature. Move vertically to the aerodrome pressure altitude.
      2.   From this point, move horizontally right to the landing weight reference line. Parallel the grid lines to the appropriate landing weight input.
      3.   Continue from this intersection to the wind component reference line. Parallel the grid line to the appropriate wind component input.
      4.   Travel horizontally right to the ground roll reference line. Either continue horizontally to the right vertical axis to read the ground roll distance or parallel the grid lines to the right vertical axis to read the landing distance.
      5.   Apply the appropriate factors to the landing distance to obtain the landing distance required.

      Example: Normal Landing
         Aerodrome Pressure Altitude   3965 ft.
         Ambient Temperature   + 25°C
         Landing Weight   3479 lb.
         Wind Component   10 Kt. Head
         Runway Slope   1% down
         Runway Surface   Grass
         Runway Condition   Wet

      Calculate Landing Distance Required

      Solution:
         Graphical Distance   1515 ft.
         Slope Correction Factor   x 1.05
         Surface Correction Factor   x 1.15
         Condition Correction Factor   x 1.15
         Regulatory Factor   x 1.43

      Landing Distance Required   =   3008 ft.

Figure 2.4 **Landing**

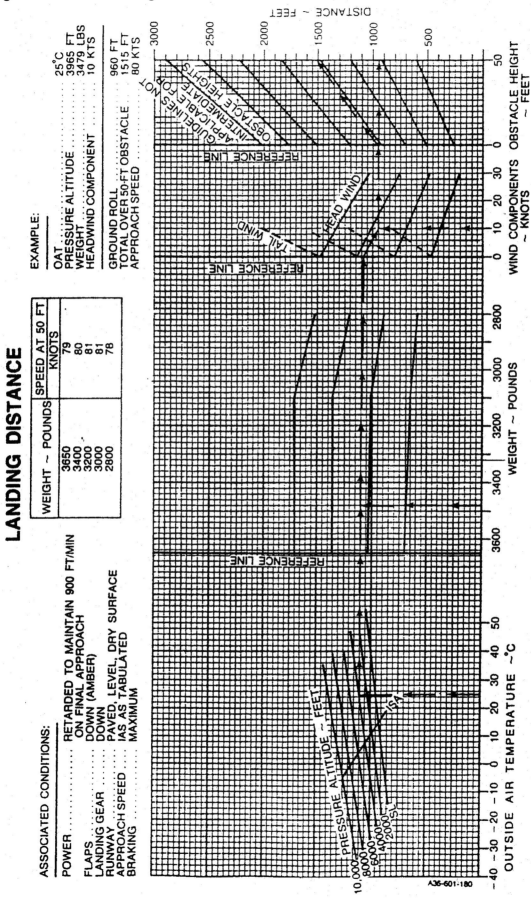

LANDING DISTANCE

EXAMPLE:

OAT .... 25°C
PRESSURE ALTITUDE .... 3965 FT
WEIGHT .... 3479 LBS
HEADWIND COMPONENT ...... 10 KTS

GROUND ROLL .... 960 FT
TOTAL OVER 50-FT OBSTACLE . 1515 FT
APPROACH SPEED .... 80 KTS

WEIGHT ~ POUNDS	SPEED AT 50 FT
	KNOTS
3650	79
3400	80
3200	81
3000	81
2800	78

ASSOCIATED CONDITIONS:

POWER .... RETARDED TO MAINTAIN 900 FT/MIN
                            ON FINAL APPROACH
FLAPS .... DOWN (AMBER)
LANDING GEAR .... DOWN
RUNWAY .... PAVED, LEVEL, DRY SURFACE
APPROACH SPEED .... IAS AS TABULATED
BRAKING .... MAXIMUM

A36-601-180

INTENTIONALLY BLANK

PERFORMANCE -AEROPLANES

SECTION III -DATA FOR MULTI-ENGINE AEROPLANE
(RECIPROCATING ENGINES)

CONTENTS

1.   GENERAL CONSIDERATIONS

1.1   The specimen aeroplane is a low wing monoplane with retractable undercarriage. It
       is powered by twin, reciprocating, engines (both of which are supercharged). These
       drive counter-rotating, constant speed propellers.

       The aeroplane, which is not certified under JAR/FCL 25, is a landplane and is
       classified in Performance Class B.

1.2   General Requirements
       This class of aeroplane includes all propeller-driven aeroplanes having 9 or less
       passenger seats and a maximum take-off weight of 5,700 Kg. or less. Performance
       accountability for engine failure, on a multi engine aeroplane in this class, need not
       be considered below a height of 300 ft.

1.3   Aeroplane Limitations
       Structural Limitations
       Maximum Take-off Mass   4.750 lb.
       Maximum Landing Mass   4,513 lb.
       Runway Crosswind Limitation
       Maximum Demonstrated Crosswind   17Kt.

2.   TAKE-OFF - MULTI-ENGINED AEROPLANES

2.1   Requirements
There are two requirements for take-off with which compliance is necessary. They are the minimum field-length and climb gradient requirements. The take off climb requirements are considered in paragraph 3.

Field-Length Requirements

a.   When no stopway or clearway is available the take-off distance x 1.25 must not exceed TORA.

b.   When a stopway and/or clearway is available the take-off distance must:
   (1)   not exceed TORA
   (2)   when multiplied by 1.3, not exceed ASDA
   (3)   when multiplied by 1.15, not exceed TODA

c.   If the runway surface is other than dry and paved the following factors must be used when determining the take-off distance in a or b. above:

SURFACE TYPE	CONDITION	FACTOR
Grass (on firm soil) Up to 20cm. Long	Dry	X 1.2
	Wet	X 1.3
Paved	Wet	X 1.0

d.   Take-off distance should be increased by 5%, for each 1% upslope. No factorisation is permitted for downslope.

NOTE: the same surface and slope correction factors should be used when calculating TOR of ASD.

2.2 Use of Take-off Graphs.

There are two sets of take-off graphs: one for a "normal" take-off with 0° flap and the other for a "maximum effort" (short field) take-off with 25° flap. Each set comprises two graphs, one for determining the take-off run and take-off distance, the other for calculating the accelerate-stop distance.

Distance Calculation

Procedure

To determine the distance used for take-off
a. Select the appropriate graph.
b. Enter the left carpet at the OAT. Travel vertically to the aerodrome pressure altitude.
c. From this point proceed horizontally right to the weight reference line. Parallel the grid lines to the appropriate take-off weight.
d. Continue horizontally right to the wind component reference line and parallel the grid lines to the wind component input
e. To read the appropriate distance:
  (1) Continue horizontally from the wind component for TOR and ASD
  (2) For take-off distance continue to the ground roll reference line then parallel the grid lines.
f. Factorise for surface and slope.

Example:

Normal Take-off	
Aerodrome Pressure Altitude	2,000 ft.
Ambient Temperature	+ 21 °
Take-off Weight	3,969 lb.
Wind Component	9 Kt. Head
Runway Slope	1.5 % Uphill
Runway Surface	Wet Grass

Calculate: Take-off Distance

Solution: Graphical Distance 1,650 ft.
Surface Factor = x 1 .3 Slope Factor x 1.075
Take-off Distance = 2,306 ft.

Weight Calculation

To calculate the field-length limited take-off weight it is necessary to apply the requirements of JAR-OPS. Only the take-off distance graph is used but the right vertical axis is entered with shortest available de-factored distance. The factors to be considered are those of slope, surface condition and regulation. Procedure and examples are shown at page 24.

Figure 3.1 **Take-off** **Normal Procedure**

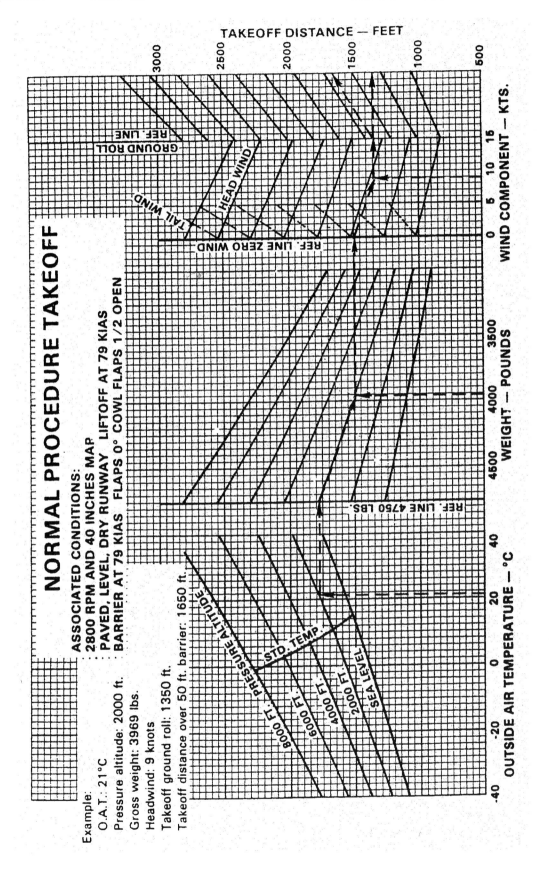

Figure 3.2   **Accelerate/ Stop Distance**   **Flaps 0°**

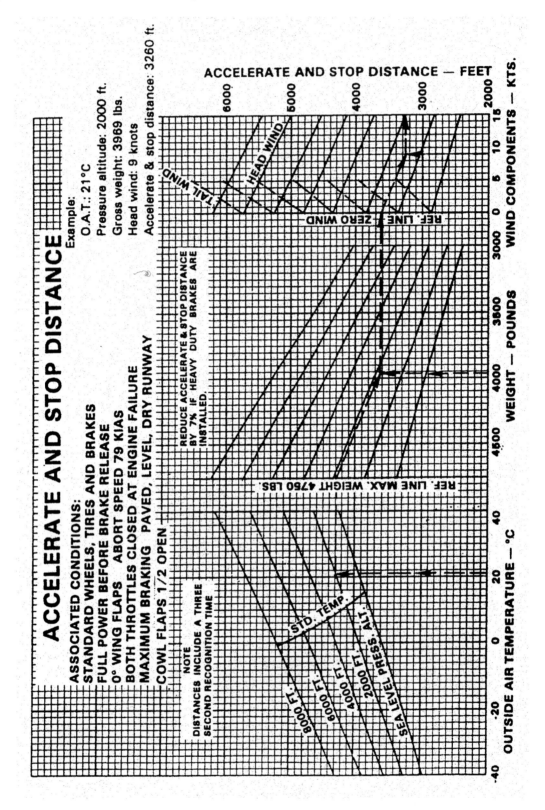

INTENTIONALLY BLANK

Weight Calculation
Procedure

1. Enter the left carpet at the ambient temperature. Move vertically to the aerodrome pressure altitude.
2. From this point, travel horizontally right to the weight reference line. Mark this position with a pencil.
3. Enter the right vertical axis at the de-factored distance. Parallel the grid lines to the ground roll reference line.
4. Now travel horizontally left to the appropriate wind component input. Parallel the grid lines to the wind component reference line.
5. From this point draw a horizontal line left through the weight grid.
6. From the position marked in 2 above, parallel the grid lines to intersect the horizontal line from 5 above.
7. At the intersection, drop vertically to the carpet to read the field length limited TOW.

Example 1: Maximum Effort Take-off (Short Field)

Aerodrome Pressure Altitude	2,000 ft.
Ambient Temperature	+ 30° C
Wind Component	5 Kt. Tail
Runway Slope	2 % Uphill
Surface Type	Grass
Surface Condition	Dry

TORA: 2.400 ft.;   ASDA: 2,500 ft.;   TODA: 2,600 ft.

Calculate the field length limited take-off weight

Solution:

	TORA	ASDA	TODA
Given Distances	2400 ft.	2500 ft.	2600 ft.
Slope Factor	1.1	1.1	1.1
Surface/Condition Factor	1.2	1.2	1.2
Regulation Factor	1.0	1.3	1.15
De-factored Distance	1818 ft.	1457 ft.	1713 ft.

Field Length Limited TOW 3820 lb. Using 1457 ft.

Example 2: Normal Take-off

Aerodrome Pressure Altitude	4,000 ft.
Ambient Temperature .	+ 20° C
Wind Component	5 Kt. Head
Runway slope	2 % down
Surface Type	Concrete
Surface Condition	Wet
TORA: 2500 ft.	No Stopway or Clearway

Calculate the field-length limited take-off weight

Solution	TORA
Given Distance	2500 ft.
Slope Factor	1 .0
Surface Condition Factor	1 .0
Regulation Factor	1 .25
De-factored Distance	2000 ft.
Field Length Limited TOW	3370 lb. Using 2000 ft.

Figure 3.3   **Take-Off**   **Maximum Effort**

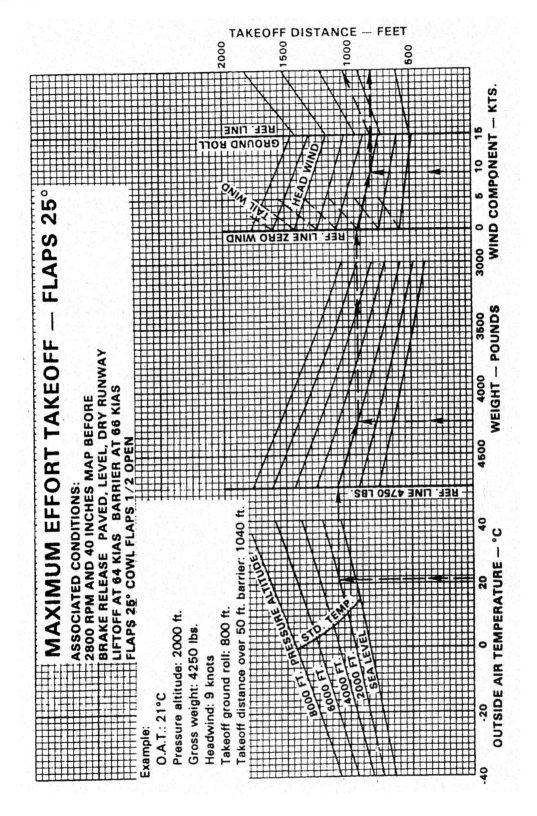

Figure 3.4　　　　**Accelerate/Stop Distance**　　　**Flaps 25°**

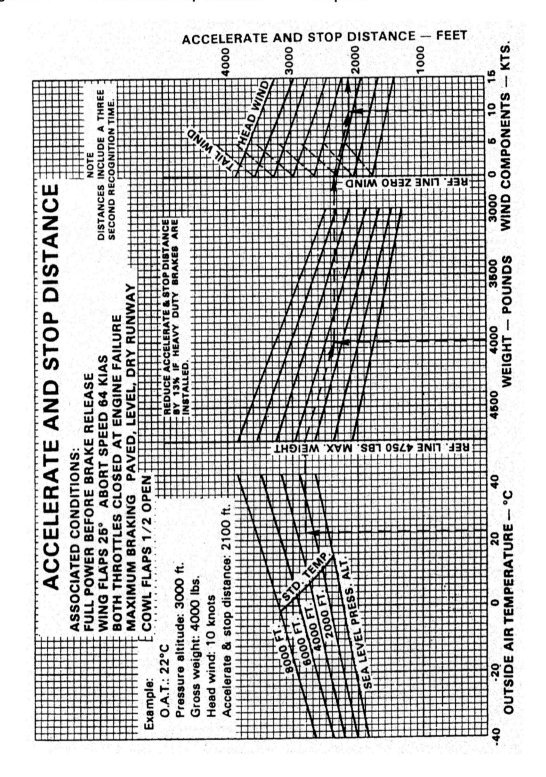

3.   TAKE-OFF CLIMB - MULTI-ENGINED AEROPLANES

3.1   Requirements
The take-off climb requirements only apply to aeroplanes with two or more engines. The take-off climb extends from 50 ft. above the surface at the end of TODR to 1500 ft. above the same surface. The maximum take-off power setting is limited to 5 minutes from the commencement of the take-off climb, at which point it must be reduced to the maximum continuous power setting.

If visual reference for obstacle avoidance is lost, it is assumed that the critical power unit becomes inoperative at this point. All obstacles encountered in the accountability area must be cleared by a vertical interval of 50 ft.

Turns are not permitted in the take-off climb before the end of the TODR and thereafter the angle of bank may not exceed 15°.

The Obstacle Domain ,
The dimensions of the obstacle accountability area are as follows: -
a.   Starting semi-width at the end of TODA of 90 m., if the wing span is less than 60 m., then (60 m. + ½ wing span) is the semi-width to be used.
b.   The area expands from the appropriate semi-width, at the rate of 0.125 x D, to the maximum semi-width where D is the horizontal distance travelled from the end of TODA or TODR if shorter.
c.   Maximum Semi-width.
d.

Change of Track Direction	0° to 15°	Over 15°
Able to Maintain Visual Guidance or same Accuracy	300m.	600 m.
All Other conditions	600 m	900 m.

Minimum Gradients of Climb

The minimum permissible gradients of climb are:
a.   All engines operating … 4% at screen height
b.   One engine inoperative
      (1)   At 400 ft. above the take-off surface level … measurably positive.
      (2)   At 1500 ft. above the take-off surface level … 0.75 %.

3.2 Use of Take-Off Climb Data
Because the graphs provided only permit the calculation of the rate of climb it is necessary to utilise the following formula to solve take-off climb problems:

$$\text{Time to Climb} = \frac{\text{Height Difference}}{\text{Rate of Climb}} \times 60 \text{ seconds}$$

$$\text{Distance to Climb} = \frac{\text{Height Difference}}{\text{Rate of Climb}} \times \frac{\text{Groundspeed N.M.}}{60}$$

$$\text{Still Air Gradient of Climb} = \frac{\text{Rate of Climb}}{\text{TAS Kt.}} \times \frac{6000}{6080} \%$$

Climb graphs
There are three graphs provided for climb calculations:
a.   Gear extended, max. take-off power (Fig. 3.5)
b.   Gear retracted, max. take-off power (Fig. 3.6)
c.   Gear retracted, max. continuous power (Fig. 3.7)

Use of Graph (Fig. 3.5)

1.   Enter the left carpet with the temperature and pressure altitude of the aerodrome.
2.   Travel horizontally to the curved reference line.
3.   From this intersection drop a vertical line to the bottom scale. Read off rate of climb.

An example is shown on the graph.

Figure 3.5   Take-Off   Climb Performance - Gear Extended

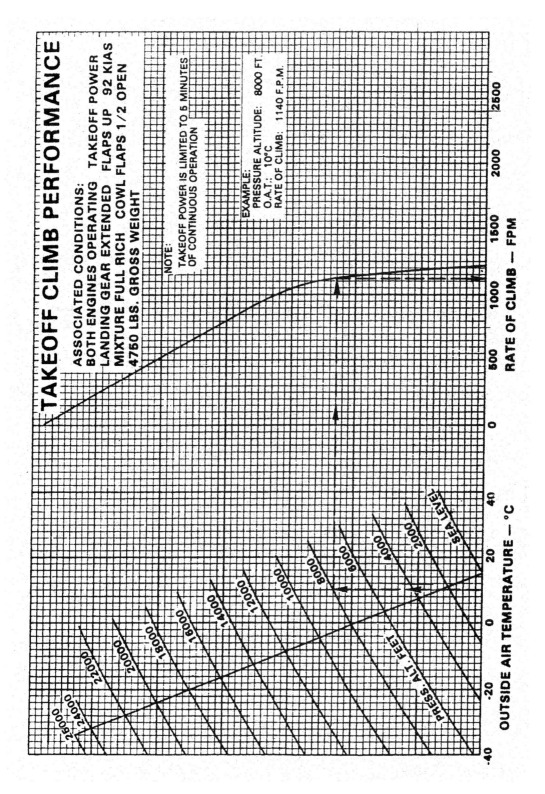

Use of Graphs (Fig. 3.6 and Fig. 3.7)
1.   Enter the left carpet. Travel vertically to the aerodrome pressure altitude.
2.   From this point, travel horizontally right to intercept the interpolated value of take-off weight.
3.   Drop vertically to read the all engine operating rate of climb.
4.   From the TOW intersection, continue horizontally right to intercept the second interpolated weight (if applicable).
5.   Drop vertically to the carpet to read the one engine inoperative rate of climb.

Example 1:

Aerodrome Preasure Altitude	10,000 ft
Ambient Temperature	+10° C
Take off Weight	4000 lb.
Gear up (Undercarriage Retracted)	
Flaps 0°;	
Climb speed	92 Kt. IAS
Cloud Base	400 ft above Reference Zero
Wind component	20 Kt. Head

Calculate the distance from the end of TODR to 1500 ft. above Reference Zero.

Solution:

All engines rate of climb at take-off power 1660 fpm
One engine inoperative rate of climb 300 fpm at take-off power
One engine inoperative rate of climb 220 fpm at MCP
Time to cloud base = $\frac{350}{1660}$ x 60 = 12.65 seconds

Time to 1500 ft. from cloud base = $\frac{1100}{300}$ x 60 = 220 secs. = 3min. 40 secs.

92 Kt. IAS = 110 Kt. TAS: G/S = 110 -20 = 90 Kt.

Distance to cloud base = $\frac{350}{1660}$ x $\frac{90}{60}$ x 1.3 = 0.42 NM

Distance cloud base to 1500 = $\frac{1100}{300}$ x $\frac{90}{60}$ = 5.5 NM

Total Distance = 0.42 + 5.5 = 5.92NM

Example 2.

Aerodrome Pressure Altitude	6000 ft.
Ambient Temperature	+ 20° C
Take-off Weight	4500 lb.
Gear Up (Undercarriage Retracted)	
Flaps 0°; Climb Speed	92 Kt. IAS.
Cloud Base	400 ft. Above Reference Zero
Wind Component	20 Kt. Tail

Obstacle in the domain at 14000 ft. from the end TODR and 600 ft. above Reference Zero

Calculate the vertical clearance of the obstacle by the aeroplane.

Solution

All engines operating rate of climb at take-off power 1500 fpm
One engine inoperative rate of climb at take-off power 260 fpm
One engine inoperative rate of climb at maximum cont. power 220 fpm

Time to cloud base = $\frac{350}{1500}$ x 60 = 14.0 secs.

Time to 1500 ft. from cloud base = $\frac{1100}{260}$ x 60 = 253.8 secs. = 4 m. 13.8.s

92 Kt. IAS = 104 Kt. TAS. G/S = 104 + 20 = 124 Kt.

Distance to cloud base = $\frac{350}{1500}$ x $\frac{124}{60}$ x 6080 x 1.3 = 3811 ft.

Distance cloud base to obstacle = 14000 -3811 = 10189 ft.

Height gain = $\frac{10189 \times 260 \times 60}{124 \times 6080}$ = 210.8 ft.

Height at obstacle = 400 + 211 =611 ft.

Clearance = 611 - 600 = 11 ft

Figure 3.6   **Take-Off Climb Performance -Gear Retracted**

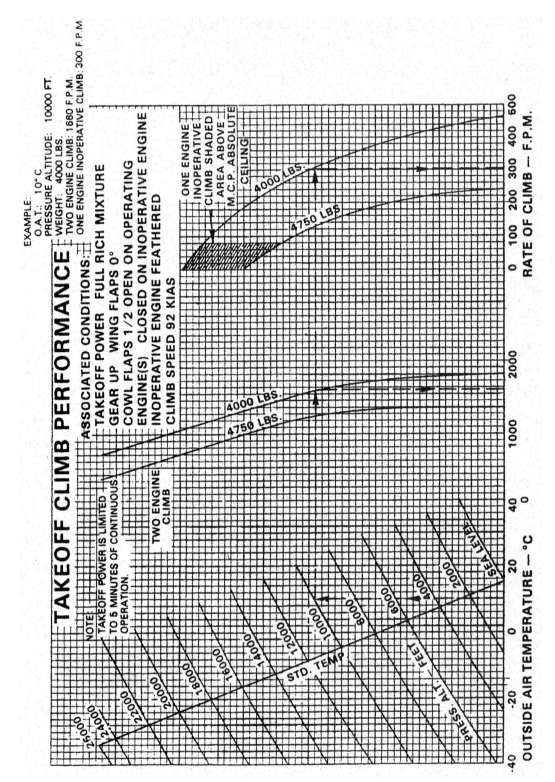

Figure 3.7   **Climb Performance -Gear Retracted**
Maximum Continuous Power

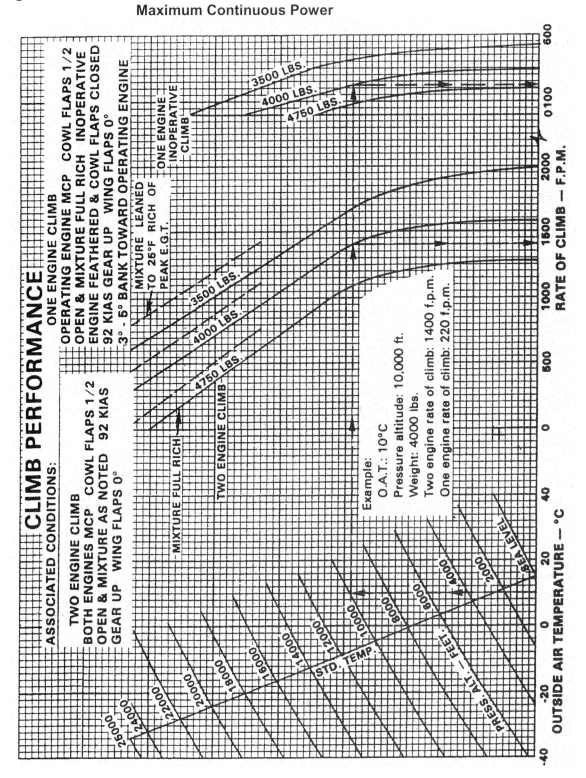

4.   EN-ROUTE

The en-route phase extends from 1500 ft. above the take-off surface level to 1000 ft. above the landing aerodrome surface level.

4.1   Requirements
        In the event of engine failure, with the remaining engine(s) set at the maximum continuous setting, the aeroplane must be able to continue flight at or above the relevant minimum safe altitude to an aerodrome at which the landing requirements can be attained.

        To show compliance:
        a.   The aeroplane may not be assumed to be flying above that altitude at which the rate of climb is 300 fpm with all engines operating.
        b.   The one engine inoperative net gradient of climb is the gross gradient of climb decreased by 0.5% or the net gradient of descent is the gross gradient of descent increased by 0.5%.

5.   LANDING. - MULTI-ENGINED AEROPLANES

5.1   Requirements
There are three requirements for landing with which compliance is necessary. They are the climb gradient requirements, in the event of either a balked landing or a missed approach, and the landing field length requirement.

Field Length Requirements
a.   The landing distance, from a screen height of 50 ft., must not exceed 70% of the landing distance available. i.e. a factor of 1.43.
b.   If the landing surface is grass up to 20 cm. long on firm soil, the landing distance should be multiplied by a factor of 1.15.
c.   If the METAR or TAF of combination of both indicate that the runway may be wet at the estimated time of arrival, the landing distance should be multiplied by a factor of 1.15.
d.   The landing distance should be increased by 5% for each 1% downslope. No allowance is permitted for upslope.
e.   The despatch rules for scheduled (planned) landing calculations are in JAR -OPS 1.550 (c).

Balked Landing Requirement
The minimum acceptable gross gradient of climb after a balked landing is 2.5%. This must be achieved with:
a.   The power developed 8 seconds after moving the power controls to the take-off position.
b.   The landing gear (undercarriage) extended.
c.   Flaps at the landing setting.
d.   Climb speed equal to V_{REF}.

Missed Approach Requirement
The minimum acceptable gross gradient of climb, after a missed approach, is 0.75% at 1500 ft. above the landing surface. This must be achieved with:
a.   The critical engine inoperative and the propeller feathered.
b.   The live engine set at maximum continuous power.
c.   The landing gear (undercarriage) retracted.
d.   The flaps retracted
e.   Climb speed not less than 1.2 V_{S1}

5.2   Balked Landing Climb Graph
        The graph provided for this purpose is constructed for the maximum landing
        weight of 4513 lb. (Fig. 3.8)

        Use of Graph
        1.   Enter left carpet at the ambient temperature. Travel vertically to the
                aerodrome pressure altitude.
        2.   From this point travel horizontally right to intercept the rate of climb
                graph line. Now drop a vertical to read the rate of climb.
        3.   Convert the rate of climb to a gradient of climb using the formula:
                Gradient of climb = $\dfrac{ROC}{TAS} \times \dfrac{6000}{6080}$

Example:
        Aerodrome Pressure Altitude   3000 ft.
        Ambient Temperature   +22° C

Solution:
        Graphical ROC = 810 fpm.
        IAS 85 Kt. = 91 Kt. TAS.
        Gradient of Climb = $\dfrac{810}{91} \times \dfrac{6000}{6080}$ = 8.78 %

Figure 3.8   Landing Climb Performance

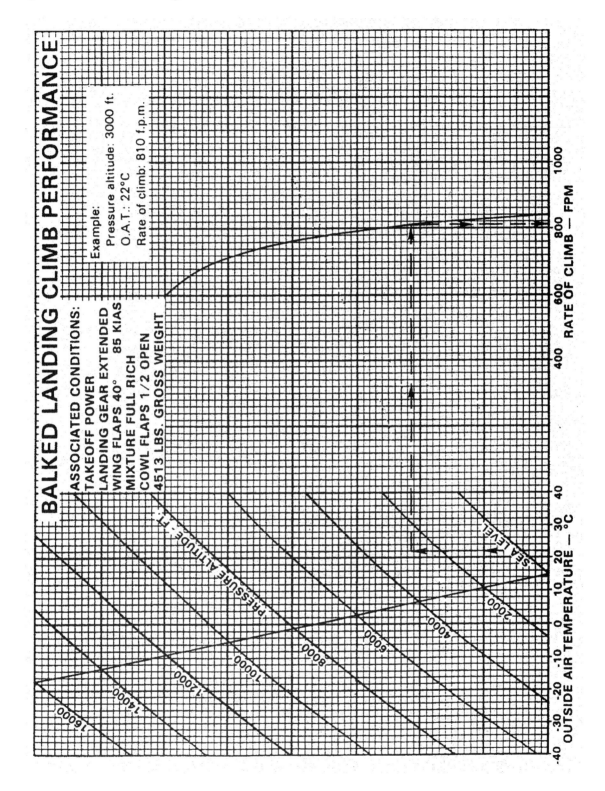

5.3   Use of Landing Field Length Graphs
       There are two landing field length graphs: one for normal landings with 40°
       landing flap (Fig. 3.9) and the other for short field landings with 40° landing
       flap. (Fig. 3.10).

       Distance Calculations
       1.   Enter left carpet at the ambient temperature. Move vertically to the
             aerodrome pressure altitude.
       2.   From this point, move horizontally right to the landing weight reference
             line. Parallel the grid lines to the appropriate landing weight input,
       3.   Continue from this intersection to the wind component reference line.
             Parallel the grid lines to the appropriate wind component input.
       4.   Travel horizontally right to the ground roll reference line. Either
             continue horizontally to the right vertical axis to read the ground roll
             distance or parallel the grid lines to the right vertical axis to read the
             landing distance.
       5.   Apply the appropriate factors to the landing distance to obtain the
             landing distance required.

       Example: Normal Landing
             Aerodrome Pressure Altitude .   3000 ft.
             Ambient Temperature   +22° C
             Landing Weight   3566 lb.
             Wind Component   10 Kt. Head
             Runway Slope   1% Down
             Runway Surface   Grass
             Runway Condition   Wet

       Calculate Landing Distance Required

       Solution:
             Geographical Distance   2260 ft.
             Slope Correction Factor   x 1.05
             Surface Correction Factor   x 1.15
             Condition Correction Factor   x 1.15
             Regulatory Factor   x 1.43

       Landing Distance Required =   4488 ft

Landing Weight Calculations

The procedure for calculating the field length limited landing weight is:

1. De-factorise the landing distance available.
2. Enter the left carpet at the ambient temperature. Move vertically to the aerodrome pressure altitude.
3. From this point, travel horizontally right to the weight reference line. Mark with a pencil.
4 Enter right vertical axis with the distance from 1 above. Parallel the grid lines to the ground roll reference line.
5. From this point, travel horizontally left to the appropriate wind component input. Parallel grid lines to the wind component reference line.
6. Now draw a line horizontally from this point through the weight grid.
7. From the pencil mark in 3 above, parallel the grid-lines to intersect the horizontal line. Drop a vertical to the carpet to read field length limited landing weight.

Example: Short Field Landing

Aerodrome Pressure Altitude	3000 ft.
Ambient Temperature	+22° C
Landing Distance Available	3733 ft.
Wind Component	10 Kt. Head
Runway Slope	1% down
Surface Type	Grass
Surface Condition	Wet

Calculate the field length limited landing weight

Solution:

Landing Distance Available	3733 ft.
Slope Correction Factor	÷1.05
Surface Type Correction Factor	÷1.15
Surface Condition Correction Factor	÷1.15
Regulatory Factor	1.43
De-factorised LDA =	1880 ft.

Field Length Limited Landing Weight   =   <u>3650 lb.</u>

Figure 3.9   **Landing Distance**   **Normal Procedure**

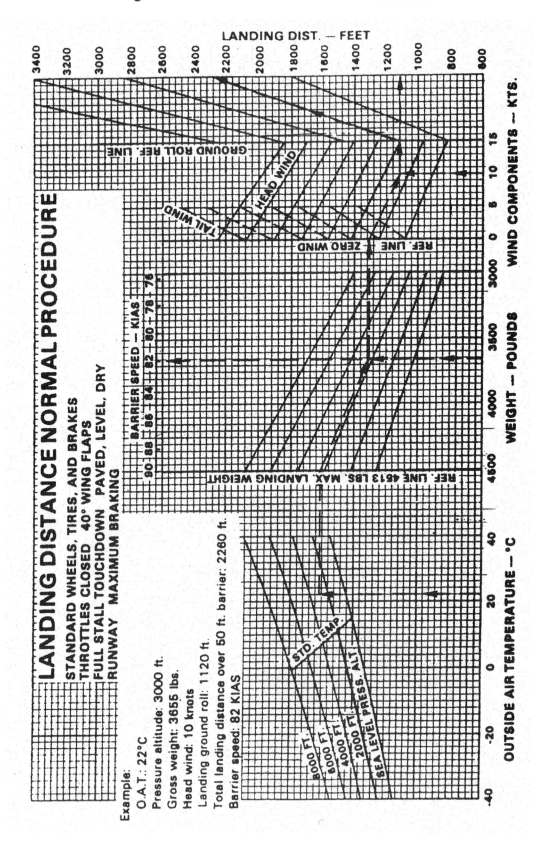

Figure 3.10   **Landing Distance**   **Short Field**

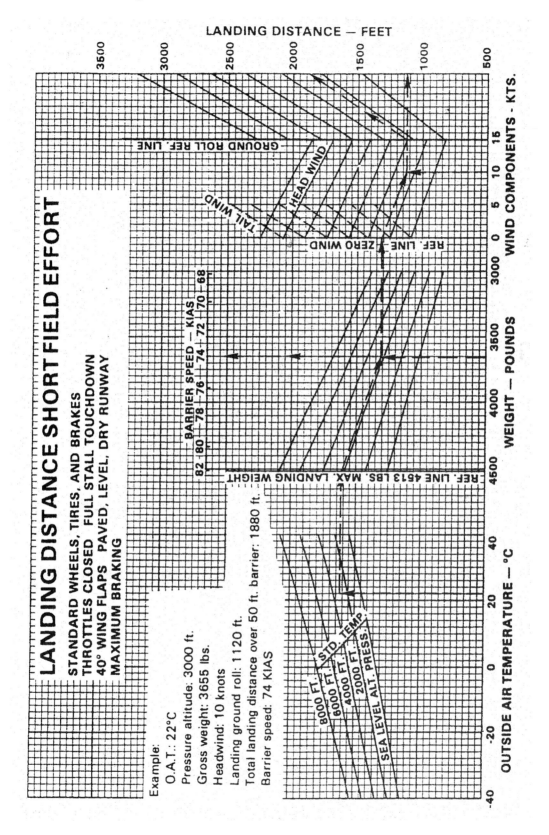

INTENTIONALLY BLANK

PERFORMANCE -AEROPLANES

SECTION IV -DATA FOR MEDIUM RANGE JET TRANSPORT

CONTENTS

1.   GENERAL CONSIDERATIONS

1.1   Performance Limitations

Performance Classification
The specimen aircraft is a landplane powered by two turbo fan engines. It is certificated in the Transport Category (Passenger) and is operated in accordance with JAR/FAR 25 Performance Class A.

Flight Over Water Speed
The true airspeed to be assumed for the purpose of compliance with legislation governing flight over water and en-route climb performance is 380 knots.

Engine Relighting
The maximum altitude to be assumed for engine relighting is 25000 feet.

Maximum Crosswind Component
The maximum crosswind component in which the aeroplane has been demonstrated to be satisfactory for take-off and landing is 33 knots. This wind speed is related to a height of 10 metres.

Standard Performance Conditions
Performance information relates to an average aeroplane of the type and the data are based on: -
a)   Certified engine thrust ratings less installation losses, airbleed and accessory losses.
b)   Full temperature accountability within operational limits except for landing distance, which is based on standard day temperatures.
c)   Trailing edge flap settings:
            •   5° or 15° for take-off,
            •   4° transition setting,
            •   22° for approach,
            •   15°, 30°, 40° for landing
        with leading edge devices in the full down position for these flap settings.
d)   Operations on smooth, hard-surfaced, runways.

1.2   Aeroplane Limitations

Mass (Weight)
Maximum structural take-off mass is 62800 Kg.
Maximum structural landing mass is 54900 Kg.
Maximum zero fuel mass is 51300 Kg.
On a given occasion, the maximum permitted take-off and landing mass may be less than the structural limits given above.

Wing Span
The wingspan of the aeroplane is 28.88 metres.

Power Plant
The engines shall not be operated continuously at maximum take-off thrust for periods exceeding 5 minutes.

Operating Limitations

Operational mass (weight) limits are determined from the following performance considerations: -
a)   TAKE-OFF FIELD LENGTHS
b)   TAKE-OFF (MASS-ALTITUDE-TEMPERATURE) LIMITS
c)   TYRE SPEED LIMITS
d)   BRAKE ENERGY LIMITS
e)   NET TAKE-OFF FLIGHT PATH
f)   EN ROUTE CLIMB PERFORMANCE
g)   LANDING (MASS-ALTITUDE-TEMPERATURE) LIM ITS
h)   LANDING FIELD LENGTHS

1.3   Additional Definitions

Temperature
Outside Air Temperature (OAT or SAT) -the free air static (ambient) temperature.
Total Air Temperature (TAT) -static air temperature plus adiabatic compression (ram) rise as indicated on the Total Air Temperature indicator.

Altitude
The altitude shown on the charts is pressure altitude. This is the height, in the International Standard Atmosphere, at which the prevailing pressure occurs. It may be obtained by setting the sub-scale of a pressure altimeter to 1013 hPa (29.92 inches or 760 mm. of mercury).

International Standard Atmosphere (ISA)
A structure of assumed conditions relating to the change of pressure, temperature and density with height in the atmosphere.

Gross Performance
The average performance that a fleet of aeroplanes should achieve if satisfactorily maintained and flown in accordance with the techniques described in the manual.

Net Performance
Net performance is the gross performance diminished to allow for various contingencies that cannot be accounted for operationally e.g., variations in piloting technique, temporary below average performance, etc. It is improbable that the net performance will not be achieved in operation, provided the aeroplane is flown in accordance with the recommended techniques.

IAS
The indicated airspeed is the reading obtained on a pitot-static airspeed indicator having no instrument calibration error. If the calibration error of the particular instrument is not known, then the actual reading may be taken to be equal to IAS because the tolerances permitted on the instrument are small.

TAS
The 'true airspeed' is the speed of the aeroplane relative to the undisturbed air.

Decision Speed -V_1
An engine failure having been promptly recognised, V_1 is the speed at which:
a)   the continued Take-off Distance Required will not exceed the Take-off Distance Available
b)   the continued Take-off Run Required will not exceed the Take-off Run Available,
c)   the Accelerate-Stop Distance will not exceed the Emergency Distance Available.
V_1 must not be less than V_{MCG}, not greater than V_R and not greater than V_{MBE}.

Maximum Brake Energy Speed -V_{MBE}
The maximum speed on the ground from which a stop can be accomplished within the energy capabilities of the brakes.

Rotation Speed -V_R
The speed at which, during the take-off, rotation is initiated.

Take-Off Safety Speed -V_2
The target speed to be attained at the screen height with one engine inoperative, and used to the point where acceleration to flap retraction speed is initiated.

Steady Initial Climb Speed -V_4
The all engines operating take-off climb speed used to the point where acceleration to flap retraction speed is initiated. V_4 should be attained by a height of 400 feet.

Air Minimum Control Speed -V_{MCA}
The minimum flight speed at which the aeroplane is controllable with a maximum of 5° bank when the critical engine suddenly becomes inoperative with the remaining engines at take-off thrust.

Approach and Landing Minimum Control Speed -V_{MCL}
The minimum speed with a wing engine inoperative where it is possible to decrease thrust to idle or increase thrust to maximum take-off without encountering dangerous flight characteristics.

Ground Minimum Control Speed -V_{MCG}
The minimum speed on the ground at which the take-off can be safely continued, when the critical engine suddenly becomes inoperative with the remaining engines at take-off thrust.

Threshold Speed -V_T
The speed at which the pilot should aim to cross the runway threshold to ensure that the scheduled landing field lengths are consistently achieved. The speeds at the threshold are: -
V_{TO} -all engines operating
V_{T1} -a critical engine inoperative
Maximum Threshold Speed -V_{TMAX}
The speed at the threshold above which the risk of exceeding the scheduled landing field length is unacceptably high. Go-around action should normally be taken if it appears that maximum threshold speed will be exceeded. This speed is 15 knots greater than the all-engines operating target threshold speed.

Elevation
The vertical distance of an object above mean sea level. This may be given in metres or feet

Height
The vertical distance between the lowest part of the aeroplane and the relevant datum.

Gross Height
The true height attained at any point in the take-off flight path using gross climb performance. Gross height is used for calculating pressure altitudes for purposes of obstacle clearance and the height at which wing flap retraction is initiated.

Net Height
The true height attained at any point in the take-off flight path using net climb performance. Net height is used to determine the net flight path that must clear all obstacles by at least 35 feet to comply with the operating regulations.

Climb Gradient
The ratio, in the same units, and expressed as a percentage as obtained from the formula: -

$$\frac{\text{Change in Height}}{\text{Horizontal Distance Travelled}} \times 100\ \%$$

The climb gradients shown on the charts are true gradients, i.e. they are derived from true (not pressure) rates of climb.

EPR
Engine Pressure Ratio. The parameter used for setting thrust.

1.4   The Determination of Wind Component

Use the graph at Fig. 4.1
a)   Calculate the relative direction of the wind to the runway. i.e.
        wind direction -runway direction or
        runway direction -wind direction.
b)   Enter graph at left vertical axis with windspeed.
c)   Follow circle until relative direction intercepted.
d)   From the intersection draw a line horizontally left to the vertical axis to read
        the along track component. Negative values are tailwinds.
e)   From the intersection drop a vertical line to intersect the horizontal axis to
        read the crosswind component.
f)   The windspeed grids have already been factorised 50% for headwinds and
        150% for tailwinds. Therefore the grids may be entered with the reported or
        calculated along track component.

Example:
W/V 330/30 Runway 02

Wind angle = 50°
Headwind = 19 Kt. Crosswind = 23 Kt. left to right.
Note this graph is for use with take-off and landing computations only

Figure 4.1   Wind Components for Take Off and Landing

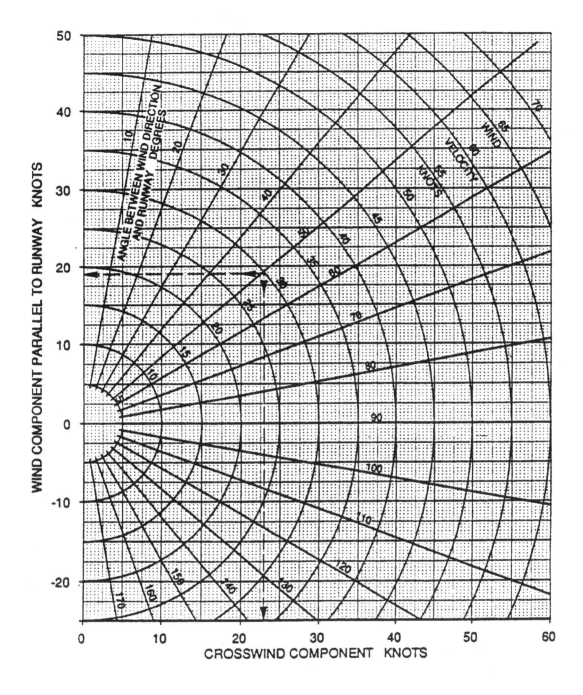

1.5   Conversion of QFE or QNH to Pressure Altitude

All altitudes in this manual refer strictly to pressure altitude. If only QFE or QNH are known then it must be used to produce a pressure altitude.

Figure 4.2   QNH TO Pressure Altitude

QNH (IN. HG.)			Correction to elevation for press. Alt. Ft.	QNH (hPa)		
28.81	to	28.91	+1000	976	to	979
28.91	to	29.02	+900	979	to	983
29.02	to	29.12	+800	983	to	986
29.12	to	29.23	+700	986	to	990
29.23	to	29.34	+600	990	to	994
29.34	to	29.44	+500	994	to	997
29.44	to	29.55	+400	997	to	1001
29.55	to	29.66	+300	1001	to	1004
29.66	to	29.76	+200	1004	to	1008
29.76	to	29.87	+100	1008	to	1012
29.87	to	29.97	0	1012	to	1015
29.97	to	30.08	- 100	1015	to	1019
30.08	to	30.19	- 200	1019	to	1022
30.19	to	30.30	- 300	1022	to	1026
30.30	to	30.41	- 400	1026	to	1030
30.41	to	30.52	- 500	1030	to	1034
30.52	to	30.63	- 600	1034	to	1037
30.63	to	30.74	- 700	1037	to	1041
30.74	to	30.85	- 800	1041	to	1045
30.85	to	30.96	- 900	1045	to	1048
30.96	to	31.07	- 1000	1048	to	1052

Example: Elevation   = 2500ft
             QNH   = 29.48 IN. HG.
             Correction   = 400 ft.
             Press Alt.   = 2900 ft.

1.6   Total Air Temperature at ISA

The cockpit temperature gauge shows the total air temperature (TAT) which is the true outside air temperature plus the rise due to ram air compression. To calculate the value of TAT while flying in ISA conditions enter the table left column at the appropriate pressure altitude and move along the line to the appropriate indicated mach number to read the ISA/TAT.
Compare the actual TAT to the tabulated ISA/TAT to obtain the temperature deviation from standard.

Figure 4.3   TAT in ISA conditions

Pressure Altitude 1000 ft.	INDICATED MACH NUMBER													
	0	.40	.50	.60	.70	.74	.78	.80	.82	.84	.86	.88	.90	.92
36 to 45	-56			-41	-35	-33	-30	-29	-27	-26	-24	-23	-21	-20
35	-54			-39	-33	-30	-28	-26	-25	-23	-22	-20	-19	-17
34	-52		-41	-36	-31	-28	-25	-24	-23	-21	-20	-18	-17	-15
33	-50		-39	-34	-29	-26	-23	-22	-20	-19	-17	-16	-14	-13
32	-48		-37	-32	-26	-24	-21	-20	-18	-17	-15	-14	-12	-10
31	-46		-35	-30	-24	-22	-19	-17	-16	-14	-13	-11	-10	-8
30	-44		-33	-28	-22	-19	-17	-15	-14	-12	-11	-9	-7	-6
29	-42		-31	-26	-20	-17	-14	-13	-11	-10	-8	-7	-5	-3
28	-40		-29	-24	-18	-15	-12	-11	-9	-8	-6	-4	-3	-1
27	-38		-27	-22	-15	-13	-10	-8	-7	-5	-4	-2	0	1
26	-37		-25	-19	-13	-11	-8	-6	-5	-3	-2	0	2	4
25	-35		-23	-17	-11	-8	-5	-4	-2	-1	1	2	4	6
24	-33	-25	-21	-15	-9	-6	-3	-2	0	1	3	5	6	8
23	-31	-23	-18	-13	-7	-4	-1	0	2	4	5	7	9	11
22	-29	-21	-16	-11	-5	-2	1	3	4	6	8	9	11	13
21	-27	-19	-14	-9	-2	0	3	5	7	8	11	12	13	
20	-25	-17	-12	-7	0	3	6	7	9	10	12	14		
19	-23	-15	-10	-5	2	5	8	9	11	13	14			
18	-21	-13	-8	-2	4	7	10	12	13	15				
17	-19	-11	-6	0	6	9	12	14	16					
16	-17	-8	-4	2	8	11	15	16						
15	-15	-6	-2	4	11	14	17	18						
14	-13	-4	0	6	13	16	19							
13	-11	-2	2	8	15	18	21							
12	-9	0	4	10	17	20								
11	-7	2	7	12	19	22								
10	-5	4	9	15	21									
9	-3	6	11	17	24									
8	-1	8	13	19	26									
7	1	10	15	21	28									
6	3	12	17	23	30									
5	5	14	19	25	32									
4	7	16	21	27										
3	9	18	23	29										
2	11	20	25	32										
1	13	22	27											

TOTAL AIR TEMPERATURE AT ISA (°C)

INTENTIONALLY BLANK

2. TAKE OFF PERFORMANCE

2.1 Field Limit

The field length limit graph (figure 4.4) accounts for runway slope, wind component, flap position, aerodrome pressure altitude and ambient temperature. It does not take into account any stopway or clearway. It is therefore a balanced field equal to the length of the take-off run available. The field length used in the graph is based on a minimum V_{MCG}. This means that, if either stopway or clearway is available, a certain amount of payload, which could have been carried, will have to be forgone.

The maximum take-off mass is determined as the lowest of:
- the structural limit
- the field length limit
- the climb limit (WAT)
- the tyre speed limit
- the brake energy limit
- the obstacle limit.

The field length requirements specified in a JAR 25 are:

a. If the take-off distance includes a clearway, the take-off run is the greatest of:
  i. All power units operating. The total of the gross distance from the start of the take-off run to the point at which V_{LOF} is reached, plus one half of the gross distance from V_{LOF} to the point at which the aeroplane reaches 35ft, all factorised by 1.15 to obtain the net TORR.
  ii. One power unit inoperative (dry runway). The horizontal distance from the brakes release point (BRP) to a point equidistant between V_{LOF} and the point the aeroplane reaches 35 ft with the critical power unit inoperative.
  iii. One power unit inoperative (wet runway). The horizontal distance from the brake release point (BRP) to V_{LOF} assuming the critical power unit inoperative at VEF.

b. The accelerate/stop distance on a wet runway is the greater of:
  i. All engines operating. The sum of the distances required to accelerate from BRP to VSTOP + 2 seconds and to decelerate from this point to a full stop on a wet hard surface.
  ii. One engine inoperative. The sum of the distances to accelerate from BRP with all engines operating (to VSTOP + 2 seconds) plus the distance to brake to a full stop on a wet hard surface with one engine inoperative.

c. The take-off distance required is the greatest of the following three distances:
  i. All engines operating. 115% of the horizontal distance travelled, with all engines operating, to reach a screen height of 35 ft.
  ii. One engine inoperative (dry runway). The horizontal distance from BRP to the point at which the aeroplane attains 35 ft., assuming the critical power unit fails at V_1 on a dry, hard surface.
  iii. One engine inoperative (wet or contaminated runway). The horizontal distance from BRP to the point at which the aeroplane attains 15 ft., assuming the critical power unit fails at V_1 on a wet or contaminated hard surface.

Method of Use Of The "Take-Off Performance Field Limit" Graph (Fig 4.4)

a) Enter the right carpet with TORA. Move vertically to the runway slope reference line.

b) Parallel the grid lines to the appropriate runway slope then continue vertically to the wind component reference line.

c) Parallel the grid lines to the appropriate wind component then continue vertically to the flap reference line.

d) If flap is 15°, parallel grid lines then draw a vertical line through the weight grid. If flap is 5° draw a vertical line from the reference line through the weight grid.

e) Enter left carpet at the aerodrome ambient temperature and proceed vertically to the aerodrome pressure altitude.

f) Proceed horizontally right to the weight grid reference line.

g) From this point interpolate the grid lines to intersect the vertical line drawn in d. above.

h) From this intersection draw a horizontal line to read the field length limited TOM.

Example

Field Length Available (TORA)	9,600 ft.
Runway slope	1% Uphill
Wind Component	20 Kt. Head
Flaps	15°
Ambient Temperature	+ 33° C
Aerodrome Pressure Altitude	2,000 ft.
Field-Length Limited TOM	63,000 Kg.

Figure 4.4   **Take Off Performance**   **Field Limit Graph**

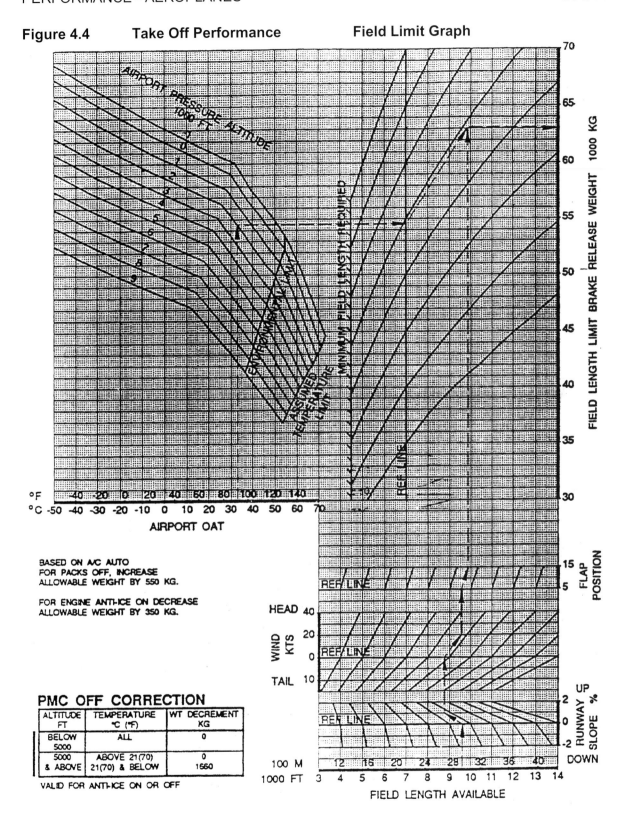

BASED ON A/C AUTO
FOR PACKS OFF, INCREASE
ALLOWABLE WEIGHT BY 550 KG.

FOR ENGINE ANTI-ICE ON DECREASE
ALLOWABLE WEIGHT BY 350 KG.

PMC OFF CORRECTION

ALTITUDE FT	TEMPERATURE °C (°F)	WT DECREMENT KG
BELOW 5000	ALL	0
5000 & ABOVE	ABOVE 21(70)	0
	21(70) & BELOW	1660

VALID FOR ANTI-ICE ON OR OFF

2.2   Take Off Climb

The graph at figure 4.5 guarantees attainment of the most severe gradient
requirement of the net flight path. It does not guarantee obstacle clearance.

Method of Use
Enter the graph at aerodrome ambient temperature
Move vertically to the aerodrome pressure altitude.
Travel horizontally left to the flap reference line and apply the appropriate setting to
read climb limit weight.
Apply corrections.

Example:
Aerodrome Pressure Altitude   2,000 ft.
OAT   + 33°C
Flaps   15°
MAT Limited TOM   53,400 Kg.

Figure 4.5   **Take Off Performance**   **Climb Limit**

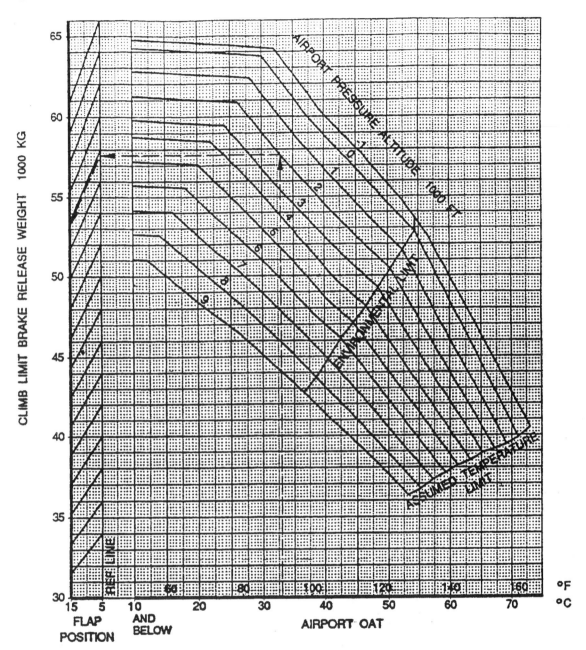

BASED ON A/C AUTO WITH APU ON OR OFF. FOR PACKS OFF, INCREASE
ALLOWABLE WEIGHT BY 900 KG.
FOR OPERATION WITH ENGINE ANTI-ICE ON SUBTRACT 190 KG WHEN AIRPORT
PRESSURE ALTITUDE IS AT OR BELOW 8000 FEET OR 530 KG WHEN AIRPORT
PRESSURE ALTITUDE IS ABOVE 8000 FEET.

PMC OFF CORRECTION

ALTITUDE FT	TEMPERATURE °C (°F)	WT DECREMENT KG
BELOW 5000	ALL	0
5000 & ABOVE	ABOVE 21 (70)	0
	21 (70) & BELOW	1860

2.3   Take Off Tyre Speed Limit

The graph at figure 4.6 presents the limitation on take-off weight for 225-mph tyres and 5° flap.

Method of Use
Enter the graph with aerodrome OAT. Proceed vertically to the aerodrome pressure altitude, then horizontally left to read the tyre speed limit.
Correct if necessary.

For 210-mph tyres and/or 15° flap, apply the correction below the graph.

Example:
OAT	+33° C
Airfield Pressure Altitude	2000 ft
Flaps	15°
Tyres	210 mph
Uncorrected limit	80400 kg.
Correction	-1500 kg.
Tyre Limit	78900 kg.

Figure 4.6　　　**Take-Off Performance**　　　**Tyre Speed Limit**

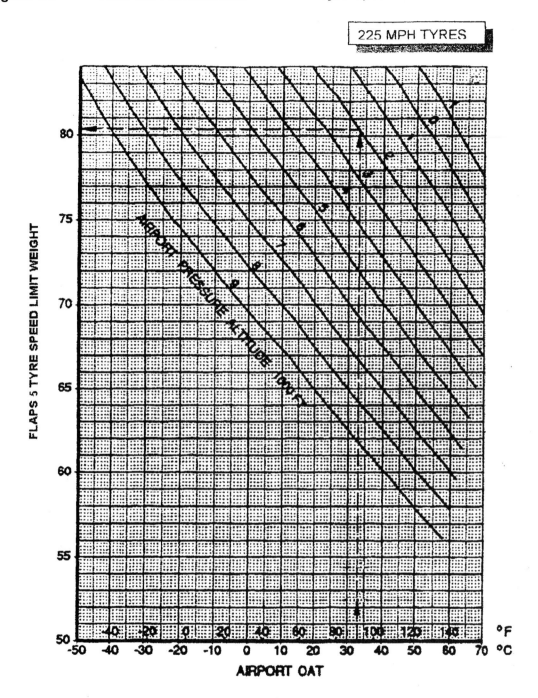

225 MPH TYRES

Increase 'flaps 5' tyre speed limit weight by 6600 kg for flaps 15°
Decrease 'flaps 5' tyre speed limit weight by 9600 kg for 210 mph tyres and flaps 5
Decrease 'flaps 5' tyre speed limit weight by 1500 kg for 210 mph tyres and flaps 15.
Increase tyre speed limit weight by 400 kg per knot of headwind.
Decrease tyre speed limit weight by 650 kg per knot of tailwind

PMC OFF CORRECTION

ALTITUDE FT	TEMPERATURE ° C	WT. DECREMENT KG
Below 5000	Above 21	250
	21 & Below	210
5000 & above	Above 21	200
	21 & Below	270

2.4   Take Off Brake Energy Limit

Figure 4.7 enables the determination of V_{MBE}.
Generally V_{MBE} will not be limiting except at hot, high aerodromes or operating with a tail wind.
Always check V_{MBE} when outside the shaded area of the top left grid or when there is a tail wind or when employing the improved climb technique. If V_1 exceeds V_{MBE}. apply the correction below the graph. Make $V_1 = V_{MBE}$ and recalculate the other V speeds for the reduced weight.

Method of Use

Enter the left vertical axis with aerodrome pressure altitude. Travel horizontally to OAT. Drop vertically to take-off weight, then horizontally right to V_{MBE}.

Example:

Take Off Weight	64000 Kg.
Airfield Pressure Altitude	5600 ft.
Ambient Temperature	-10° C
Runway Slope	1.5% Uphill
Wind Component	10 Kt. Head
PMC 'ON'	

V_{MBE} = 165 + 3 + 3=171 Kt.

Figure 4.7   Take-Off Brake Energy Limit

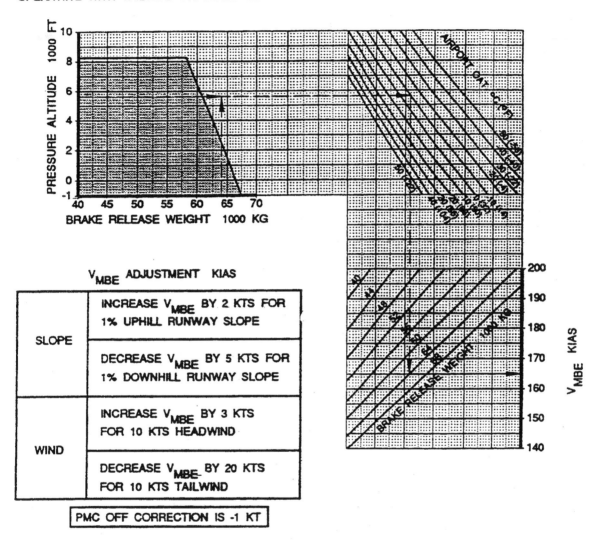

ALL FLAPS

CHECK V_{MBE} WHEN OUTSIDE SHADED AREA OR WHEN
OPERATING WITH TAILWIND OR IMPROVED CLIMB.

V_{MBE} ADJUSTMENT KIAS

SLOPE	INCREASE V_{MBE} BY 2 KTS FOR 1% UPHILL RUNWAY SLOPE
	DECREASE V_{MBE} BY 5 KTS FOR 1% DOWNHILL RUNWAY SLOPE
WIND	INCREASE V_{MBE} BY 3 KTS FOR 10 KTS HEADWIND
	DECREASE V_{MBE} BY 20 KTS FOR 10 KTS TAILWIND

PMC OFF CORRECTION IS -1 KT

ADJUST V_{MBE} FOR SLOPE AND WIND.

NORMAL TAKEOFF: DECREASE BRAKE RELEASE WEIGHT BY 300 KG FOR EACH
KNOT V_1 EXCEEDS V_{MBE}. DETERMINE NORMAL V_1, V_R, V_2 SPEEDS FOR LOWER
BRAKE RELEASE WEIGHT.

IMPROVED CLIMB TAKEOFF: DECREASE CLIMB WEIGHT IMPROVEMENT BY 160 KG
FOR EACH KNOT V_1 EXCEEDS V_{MBE}. DETERMINE V_1, V_R, V_2 SPEED INCREMENTS
FOR THE LOWER CLIMB WEIGHT IMPROVEMENT.

2.5   V Speeds and % N_1 Values

The V speeds quoted in tables Figures 4.8 and 4.9 are those for a balanced field at which TORA = ASDA = TODA. The runway is assumed to be hard, level and in still air. Sub-tables are provided to enable the speeds to be corrected for the effects of slope and wind. V_1 can be further adjusted to account for any clearway or stopway by using the following table. This table must not be used if the stopway and/or clearway were used in the determination of the field length limited take-off weight.

2.5.1 V_1 Adjustments

Clearway Minus Stopway Ft.	Normal V_1 KIAS			
	100	120	140	160
800	-	-	-3	-2
600	-	-3	-2	-1
400	-4	-3	-2	-1
200	-2	-1	-1	0
0	0	0	0	0
-400	1	1	1	1
-800	1	1	1	1

Maximum Allowable clearway

Field Length Ft.	Max. Allowable Clearway for V_1 Reduction Ft.
4000	400
6000	500
8000	550
10000	600
12000	700
14000	750

In the absence of more precise details the above table should be used as a guide to the maximum allowable clearway permitted.
In no circumstances may V_1 be less than the minimum V_{MCG} nor may it exceed V_R or V_{MBE}.
If V_1 is less than the minimum V_{MCG} then make V_1 = minimum V_{MCG}.
If V_R is less than the minimum V_{MCG} then make V_1 = V_R = min V_{MCG}.
If such is the case the V_2 = V_{MCG} plus the difference between the original V_R and V_2.
Correction to the take-off weight is unnecessary provided the TORA exceeds the TORR.

2.5.2   The Calculation of V Speeds

To calculate the V speeds use the tables (at figure 4.8 or 4.9) as appropriate in the following manner:

i.   Select the tables appropriate to the flap setting.

ii.   Enter the density sub-graph (below) with pressure altitude and ambient temperature to determine which of the columns of the tables should be used.

iii. Enter the V speed tables at the actual take-off weight. Extract V_1, V_R and V_2.

iv.   If necessary to correct V_1 for slope and/or wind component, enter the table at the top of the appropriate figure (4.8 or 4.9) take-off weight and interpolate the correction necessary.

v.   Apply the corrections to V_1.

vi.   Enter the lower table to determine the minimum V_{MCG} by entering the left column at the ambient temperature and then proceeding right along the line to the appropriate aerodrome pressure altitude (interpolating if necessary). Extract V_{MCG}.

vii. Compare V_1 with minimum V_{MCG}. If V_1 is less make V_1 = min V_{MCG}

viii. Check TORA exceeds TORR. If it does not, the take-off weight must be reduced.

DENSITY SUB GRAPH

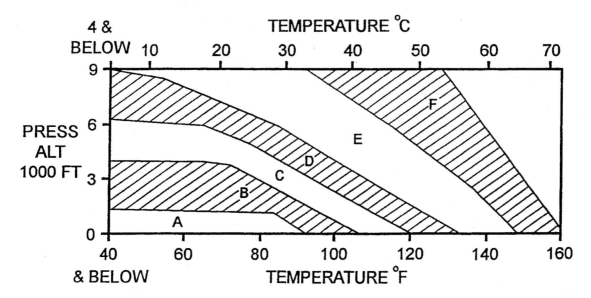

2.5.3   Stabiliser Trim Setting

To determine the take-off stabiliser trim setting select the appropriate table and use the actual take-off weight and % MAC centre of gravity to read or calculate the appropriate setting.

Figure 4.8   Take-Off Speeds

FLAPS 5°

PMC ON

	Slope/Wind V$_1$ adjustment					
Weight	Slope %			Wind Kt.		
1000 Kg.	-2	0	2	-15	0	40
70	-3	0	4	-3	0	1
60	-2	0	2	-3	0	1
50	-2	0	1	-4	0	1
40	-2	0	1	-4	0	1

* V$_1$ not to exceed V$_R$

WT 1000 Kg.	A			B			C		
	V$_1$	V$_R$	V$_2$	V$_1$	V$_R$	V$_2$	V$_1$	V$_R$	V$_2$
70	158	163	168	158	164	169			
65	151	155	161	152	156	162	153	157	162
60	144	148	155	145	148	155	146	149	155
55	137	139	149	138	140	149	138	141	148
50	129	131	142	130	132	142	131	133	142
45	121	123	136	122	124	135	122	125	135
40	113	114	130	113	116	129	113	116	128

WT 1000 Kg.	D			E			F		
	V$_1$	V$_R$	V$_2$	V$_1$	V$_R$	V$_2$	V$_1$	V$_R$	V$_2$
70									
65									
60									
55	140	143	148						
50	132	134	141	133	135	141			
45	124	126	135	125	127	134	128	128	134
40	113	117	128	114	118	127	119	120	126

In shaded area check minimum V$_1$ (MCG) for actual temp.

Minimum V$_1$ (MCG)

Actual OAT		Press. Alt. X 1000 Ft.				
°C	°F	0	2	4	6	8
55	131	104				
50	122	107	103			
40	104	111	107	103	99	94
30	86	116	111	107	104	98
20	68	116	113	111	107	102
10	50	116	113	111	108	104
-50	-58	118	115	112	109	105

For A/C packs 'off' increase V1 (MCG) by 2 knots

FLAPS 5   STABILISER TRIM SETTING

CG. % MAC	6	10	14	18	22	26	30
Stab. Trim	5 ½	5	4 ½	3 ¾	3 ¼	2 ¾	2 ¼

For weights at or below 45350 kg subtract ½ unit
For weights at or above 61250 kg add ½ unit
Stab trim settings must be between 1 and 5 ¾ units

Figure 4.9   TAKE-OFF SPEEDS

Weight	Slope/Wind V$_1$ Adjustment					
	Slope %			Wind Kt.		
1000 Kg	-2	0	2	-15	0	40
70	-3	0	4	-3	0	1
60	-2	0	2	-3	0	1
50	-2	0	1	-4	0	1
40	-2	0	1	-4	0	1

•   V$_1$ not to exceed V$_R$

Wt. 1000 Kg.	A			B			C		
	V$_1$	V$_R$	V$_2$	V$_1$	V$_R$	V$_2$	V$_1$	V$_R$	V$_2$
70	150	152	158						
65	143	145	152	145	146	151			
60	137	139	146	138	140	146	139	140	146
55	130	131	141	131	132	140	131	133	140
50	122	124	135	123	125	134	124	125	134
45	114	116	128	116	117	128	116	118	128
40	104	107	122	104	109	122	104	110	122

Wt. 1000 Kg.	D			E			F		
	V$_1$	V$_R$	V$_2$	V$_1$	V$_R$	V$_2$	V$_1$	V$_R$	V$_2$
70									
65									
60									
55	134	134	140						
50	126	126	134	127	128	133			
45	118	119	127	120	120	126	120	121	127
40	109	111	121	111	112	120	112	113	120

In shaded area check minimum V$_1$ (MCG) for actual temp.

Minimum V$_1$ (MCG)

OAT		Pressure Alt. X 1000 Ft.				
°C	°F	0	2	4	6	8
55	131	104				
50	122	107	103			
40	104	111	107	103	99	94
30	86	116	111	107	104	98
20	68	116	113	111	107	102
10	50	116	113	111	108	104
-50	-58	118	115	112	109	105

For A/C packs 'off' ADD 2 knots

FLAPS 15   STABILISER TRIM SETTING

CG. % MAC	6	10	14	18	22	26	30
Stab. Trim	5	4 ¼	3 ¾	3	2 ½	1 ¾	1

For weights at or below 45350 kg subtract ½ unit
For weights at or above 61250 kg add ½ unit
Stab trim settings must be between 1 and 5 ¾ units

2.5.4   % N_1 Values

All % N_1 tables may be used for either engine anti-icing 'on' and 'off' configurations. Correction is necessary if the air conditioning packs are off.

To determine the % N_1 values use the following procedures:
- i.   Select the table appropriate to either PMC on or off.
- ii.   Select the table that is appropriate to the phase of flight (take-off, climb or go-around).
- iii.   Enter the left column of the table with either aerodrome ambient temperature or TAT as appropriate. Read % N_1 in the aerodrome pressure altitude column.

Figure 4.10   Expanded Maximum Take-off % N$_1$

Airport OAT		Pressure Altitude Ft.									
°C	°F	-1000	0	1000	2000	3000	4000	5000	6000	7000	8000
54	129	93.3	94.1	93.6							
52	126	93.6	94.2	94.2	93.7						
50	122	93.8	94.3	94.3	94.3	93.9					
48	118	94.0	94.5	94.4	94.4	94.4	94.1				
46	115	94.1	94.7	94.6	94.5	94.5	94.6	94.4			
44	111	94.3	94.8	94.8	94.7	94.7	94.7	94.8	94.6		
42	108	94.5	95.0	95.0	94.9	94.9	94.8	94.9	95.0	94.8	
40	104	94.6	95.2	95.2	95.1	95.0	95.1	95.1	95.2	95.1	94.9
38	100	94.8	95.3	95.4	95.3	95.2	95.3	95.3	95.4	95.3	95.2
36	97	95.1	95.5	95.5	95.5	95.4	95.6	95.6	95.6	95.5	95.4
34	93	95.3	95.7	95.7	95.7	95.6	95.8	95.8	95.8	95.7	95.6
32	90	95.5	95.9	95.9	95.8	95.8	96.0	96.0	95.9	95.9	95.8
30	86	95.2	96.1	96.1	96.0	96.0	96.3	96.2	96.1	96.0	96.0
28	82	94.9	95.8	96.3	96.2	96.2	96.5	96.4	96.3	96.2	96.1
26	79	94.6	95.5	96.0	96.4	96.4	96.6	96.5	96.5	96.4	96.3
24	75	94.2	95.2	95.6	96.1	96.5	96.8	96.7	96.7	96.6	96.5
22	72	93.9	94.8	95.3	95.7	96.2	96.9	96.9	96.9	96.8	96.6
20	68	93.6	94.5	95.0	95.4	95.9	96.6	97.1	97.1	97.0	96.9
18	64	93.3	94.2	94.7	95.1	95.6	96.3	96.8	97.3	97.2	97.1
16	61	93.0	93.9	94.3	94.8	95.2	96.0	96.4	96.9	97.4	97.3
14	57	92.6	93.5	94.0	94.4	94.9	95.6	96.1	96.6	97.0	97.5
12	54	92.3	93.2	93.7	94.1	94.6	95.3	95.8	96.3	96.7	97.2
10	50	92.0	92.9	93.4	93.8	94.2	95.0	95.4	95.9	96.4	96.8
8	46	91.7	92.6	93.0	93.4	93.9	94.6	95.1	95.6	96.0	96.5
6	43	91.3	92.2	92.7	93.1	93.6	94.3	94.7	95.3	95.7	96.2
4	39	91.0	91.9	92.4	92.8	93.2	93.9	94.4	94.9	95.3	95.8
2	36	90.7	91.6	92.0	92.4	92.9	93.6	94.1	94.6	95.0	95.5
0	32	90.4	91.2	91.7	92.1	92.6	93.3	93.7	94.2	94.7	95.1
-2	28	90.0	90.9	91.4	91.8	92.2	92.9	93.4	93.9	94.3	94.8
-4	25	89.7	90.6	91.0	91.4	91.9	92.6	93.0	93.5	94.0	94.4
-6	21	89.4	90.2	90.7	91.1	91.5	92.2	92.7	93.2	93.6	94.1
-8	18	89.0	89.9	90.3	90.7	91.2	91.9	92.3	92.8	93.3	93.7
-10	14	88.7	89.6	90.0	90.4	90.8	91.5	92.0	92.5	92.9	93.4
-12	10	88.3	89.2	89.7	90.0	90.5	91.2	91.6	92.1	92.5	93.0
-14	7	88.0	88.9	89.3	89.7	90.2	90.8	91.3	91.8	92.2	92.6
-16	3	87.7	88.5	89.0	89.4	89.8	90.5	90.9	91.4	91.8	92.3
-18	0	87.3	88.2	88.6	89.0	89.5	90.1	90.6	91.1	91.5	91.9
-20	-4	87.0	87.8	88.3	88.7	89.1	89.8	90.2	90.7	91.1	91.6
-22	-8	86.6	87.5	87.9	88.3	88.7	89.4	89.9	90.3	90.8	91.2
-24	-11	86.3	87.1	87.6	88.0	88.4	89.1	89.5	90.0	90.4	90.8
-26	-15	85.9	86.8	87.2	87.6	88.0	88.7	89.1	89.6	90.0	90.5
-28	-18	85.6	86.4	86.9	87.2	87.7	88.4	88.8	89.3	89.7	90.1
-30	-22	85.2	86.0	86.5	86.9	87.3	88.0	88.4	88.9	89.3	89.7

Valid for PMC 'on', A/C 'auto', engine anti-ice 'on' or 'off'.
For A/C off Add 1.0% N$_1$
Do not operate engine anti-ice "on" at airport OAT above 10°C (50° F).

Figure 4.11　　Max T/O & Max Climb　　% N₁ values

PMC ON

MAX TAKE-OFF % N₁

AIRPORT OAT		VALID FOR 2 PACKS ON (AUTO) ENGINE A/I ON OR OFF								
		AIRPORT PRESSURE ALTITUDE FT.								
°C	°F	0	1000	2000	3000	4000	5000	6000	7000	8000
55	131	93.8	93.8	93.8						
50	122	94.3	94.3	94.3	93.9	93.6				
45	113	94.7	94.7	94.6	94.6	94.6	94.7	94.4	94.2	
40	104	95.2	95.2	95.1	95.0	95.1	95.1	95.2	95.1	94.9
35	95	95.6	95.6	95.6	95.5	95.7	95.7	95.7	95.6	95.5
30	86	96.1	96.1	96.0	96.0	96.3	96.2	96.1	96.0	96.0
25	77	95.3	95.8	96.2	96.5	96.7	96.6	96.6	96.5	96.4
20	68	94.5	95.0	95.4	95.9	96.6	97.1	97.1	97.0	96.9
15	59	93.7	94.2	94.6	95.1	95.8	96.3	96.8	97.2	97.5
10	50	92.9	93.4	93.8	94.2	95.0	95.4	95.9	96.4	96.8
5	41	92.1	92.5	92.9	93.4	94.1	94.6	95.1	95.5	96.0
0	32	91.2	91.7	92.1	92.6	93.3	93.7	94.2	94.7	95.1
-10	14	89.6	90.0	90.4	90.8	91.5	92.0	92.5	92.9	93.4
-20	-4	87.8	88.3	88.7	89.1	89.8	90.2	90.7	91.1	91.6
-30	-22	86.0	86.5	86.9	87.3	88.0	88.4	88.9	89.3	89.7
-40	-40	84.3	84.7	85.1	85.5	86.2	86.6	87.1	87.4	87.9
-50	-58	82.5	82.9	83.2	83.7	84.3	84.7	85.2	85.6	86.0

Do not operate engine anti-ice "on" at airport oat above 10° C (50° F)

% N₁ BLEED ADJUSTMENT
Configuration
A/C packs off　+ 1.0

MAX CLIMB % N₁　　250/280/0.74M

	VALID FOR 2 PACKS ON (AUTO) ENGINE A/I OFF								
	Pressure Altitude 1000 Ft.								
TAT °C	0	5	10	15	20	25	30	35	37
50	90.9	91.1	92.5						
40	92.0	92.2	93.6	93.3	93.6				
30	92.2	93.2	93.5	94.2	94.6	94.7			
20	90.6	92.8	94.3	95.0	95.4	95.6	95.7		
10	89.1	91.2	93.1	95.1	96.1	96.4	96.6	96.6	96.6
0	87.5	89.6	91.5	93.4	95.5	97.2	97.5	97.5	97.5
-10	85.9	87.9	89.8	91.7	93.7	95.9	97.9	98.4	98.4
-20	84.2	86.3	88.1	90.0	91.9	94.0	96.0	99.0	99.3
-30	82.5	84.5	86.3	88.2	90.1	92.1	94.1	97.0	97.6
-40	80.8	82.8	84.5	86.3	88.2	90.2	92.2	95.0	95.6
-50	79.1	81.0	82.7	84.5	86.3	88.3	90.2	92.9	93.5
A/C Packs Off	+ 0.5	+ 0.5	+ 0.6	+ 0.7	+ 0.8	+ 0.8	+ 0.9	+ 0.9	+ 0.9
A/C Packs High	- 0.3	- 0.3	- 0.4	- 0.4	- 0.4	- 0.4	- 0.5	- 0.6	- 0.6
Engine A/I On	- 0.7	- 0.8	- 0.9	-1.0	- 1.0	- 1.0	- 1.0	- 1.0	- 1.0
Wing A/I On	- 1.2	- 1.2	- 1.3	- 1.4	- 1.6	- 1.7	- 1.8	- 2.0	- 2.0

% N₁ Corr.

Figure 4.12   **Max Go-Around**   **% N₁ values**

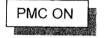

PMC ON

MAX GO-AROUND % N₁

AIRPORT OAT		TAT	VALID FOR TWO PACKS ON (AUTO) ENGINE A/I ON OR OFF								
			PRESSURE ALTITUDE FT.								
°C	°F	°C	0	1000	2000	3000	4000	5000	6000	7000	8000
55	131	58	93.9	93.9	93.9						
50	122	53	94.2	94.2	94.2	94.2	94.2				
45	113	48	94.7	94.6	94.6	94.6	94.6	94.7	94.8	94.6	
40	104	43	95.1	95.1	95.1	95.0	95.1	95.1	95.2	95.1	95.0
35	95	38	95.5	95.6	95.5	95.5	95.7	95.7	95.7	95.6	95.6
30	86	33	96.1	96.1	96.0	96.0	96.3	96.2	96.2	96.1	96.0
25	77	28	95.3	95.8	96.2	96.5	96.7	96.7	96.6	96.5	96.5
20	68	23	94.5	95.0	95.4	95.9	96.6	97.1	97.2	97.0	96.9
15	59	18	93.7	94.1	94.6	95.1	95.8	96.3	96.8	97.3	97.5
10	50	13	92.8	93.3	93.7	94.2	95.0	95.4	96.0	96.6	96.9
5	41	8	92.0	92.5	92.9	93.4	94.1	94.6	95.1	95.6	96.1
0	32	3	91.2	91.7	92.1	92.6	93.3	93.7	94.3	94.7	95.2
-10	14	-8	89.5	90.0	90.4	90.8	91.5	92.0	92.5	93.0	93.4
-20	-4	-18	87.8	88.2	88.6	89.1	89.8	90.3	90.8	91.2	91.6
-30	-22	-28	86.0	86.5	86.9	87.3	88.0	88.5	89.0	89.4	89.8
-40	-40	-38	84.2	84.7	85.1	85.5	86.2	86.6	87.1	87.5	87.9
-50	-58	-48	82.4	82.8	83.2	83.7	84.3	84.7	85.2	85.6	86.0

% N₁ Bleed Adjustments		
	TAT °C	
Configuration	- 60	+ 60
A/C Packs Off	+ 0.8	+ 1.0
A/C Packs High	- 0.3	- 0.3
Wing A/I All Engines	- 1.3	- 1.6
1 Eng. Inop.	- 2.3	- 2.7

Do not operate engine anti-ice "on" at total air temperatures above 10° C (50 ° F)

Figure 4.13   % N_1 Values

PMC - OFF

MAX TAKE-OFF % N_1 A/C Packs on (Auto)

Airport OAT		VALID FOR ENGINE A/I ON OR OFF								
		AIRPORT PRESSURE ALTITUDE FT.								
°C	°F	0	1000	2000	3000	4000	5000	6000	7000	8000
55	131	94.9	94.9	94.9						
50	122	95.4	95.4	95.4	95.4	95.4				
45	113	95.9	95.9	95.9	95.9	95.9	95.9	95.9	95.9	
40	104	96.3	96.3	96.3	96.3	96.3	96.3	96.4	96.4	96.4
35	95	96.8	96.8	96.8	96.8	96.8	96.8	96.8	96.8	96.8
30	86	96.6	96.8	97.2	97.2	97.2	97.2	97.2	97.1	97.1
25	77	95.8	96.0	96.5	97.0	97.4	97.6	97.5	97.5	97.4
20	68	95.0	95.2	95.7	96.2	96.5	96.9	97.1	97.1	97.1
15	59	94.1	94.4	94.8	95.4	95.7	96.3	96.3	96.3	96.3
10	50	93.3	93.5	94.0	94.6	94.9	95.2	95.5	95.6	95.7
5	41	92.5	92.7	93.2	93.7	94.0	94.4	94.7	94.8	94.9
0	32	91.7	91.9	92.3	92.9	93.2	93.5	93.8	93.9	94.0
-10	14	90.0	90.2	90.6	91.2	91.5	91.8	92.1	92.2	92.3
-20	-4	88.2	88.4	88.9	89.4	89.7	90.0	90.3	90.4	90.5
-30	-22	86.5	86.7	87.1	87.6	87.9	88.2	88.5	88.6	88.7
-40	-40	84.7	84.9	85.3	85.8	86.1	86.4	86.7	86.8	86.9
-50	-58	82.8	83.0	83.5	83.9	84.2	84.5	84.8	84.9	85.0

Do not operate engine anti-ice "on" at airport OAT above 10° C (50° F)

MAX TAKE-OFF % N_1 A/C Packs Off

Airport OAT		VALID FOR ENGINE A/I ON OR OFF								
		AIRPORT PRESSURE ALTITUDE FT.								
°C	°F	0	1000	2000	3000	4000	5000	6000	7000	8000
55	131	95.9	95.9	95.9						
50	122	96.4	96.4	96.4	96.4	96.4				
45	113	96.8	96.8	96.8	96.8	96.8	96.8	96.8	96.9	
40	104	97.3	97.3	97.3	97.3	97.3	97.3	97.3	97.3	97.3
35	95	97.7	97.7	97.7	97.7	97.7	97.7	97.7	97.7	97.7
30	86	97.5	97.7	98.1	98.1	98.1	98.1	98.1	98.1	98.0
25	77	96.7	96.9	97.4	97.9	97.9	97.9	97.9	97.9	97.9
20	68	95.9	96.1	96.6	97.1	97.1	97.1	97.1	97.1	97.1
15	59	95.0	95.3	95.7	96.3	96.3	96.3	96.3	96.3	96.3
10	50	94.2	94.4	94.9	95.6	95.8	96.1	96.3	96.3	96.3
5	41	93.4	93.6	94.1	94.6	94.9	95.2	95.4	95.4	95.4
0	32	92.5	92.7	93.2	93.8	94.1	94.4	94.5	94.5	94.5
-10	14	90.8	91.0	91.5	92.0	92.3	92.6	92.8	92.8	92.8
-20	-4	89.1	89.3	89.7	90.3	90.6	90.9	91.0	91.0	91.0
-30	-22	87.3	87.5	88.0	88.5	88.8	89.1	89.2	89.2	89.2
-40	-40	85.5	85.7	86.1	86.6	86.9	87.2	87.3	87.4	87.4
-50	-58	83.6	83.8	84.3	84.7	85.0	85.3	85.5	85.5	85.5

Note - For maximum climb and go-around use PMC 'on' % N_1

TAKE-OFF SPEEDS ADJUSTMENT

ALTITUDE FT.	TEMPERATURE °C (°F)	SPEED ADJUSTMENT KIAS	
		V_1 (MCG)	V_1 & V_R
Below 5000	Above 21 (70)	+6	0
	21 (70) & Below	+4	0
5000 & Above	Above 21 (70)	+6	0
	21 (70) & Below	+4	+1

2.6   Contaminated Runway Take-Off Calculations

1.   These calculations assume an engine failure at V_1
2.   Contaminated runway take-offs are prohibited for:
   a)   variable or reduced thrust take-offs
   b)   contaminant depths exceeding 13mm (0.5 ins) due to spray impingement damage.

The Determination of Take-Off Mass

i..   Calculate the normal limiting take-off mass for a dry runway i.e., field length limit, climb limit or obstacle limit.
ii.   Select the table(s) appropriate to the depth of contaminant (interpolating if necessary)
iii.   Enter the left column of the left table at the normal limiting take-off mass, travel right to the aerodrome pressure altitude column. Interpolate for mass and pressure altitude if necessary. Extract the mass reduction. Calculate maximum take-off mass.
iv.   If in the shaded area, proceed to the right table. Enter the left column with the take-off run available (TORA), move right to the appropriate aerodrome pressure altitude column. Interpolate as necessary. Extract the maximum permissible take-off Mass. Make $V_1 = V_{MCG}$.
v.   The lower of the two values from (iii) and (iv) above is the maximum take off Mass for a contaminated runway.
vi.   Calculate the V speeds for the actual take-off mass.
vii.   If not in the shaded area in (iii) above, then re-enter the left table at the actual to determine the V_1 reduction to be made.
viii. Apply the reduction to V_1.

Example:
Given: -

Aerodrome Pressure Altitude	2000 ft.;
OAT	+2° C;
TORA	5800 ft.;
Runway Slope	2% Down
Wind Component	10 Kt. Head
TOM	50000 Kg.
PMC	on
Runway condition	10 mm. Slush

Calculate contaminated R/W TOM and V speeds.

Mass & V_1 interpolation for 50,000 Kg @ 2000 ft = -7.45 Revised
Mass = 42550Kg.
Limiting Mass interpolation for 5800 ft. TORA @ 2000 ft. Revised
Mass = 44780 Kg
V speeds = V_1 116 -2 = 114 Kt. V_R 118 Kt. V_2 133 Kt V_{MCG} = 114 Kt

4.14   **Advisory Information**   **Contaminated Runways**

 **ALL FLAPS**   ONE ENGINE INOPERATIVE – A/C AUTO OR OFF

0.08 INCH (2MM) SLUSH/STANDING WATER DEPTH

Mass x 1000kg		Mass and V$_1$ Reductions		
	Press Alt Ft	0	4000	8000
40	1000Kg	2.9	3.4	4.0
	KIAS	22	21	19
44	1000Kg	3.7	4.2	4.9
	KIAS	22	21	18
48	1000Kg	4.3	5.0	5.8
	KIAS	21	19	17
52	1000Kg	4.9	5.7	6.5
	KIAS	20	18	15
56	1000Kg	5.6	6.3	7.0
	KIAS	18	16	14
60	1000Kg	6.1	6.8	7.3
	KIAS	16	15	12
64	1000Kg	6.6	7.2	7.6
	KIAS	15	13	10
68	1000Kg	6.9	7.5	8.2
	KIAS	13	11	8

Field Length Available Ft	V$_1$ = V$_{MCG}$ Limit Mass 1000Kg		
	Pressure Altitude (Ft)		
	0	4000	8000
5600	39	-	-
5800	42	34	-
6000	45	37	-
6200	49	39	-
6400	52	42	-
6600	55	45	36
6800	59	47	39
7000	62	50	41
7200	65	53	43
7400	69	56	46
7600	-	59	48
7800	-	62	51
8000	-	65	53
8200	-	68	56
8400	-	71	58
8600	-	-	61
8800	-	-	64
9000	-	-	66

0.25 INCH (6MM) SLUSH/STANDING WATER DEPTH

Mass x 1000kg		Mass and V$_1$ Reductions		
	Press Alt Ft	0	4000	8000
40	1000Kg	3.4	4.0	4.9
	KIAS	19	16	12
44	1000Kg	4.3	5.3	6.3
	KIAS	18	14	10
48	1000Kg	5.3	6.3	7.5
	KIAS	16	12	9
52	1000Kg	6.2	7.2	8.4
	KIAS	13	10	8
56	1000Kg	7.0	8.1	9.1
	KIAS	11	8	7
60	1000Kg	7.9	8.8	9.5
	KIAS	8	7	6
64	1000Kg	8.7	9.4	9.7
	KIAS	7	5	5
68	1000Kg	9.4	9.9	9.7
	KIAS	5	5	5

Field Length Available Ft	V$_1$ = V$_{MCG}$ Limit Mass 1000Kg		
	Pressure Altitude (Ft)		
	0	4000	8000
5400	40	-	-
5600	43	35	-
5800	47	38	-
6000	50	41	-
6200	53	43	36
6400	56	46	38
6600	59	48	41
6800	63	51	43
7000	66	54	45
7200	70	57	48
7400	-	60	50
7600	-	63	52
7800	-	65	55
8000	-	68	57
8200	-	-	59
8400	-	-	62
8600	-	-	64
8800	-	-	67

4.14 (Cont.)   Advisory Information   Contaminated Runways

0.5 INCH (13MM) SLUSH/STANDING WATER DEPTH

Mass x 1000kg	Press Alt Ft	Mass and V₁ Reductions		
		0	4000	8000
40	1000Kg	4.2	5.0	6.5
	KIAS	13	6	0
44	1000Kg	5.5	6.8	8.5
	KIAS	9	2	0
48	1000Kg	6.8	8.8	10.3
	KIAS	4	0	0
52	1000Kg	8.2	10.2	11.8
	KIAS	1	0	0
56	1000Kg	9.5	11.4	12.7
	KIAS	0	0	0
60	1000Kg	11.5	12.5	13.1
	KIAS	0	0	0
64	1000Kg	15.0	13.2	13.0
	KIAS	0	0	0
68	1000Kg	13.8	13.8	12.4
	KIAS	0	0	1

Field Length Available Ft	$V_1 = V_{MCG}$ Limit Mass 1000Kg		
	Pressure Altitude (Ft)		
	0	4000	8000
5000	38	-	-
5200	41	35	-
5400	44	37	-
5600	48	40	-
5800	51	42	36
6000	54	45	38
6200	56	47	40
6400	59	50	42
6600	62	52	44
6800	65	55	47
7000	68	57	49
7200	70	60	51
7400	-	62	53
7600	-	64	55
7800	-	67	58
8000	-	69	60
8200	-	-	62
8400	-	-	65

Instructions:

T.O. Mass

1. Enter Mass (weight) and V_1 reductions table for slush depth with dry field obstacle limit. to obtain slush mass (weight) decrement.
2. If in a shaded area, find V_1 (MCG) limited mass (weight) for available field length and pressure altitude
3. Max allowable slush weight is lesser of weights from I and 2.

T.O. Speeds

1. Obtain V_1 , V_R, V_2 for actual weight from QRH
2. If V_1 (MCG) limited, set $V_1 = V_1$ (MCG).
   If not V_1 (MCG) limited, enter weight and V_1 reductions table with actual weight to obtain V_1 speed adjustment.
   If adjusted V_1 is less than V_1 (MCG) set $V_1 = V_1$ (MCG).

2.7   Increased V_2 Take-Off

If the maximum take-off Mass is limited by the minimum acceptable climb gradients of the net flight path, and there is a large excess of available field length over that which is required, then it is possible to improve the take-off weight and still attain the minimum climb gradient requirement (subject to the limitations of the tyre speed and field length limitations). This is done by holding the aeroplane on the ground until it reaches an increased V_R and climbing at an increased V_2, which equates to Vx, the speed that will attain the maximum climb gradient.

Method of Use
- i.   Select. the set of graphs appropriate to the flap setting on the "improved climb performance field length limit" graph (Fig. 4.15).
- ii.   Enter the carpet of the left graph with the value of the field length limit weight minus the climb limit weight. Travel vertically up to the normal 'climb limit.' weight line.
- iii.   From this intersection move horizontally left to the vertical axis to read the climb weight improvement and horizontally right to the vertical axis to read the increase to apply to V_1.
- iv.   Continue horizontally right to the reference line of the right graph. From this point interpolate the grid lines to reach a vertical input from the carpet of the right graph of the normal climb limit weight.
- v.   From this intersection, travel horizontally right to the vertical axis to read the increase to apply to V_R and V_2.
- vi.   Repeat this process in the improved climb performance tyre speed limit graph (Fig. 4.16) except the initial entry point is the tyre limit weight minus the climb limit weight.
- vii.   The lower of the two weight increases is that which must be used together with its associated speed increases.
- viii.   Add the weight. increase to the normal climb weight limit.
- ix.   Determine the V speeds for this increased weight
- x.   Apply the speed increases to the appropriate speeds. Check V_{MBE}.

Figure 4.15   Improved Climb Performance   Field Length Limit

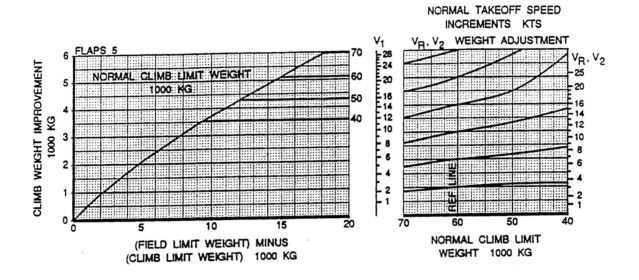

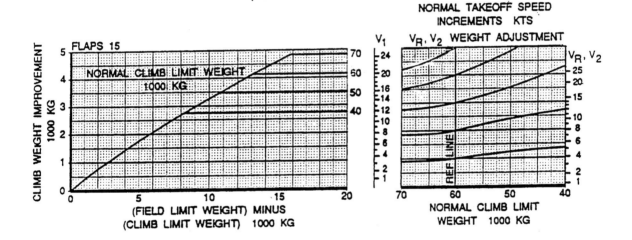

USE SMALLER OF IMPROVED CLIMB WT. (FIELD LENGTH LIMITS) OR. (TIRE SPEED LIMITS). APPLY SPEED INCREMENTS TO NORMAL V_1 V_R V_2 FOR ACTUAL TAKEOFF WT. CHECK BRAKE ENERGY LIMITS.

Figure 4.16   Improved Climb Performance   Tyre Speed Limit

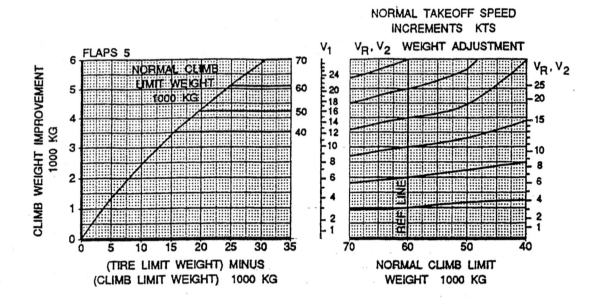

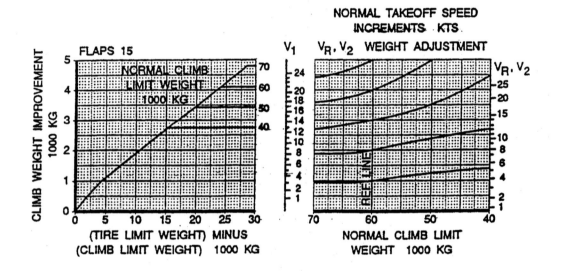

USE SMALLER OF IMPROVED CLIMB WT. (FIELD LENGTH LIMITS) OR (TIRE SPEED LIMITS).
APPLY SPEED INCREMENTS TO NORMAL V_1, V_R, V_2 FOR ACTUAL TAKEOFF WT.
CHECK BRAKE ENERGY LIMITS.

2.8   Reduced Thrust Take-Off

The reduced thrust take-off procedure is referred to by a number of different names such as the "Variable Thrust Take-off' or the "Assumed Temperature Take-off". It is a technique employed to preserve engine life or reduce the noise generated at take-off. This technique can only be used when the available distance greatly exceeds that which is required. The maximum reduction in thrust permitted is 25% of that required for a normal take-off.

Restrictions
A reduced thrust take-off is not permitted with:
- icy or very slippery runways
- contaminated runways
- anti-skid unserviceable
- reverse thrust unserviceable
- increased V_2 procedure
- with the PMC off.

Calculation Procedure.
It is first necessary to determine the most limiting performance condition. The only common parameter to enable comparison is that of temperature. Thus the maximum permissible temperature must be calculated for the actual take off weight from each of the following:
- field limit graph
- climb limit graph
- tyre speed limit graph
- obstacle limit graph

From these temperatures, select. the lowest and ensure that it does not exceed the environmental limit. If it does, then the environmental limit becomes the assumed temperature.

Chart. Procedure

a) Calculate the maximum assumed temperature from the top table. Enter the left column with the actual ambient temperature and read the maximum temperature in the column appropriate to the aerodrome pressure altitude.

b) From the centre table, bottom line, determine the minimum assumed temperature for the aerodrome pressure altitude.

c) From the same table, for the assumed temperature to be used, determine the maximum take-off % N_1. Add 1.0% N_1 if air conditioning packs are off. The assumed temperature used must neither exceed the maximum from (a) above or be below the minimum from (b) above.

d) Enter the left column of the bottom table with assumed temperature minus ambient temperature. Travel right along the line to the column appropriate to the ambient temperature, interpolating if necessary. Read the % N_1 adjustment to be subtracted from that found at (c) above.

e) Subtract the value determined at. (d) from that. at (c) to determine the % N_1 to be set at take-off.

Figure 4.17   Assumed Temperature Reduced Thrust

> **PMC - ON**

Assumed temp. %N_1 = Max. Take-off % N_1 minus % N_1 adjustment

Maximum Assumed Temperature*

OAT °C	Press. Alt. 1000 Ft.								
	0	1	2	3	4	5	6	7	8
55	71	71							
50	69	68	68	69	70				
45	67	66	66	67	67	67	68	70	
40	65	64	64	64	64	64	64	66	68
35	63	62	62	62	61	61	62	63	64
30	61	60	60	59	59	59	59	60	61
25	61	59	58	57	56	56	56	57	58
20	61	59	58	57	55	53	54	54	55
15 & below	61	59	58	57	55	53	53	52	52

OAT °F	Press Alt. 1000 Ft.								
	0	1	2	3	4	5	6	7	8
130	159	159							
120	155	154	154	155	157				
110	151	149	149	150	151	151	152	155	
100	148	145	145	145	145	147	145	149	151
90	143	141	141	140	140	140	140	143	144
80	142	139	138	136	135	135	135	137	138
70	142	138	136	135	131	129	129	130	132
60 & below	142	138	136	135	131	127	127	126	126

*Based on 25% Take-off Thrust Reduction

Figure 4.17   Assumed Temperature Reduced Thrust (Continued)

PMC - ON

Max. Take-off % N_1

Valid for 2 packs on (auto) engine A/I on or off					For A/C off add 1.0 % N_1				
Assumed Temp		Airport Pressure Altitude Ft.								
°C	°F	0	1000	2000	3000	4000	5000	6000	7000	8000
75	167	85.4	85.4							
70	158	87.6	87.4	87.4	87.6					
65	149	89.7	89.4	89.2	89.2	89.2	89.3	89.5	89.9	90.4
60	140	91.8	91.3	91.0	90.8	90.7	90.7	90.8	91.1	91.4
55	131	93.8	93.2	92.7	92.4	92.1	92.1	92.0	92.1	92.3
50	122	94.3	94.3	94.3	93.9	93.6	93.4	93.2	93.2	93.2
45	113	94.7	94.7	94.6	94.6	94.6	94.7	94.4	94.2	94.0
40	104	95.2	95.2	95.1	95.0	95.1	95.1	95.2	95.1	94.9
35	95	95.6	95.6	95.6	95.5	95.7	95.7	95.7	95.6	95.5
30	86	96.1	96.1	96.0	96.0	96.3	96.2	96.1	96.0	96.0
25	77		96.6	96.5	96.5	96.7	96.6	96.6	96.5	96.4
20	68					97.1	97.1	97.1	97.0	96.9
15	59							97.6	97.5	97.5
Minimum Assumed Temp. °C (°F)		30 (86)	28 (82)	26 (79)	24 (75)	22 (72)	20 (68)	18 (64)	16 (61)	15 (59)

% N_1 Adjustment for Temperature Difference

Assumed Temp. Minus OAT		Outside Air Temperature													
°C	°F	°C	-40	-20	0	5	10	15	20	25	30	35	40	45	50	55
		°F	-40	-4	32	41	50	59	68	77	86	95	104	113	122	131
10	18						1.6	1.6	1.6	1.5	1.5	1.5	1.5	1.4	1.4	1.3
20	36				3.3	3.2	3.2	3.1	3.0	3.0	2.9	2.8	2.8	2.7	2.5	2.3
30	54				4.8	4.8	4.6	4.5	4.4	4.3	4.0	3.8	3.6	3.6	3.6	3.6
40	72			6.0	6.2	6.1	6.0	5.8	5.7	5.2	5.0	5.0				
50	90			8.2	7.5	7.3	7.2	6.6	6.5							
60	108		10.4	9.5	8.7	8.1	7.9									
70	126		11.8	10.7	9.3											
80	144		13.0	11.8	10.1											
90	162		14.0	12.4												
100	180		15.0	12.8												
110	198		15.4													

2.9   Anti-Skid Inoperative (Simplified method)

Because the accelerate/stop distance will be adversely affected if the anti-skid is inoperative V_1 has to be reduced to comply with the take-off requirements. This will increase the take-off distance required beyond that which would normally be required and thus decrease the distance to any obstacles encountered after take-off.

Simplified Calculation
I.   Decrease the normal runway/obstacle limited take-off weight by 7700 Kg.
II.   Recalculate the V speeds for this reduced weight.
III. Further reduce V_1 by the amount shown in the table at figure 4.18 below.

Figure 4.18 V_1 Decrements

Anti-skid Inoperative V_1 Decrements	
Field length Ft	V_1 Reduction Kt
6000	28
8000	21
10000	17
12000	14
14000	11

IV.   If the actual take-off weight is already less than the anti-skid inoperative limited take-off weight ensure V_1 does not exceed the anti-skid operative V_1.
V.   If V_1 is less than V_{MCG} (see fig.4. 19) set $V_1 = V_{MCG}$, if ASDA exceeds 7900 ft.
VI.   Always ensure V_1 is not less than V_{MCG} shown in figure 4.19 below.

Figure 4.19   Minimum $V_{1 (MCG)}$

Max Take-off Thrust   Minimum $V_{1 (MCG)}$

OAT		Pressure Altitude (Ft)					
°C	°F	0	2000	4000	6000	8000	9000
55	131	105					
50	122	107	103				
45	113	109	105	101	96		
40	104	111	107	103	99	94	
35	95	113	109	105	101	96	94
30	86	116	111	107	104	98	96
25	77	116	113	109	105	100	98
20	68	116	113	111	107	102	100
15	59	116	113	111	108	104	102
-50	-58	118	115	112	109	105	104

For Packs 'OFF' add 2 Kt

3.   OBSTACLE CLEARANCE

These graphs are provided for Flaps 5° and Flaps 15° (Figures 4.20 and 4.21). They provide a rapid means of obtaining the value of obstacle clearance after take-off. They are intended for use when a detailed airport analysis is not available. Detailed analysis for the specific case from the aeroplane flight manual may result in a less restrictive weight and can account for the non-use of the air conditioning packs.

These graphs are not valid for NC packs off or for take-offs using the improved climb technique.

Obstacle Limit Weight Determination

i.   Select graph appropriate to the flap setting.
ii.   Adjust the obstacle height to account for runway slope.
iii.   Enter the bottom left vertical axis at the adjusted obstacle height.
iv.   Travel horizontally right to intersect the horizontal distance of the obstacle measured from the brake release point.
v.   From this intersection, move vertically up to the ambient temperature reference line, then parallel the grid lines to the appropriate temperature;
vi.   Continue vertically to the aerodrome pressure altitude reference line. Parallel the grid lines to the appropriate pressure altitude before continuing vertically to the wind component reference line.
vii.   Parallel the grid lines from this point to the value of the wind component then continue vertically to read the obstacle limited takeoff weight.

Example:

Flap Setting	5°
Aerodrome Pressure Altitude	1,000 ft.
Ambient Temperature	+ 37° C
Wind Component.	20 Kt. Head
Runway Slope	2% down
Obstacle Distance from BRP	18,000 ft.
Obstacle Pressure Altitude	1,160 ft.
Take-off Distance Required	10,000 ft.
Obstacle Height =	1160 -[1000 -(10000 x 2%)] = 360 ft.
Obstacle Limited TOW =	51,700 Kg.

Figure 4.20   **Obstacle Limits**   **Flaps 5°**

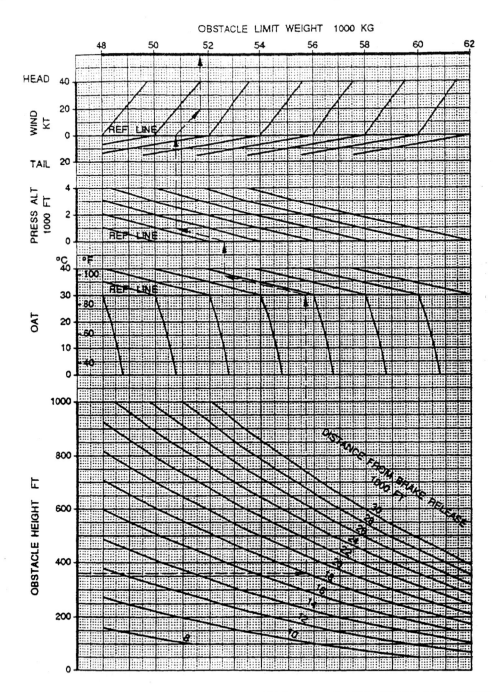

PMC OFF CORRECTION

TEMPERATURE °C	WEIGHT DECREMENT KG
21.1 AND BELOW	4970
ABOVE 21.1	4420

OBSTACLE HEIGHT MUST BE CALCULATED FROM THE
LOWEST POINT OF THE RUNWAY TO CONSERVATIVELY
ACCOUNT FOR RUNWAY SLOPE.

Figure 4.21   Obstacle Limits   Flaps l5°

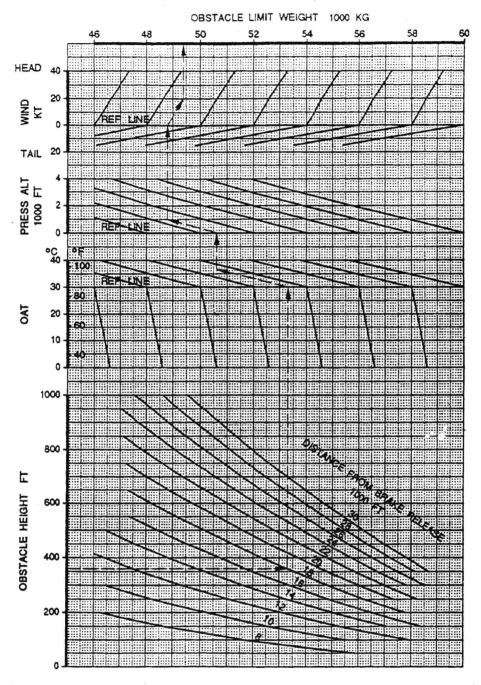

PMC OFF CORRECTION

TEMPERATURE °C	WEIGHT DECREMENT KG
21.1 AND BELOW	4970
ABOVE 21.1	4420

OBSTACLE HEIGHT MUST BE CALCULATED FROM THE
LOWEST POINT OF THE RUNWAY TO CONSERVATIVELY
ACCOUNT FOR RUNWAY SLOPE.

4.   EN-ROUTE

4.1   Maximum % N₁ Value

In the event of an engine failure during the cruise, it will generally be necessary to reduce speed and descend to a lower altitude. This is accomplished by setting maximum continuous thrust on the remaining live engine and allowing the speed to bleed off, while maintaining altitude, to the optimum drift-down speed. One engine inoperative information is based upon one pack operating with the A/C switch on "auto" or "high".

The initial maximum continuous $\%N_1$ setting, with normal engine bleed for air conditioning, one pack on, following engine failure in the cruise for 0.74 mach, may be determined from the upper table of figure 4.22. The drift down speed and level off altitude (stabilising altitude) may be determined from the lower table of the same page, for specific weights and temperature deviation. This is the gross level-off altitude.

4.2   Level Off Altitude

For performance planning purposes the level off altitude should be determined from figure 4.23. This is based on the net one engine inoperative performance (i.e., gross gradient -1.1%).

Figure 4.22   Driftdown

OPTIMUM DRIFTDOWN SPEED   ONE ENGINE INOPERATIVE

INITIAL MAXIMUM CONTINUOUS $\%N_1$ 0.74 MACH   A/C AUTO (HIGH)

Pressure Altitude Ft	TAT °C												
	-50	-40	-30	-25	-20	-15	-10	-5	0	5	10	15	20
37000	93.0	95.0	97.0	98.0	98.7	98.3	98.0	97.6	97.0	96.5	96.0	95.7	95.1
35000	92.4	94.5	96.5	97.5	98.4	98.3	98.0	97.6	97.0	96.5	96.0	95.7	95.1
33000	91.5	93.5	95.5	96.5	97.4	98.4	98.1	97.7	97.1	96.6	96.1	95.8	95.2
31000	90.3	92.3	94.3	95.3	96.2	97.1	98.1	97.7	97.1	96.6	96.1	95.8	95.2
29000	89.3	91.3	93.2	94.1	95.1	96.0	96.9	97.7	97.1	96.6	96.1	95.8	95.2
27000	88.1	90.1	92.0	92.9	93.9	94.8	95.7	96.6	97.1	96.6	96.1	95.8	95.2

DRIFTDOWN SPEED/LEVEL OFF

Weight 1000 Kg.		Optimum Driftdown Speed KIAS	Level Off Altitude Ft.		
Start Driftdown	Level Off		ISA + 10° C & below	ISA + 15° C	ISA + 20° C
70	67	245	14200	12900	11400
65	62	237	16700	15500	14200
60	57	228	19200	18200	17000
55	52	218	21900	20900	19800
50	48	209	24800	23800	22800
45	43	198	27700	26900	26000
40	38	187	30700	30100	29300
35	33	175	33900	33400	32700

Figure 4.23   Net Level-Off Altitude

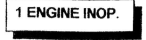

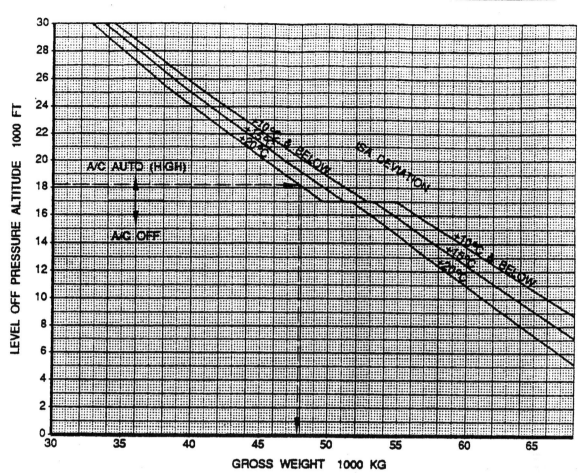

BLEED CONFIGURATION	WEIGHT ADJUSTMENT KG
ENG A/I ON	-1950
ENG AND WING A/I ON	-5650
A/C AUTO (HIGH) BELOW 17000 FT	-2500

Figure 4.24   Driftdown Profiles   Net Flight Path

1 ENGINE INOP.

A/C AUTO (HIGH)
PRESSURE ALTITUDE 37000 FT

MAX CONTINUOUS THRUST LIMITS

BLEED CONFIGURATION	WEIGHT ADJUSTMENT KG
ENG A/I ON	+1950
ENG AND WING A/I ON	+5650
A/C OFF (BELOW 17000 FT)	-1750

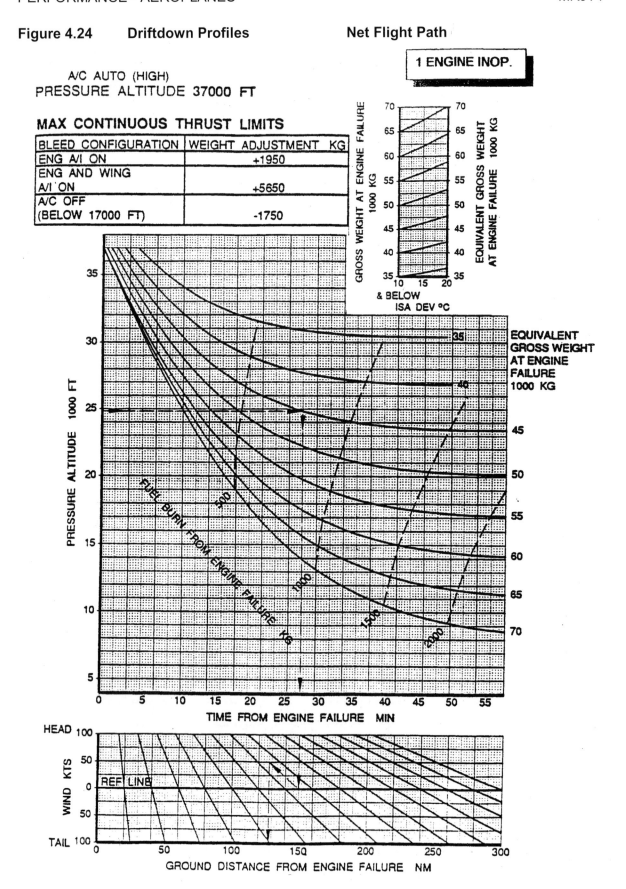

Figure 4.25   Driftdown Profiles   Net Flight Path

PRESSURE ALTITUDE 33000 FT TO 35000 FT

MAX CONTINUOUS THRUST LIMITS

BLEED CONFIGURATION	WEIGHT ADJUSTMENT KG
ENG A/I ON	+1950
ENG AND WING A/I ON	+5650
A/C OFF (BELOW 17000 FT)	-1750

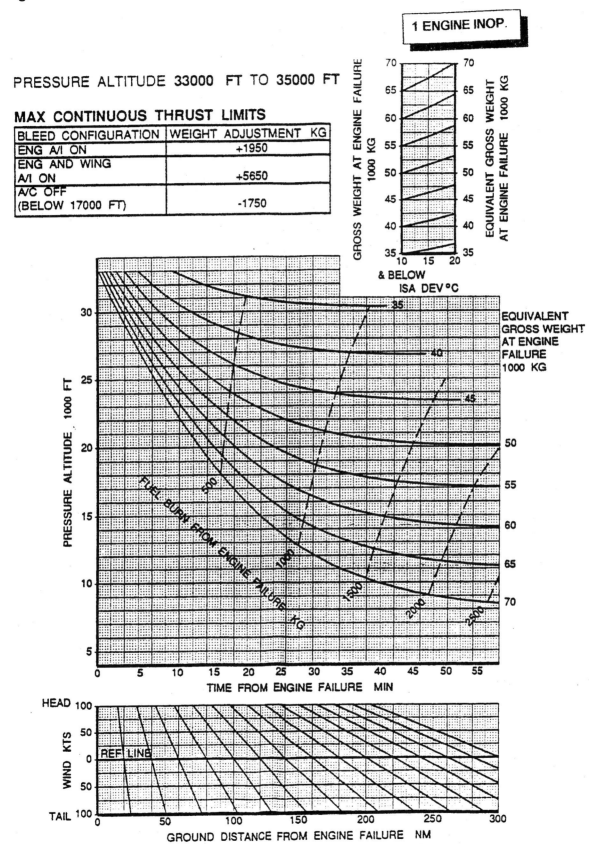

Figure 4.26   Driftdown Profiles   Net Flight Path

1 ENGINE INOP.

PRESSURE ALTITUDE 29000 FT TO 31000 FT

MAX CONTINUOUS THRUST LIMITS

BLEED CONFIGURATION	WEIGHT ADJUSTMENT KG
ENG A/I ON	+1950
ENG AND WING A/I ON	+5650
A/C OFF (BELOW 17000 FT)	-1750

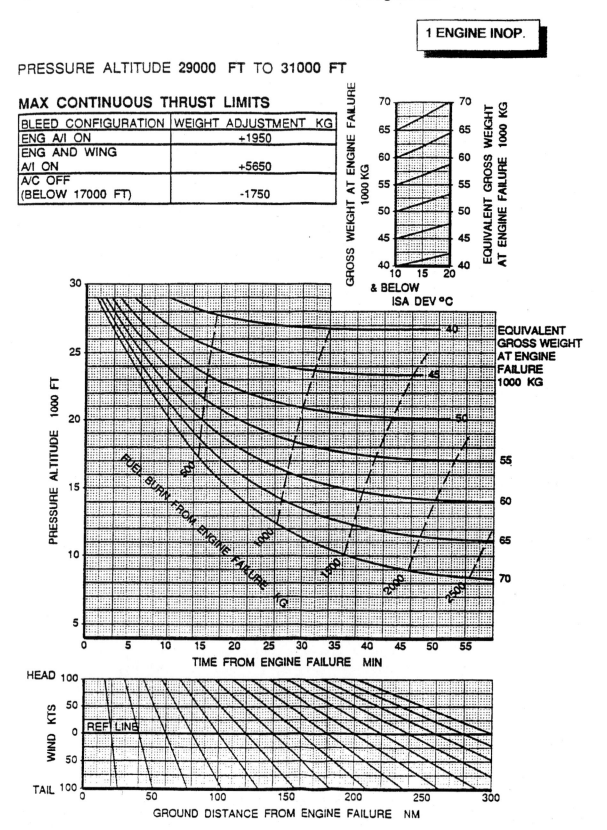

Figure 4.27   **Driftdown Profiles**   **Net Flight Path**

1 ENGINE INOP.

PRESSURE ALTITUDE **25000** FT TO **27000** FT

MAX CONTINUOUS THRUST LIMITS

BLEED CONFIGURATION	WEIGHT ADJUSTMENT KG
ENG A/I ON	+1950
ENG AND WING A/I ON	+5650
A/C OFF (BELOW 17000 FT)	-1750

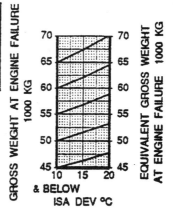

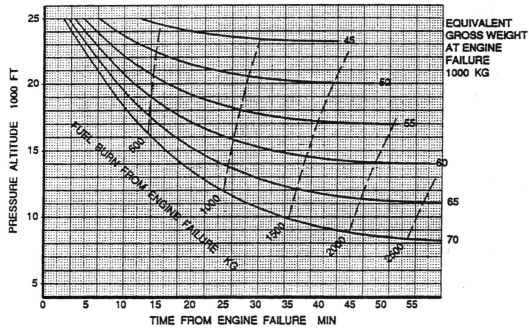

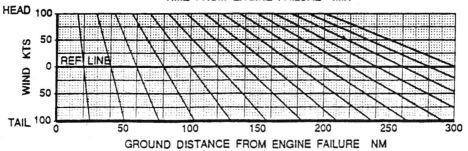

5.   LANDING

5.1   Landing Performance

The landing performance calculations are divided in two elements.
a)   The field length limited landing weight can be determined from figure 4.28.
b)   The landing "climb limit" (figure 4.29). This weight ensures the minimum
        permissible gradient is obtained and should be corrected in accordance with
        the statements beneath the graph.
c)   The maximum landing weight is the lower of a) and b).

5.2   Quick Turnaround Limit

The maximum permissible landing weight for a quick turnaround can be determined
from Figure 4.30. The weights are tabulated for aerodrome pressure altitude and
ambient temperature and should be adjusted in accordance with the statement below
the tables for runway slope and wind component. If the landing weight exceeds this
value then after 53 minutes check the wheel thermal plugs have not melted before
commencing a take-off.

Figure 4.28   **Landing Performance**   **Field Length Limit**

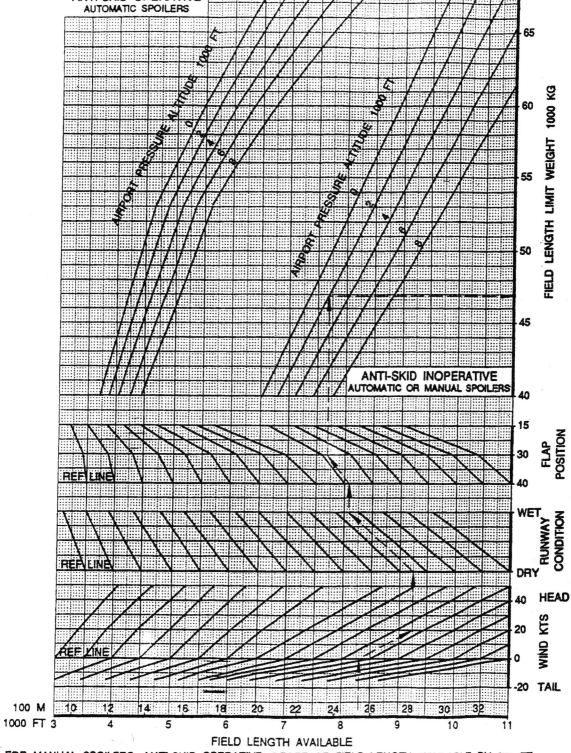

FOR MANUAL SPOILERS, ANTI-SKID OPERATIVE, DECREASE FIELD LENGTH AVAILABLE BY 650 FT.
STRUCTURAL WEIGHT LIMITS MUST BE OBSERVED.

Figure 4.29   **Landing Performance**   **Climb Limit**

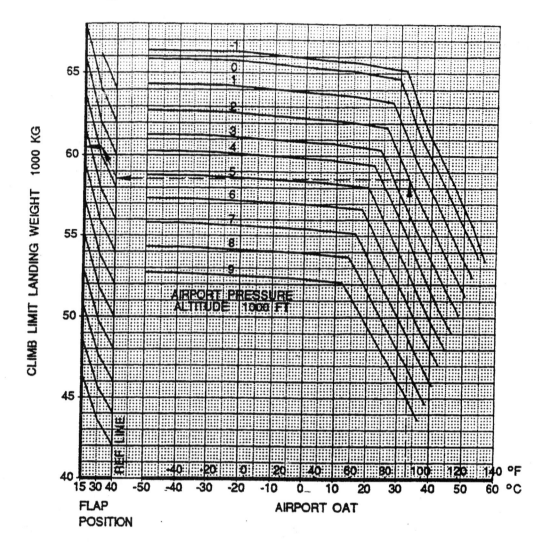

BASED ON A/C AUTO. FOR PACKS OFF INCREASE:
THE FLAPS 40 ALLOWABLE WEIGHT BY 1250 KG,
THE FLAPS 30 ALLOWABLE WEIGHT BY 1310 KG OR
THE FLAPS 15 ALLOWABLE WEIGHT BY 1440 KG.

IF OPERATING IN ICING CONDITIONS DURING ANY PART OF THE FLIGHT WHEN THE
FORECAST LANDING TEMPERATURE IS BELOW 8°C (46°F),
REDUCE THE FLAPS 40 CLIMB LIMIT WEIGHT BY 4830 KG,
REDUCE THE FLAPS 30 CLIMB LIMIT WEIGHT BY 4730 KG OR
REDUCE THE FLAPS 15 CLIMB LIMIT WEIGHT BY 4960 KG

FOR ANTI-ICE OPERATION, DECREASE ALLOWABLE WEIGHT BY THE AMOUNT SHOWN IN
THE FOLLOWING TABLE

* ANTI-ICE OPERATION DECREMENT KG		
FLAPS	ENG	ENG & WING
15	650	5800
30	600	5350
40	550	5250

*ANTI-ICE BLEED SHOULD NOT BE USED ABOVE 10°C (50°F)

Figure 4.30 Quick Turnaround Limit

After landing at weight exceeding those shown below adjusted for slope and wind, wait at least 53 minutes and then check that wheel thermal plugs have not melted before executing a take-off.

FLAPS 15

Airport Pressure Altitude Ft		Maximum Quick Turnaround weight 1000 Kg										
		Airport OAT										
	°F	-60	-40	-20	0	20	40	60	80	100	120	130
	°C	-51	-40	-29	-18	-7	4	16	27	38	49	54
-1000		59	57	56	54	53	52	51	50	49	49	48
0		58	56	55	54	52	51	50	49	49	48	47
1000		56	55	54	53	51	50	49	49	48	47	46
2000		55	54	53	52	50	49	49	48	47	46	
3000		54	53	52	51	49	49	48	47	46	45	
4000		54	52	51	50	49	48	47	46	45		
5000		53	51	50	49	48	47	46	45	44		
6000		51	50	49	48	47	46	45	44	44		
7000		50	49	48	47	46	45	44	44	43		
8000		49	49	47	46	45	44	44	43	42		
9000		49	48	46	45	44	44	43	42	41		

FLAPS 30

Airport Pressure Altitude Ft		Maximum Quick Turnaround weight 1000 Kg										
		Airport OAT										
	°F	-60	-40	-20	0	20	40	60	80	100	120	130
	°C	-51	-40	-29	-18	-7	4	16	27	38	49	54
-1000		66	64	62	61	59	58	57	56	55	54	54
0		64	63	61	60	59	57	56	55	54	53	53
1000		63	62	60	59	58	56	55	54	53	52	52
2000		62	60	59	58	56	55	54	53	52	51	
3000		61	59	58	57	55	54	53	52	51	50	
4000		60	58	57	55	54	53	52	51	50		
5000		59	57	56	54	54	52	51	50	49		
6000		58	56	55	54	52	51	50	49	49		
7000		56	55	54	53	51	50	49	49	48		
8000		55	54	53	51	50	49	49	48	47		
9000		54	53	52	50	49	49	48	47	46		

FLAPS 40

Airport Pressure Altitude Ft		Maximum Quick Turnaround weight 1000 Kg										
		Airport OAT										
	°F	-60	-40	-20	0	20	40	60	80	100	120	130
	°C	-51	-40	-29	-18	-7	4	16	27	38	49	54
-1000		68	67	65	64	62	60	59	58	57	56	55
0		67	65	64	62	61	59	58	57	56	55	54
1000		66	64	63	61	60	59	57	56	55	54	54
2000		64	63	61	60	59	57	56	55	54	53	
3000		64	62	60	59	58	56	55	54	53	52	
4000		62	60	59	58	56	55	54	53	52	51	
5000		61	59	58	57	55	54	53	52	51		
6000		60	58	57	55	54	53	52	51	50		
7000		59	57	56	54	53	52	51	50	49		
8000		58	56	54	54	52	51	50	49	49		
9000		56	55	54	53	51	50	49	49	48		

Add   350 Kg per 1% uphill slope
Subtract 1150 Kg per 1% downhill slope
Add   1100 Kg per 10 Kt. headwind
Subtract 7450 Kg per 10 Kt. tailwind.

5.3   Brake Cooling Schedule

The graph at Figure 4.31 provides advisory information to enable the operator to avoid brake overheats problems. The chart enables due allowance to be made for a single stop and, by using the graph as indicated, provides advice on the procedure to be adopted and the minimum cooling time. Separate sub-graphs are provided for determining the stop distance with manual braking. To use the chart, enter the top left vertical axis at the gross weight of the aeroplane.

Move horizontally right to the approach speed minus 3 Kt. Travel vertically down applying the aerodrome pressure altitude and ambient temperature as appropriate, then allow for the type of braking to be used. To the resultant brake energy per brake, add one million foot/pounds for each taxi mile. From this point on the horizontal axis, drop vertically through the table to determine the advised procedure and recommended cooling time.

In this procedure, allowance for wind component should be made by adjusting the "brakes on" speed by 50% of any headwind or 150% of any tailwind.

Figure 4.31   Brake Cooling Schedule

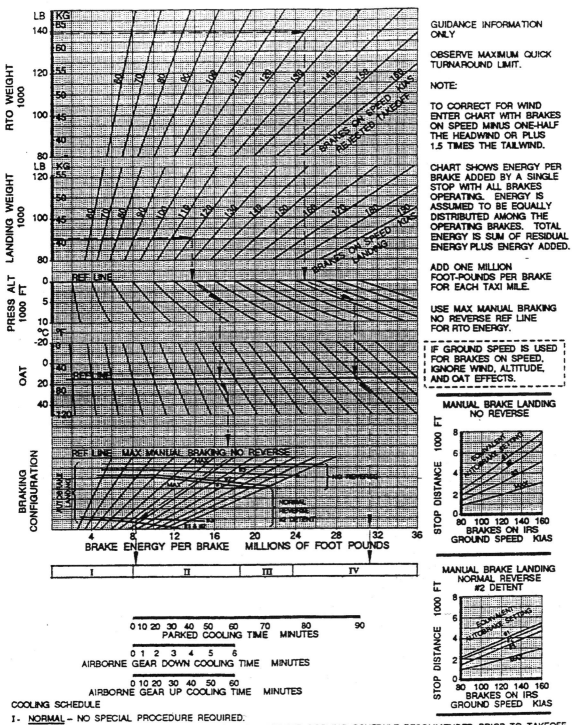

GUIDANCE INFORMATION ONLY

OBSERVE MAXIMUM QUICK TURNAROUND LIMIT.

NOTE:

TO CORRECT FOR WIND ENTER CHART WITH BRAKES ON SPEED MINUS ONE-HALF THE HEADWIND OR PLUS 1.5 TIMES THE TAILWIND.

CHART SHOWS ENERGY PER BRAKE ADDED BY A SINGLE STOP WITH ALL BRAKES OPERATING. ENERGY IS ASSUMED TO BE EQUALLY DISTRIBUTED AMONG THE OPERATING BRAKES. TOTAL ENERGY IS SUM OF RESIDUAL ENERGY PLUS ENERGY ADDED.

ADD ONE MILLION FOOT-POUNDS PER BRAKE FOR EACH TAXI MILE.

USE MAX MANUAL BRAKING NO REVERSE REF LINE FOR RTO ENERGY.

IF GROUND SPEED IS USED FOR BRAKES ON SPEED, IGNORE WIND, ALTITUDE, AND OAT EFFECTS.

MANUAL BRAKE LANDING NO REVERSE

MANUAL BRAKE LANDING NORMAL REVERSE #2 DETENT

COOLING SCHEDULE

I - NORMAL – NO SPECIAL PROCEDURE REQUIRED.
II - COOLING RECOMMENDED – COOL AS SCHEDULED. GROUND COOLING SCHEDULE RECOMMENDED PRIOR TO TAKEOFF.
III - CAUTION – WHEEL FUSE PLUGS MAY MELT. DELAY TAKEOFF AND INSPECT AFTER 30 MINUTES. AFTER TAKEOFF, EXTEND GEAR SOON FOR AT LEAST 7 MINUTES.
IV - FUSE PLUG MELT – NO REVERSE THRUST OR BRAKING CONFIG CREDIT ALLOWED IN THIS AREA. CLEAR RUNWAY IMMEDIATELY. DO NOT SET PARKING BRAKE. DO NOT APPROACH GEAR OR ATTEMPT TAXI OR SUBSEQUENT TAKEOFF BEFORE WAITING MANDATORY TIME SHOWN ON QUICK TURNAROUND LIMIT CHART. ALERT FIRE EQUIPMENT.
   * CATEGORY A AND C BRAKES ONLY

INTENTIONALLY BLANK

AIR LAW AND AIR TRAFFIC PROCEDURES

AERODROMES

1   INTRODUCTION

Both the CAA and JAA syllabi require a knowledge of aerodromes. The specific syllabus content is covered in the course notes, however, the main information reference concerning aerodromes is the AD section of the AIP. Each aerodrome is different. Whilst standards can be applied to procedures and to the methods by which facilities are provided, the aerodrome directory (AD) entry contains information, in a standard layout format, specifically applicable to each aerodrome.

2   AIP ENTRY

Each AIP aerodrome entry contains as much information about an aerodrome as is necessary to comply with the ICAO standard for the publication of such data and also to allow the safe and expeditious flow of air traffic using the aerodrome facility. The following is the complete list of headings under which data is provided for aerodromes:

Section 1   **Aerodrome Information**

1	Aerodrome name and aerodrome chart (Non ICAO)
2	Geographic and administrative data
3	Operating hours
4	Handling services and facilities
5	Passenger facilities
6	Rescue and fire fighting services
7	Seasonal availability (clearing)
8	Aprons, taxiways and check locations
9	Surface movement guidance and control system markings
10	Aerodrome obstacles
11	Meteorological information provided
12	Runway physical characteristics
13	Declared distances
14	Approach and runway lighting
15	Other lighting and secondary power supply
16	Helicopter landing areas
17	ATS airspace
18	ATS communications facilities
19	Radio navigation and landing aids
20	Local traffic regulations
21	Noise abatement procedures
22	Flight procedures
23	Additional information
24	List of charts related to the aerodrome

Section 2   Charts

1   Aerodrome charts (ICAO); parking/docking;
2   CTR local area flying and entry/exit routes;
      CTR helicopter routes
3   Helicopter crossing operations
4   Radar vectoring area (RVA)
5   Standard instrument departure routes (SIDs)
6   Standard arrival routes (STARs)
7   Instrument approach procedure charts (plates)

3   Examples

The easiest way to become familiar with the aerodrome data available in the AIP is to study a couple of examples. The following pages are a direct extract from the AIP aerodrome directory for London Heathrow and Oxford Kidlington.

It is stressed that under no circumstances must the information contained in these pages and the associated charts be used for navigation purposes. The AIP is a regularly amended document and these entries are copied and provided for instructional purposes only.

EGLL AD 2.2 – AERODROME GEOGRAPHICAL AND ADMINISTRATIVE DATA

1	ARP Co-ordinates and site at AD:	**Lat:** 512839N **Long:** 0002741W Mid point of Runway 09L/27R.
2	Direction and distance from city:	12 nm W of London.
3	Elevation/Reference temperature:	80 ft – 22°C.
4	Magnetic Variation/Annual Change:	W3.5° (2000) – 0.12° decreasing.
5	AD Administration: Address:	Heathrow Airport Limited. London (Heathrow) Airport, Hounslow, Middlesex UB3 5AP.
	Telephone: Fax: Telex:	020-8759 4321 (HAL). 020-8745 3328 (NATS Ltd). 020-8745 4290 (HAL). 020-8745 3491/3492 (NATS Ltd/FBU). 934892 (HAL). 22807 (NATS Ltd).
6	Type of Traffic Permitted (IFR/VFR):	IFR/SVFR.

EGLL AD 2.3 – OPERATIONAL HOURS

1	AD:	H24.
2	Customs and Immigration:	H24.
3	Health and Sanitation:	
4	AIS Briefing Office:	
5	ATS Reporting Office (ARO):	H24.
6	MET Briefing Office:	
7	ATS:	H24.
8	Fuelling:	H24.
9	Handling:	H24.
10	Security:	H24.
11	De-icing:	H24.
12	Remarks:	Refer to AD 2.20 item 1.

EGLL AD 2.4 – HANDLING SERVICES AND FACILITIES

1	Cargo handling facilities:	Full. Nearest railway siding, Feltham 2.6 nm.
2	Fuel/oil types:	AVTUR JET A-1. Oils; Various by arrangement with fuel companies.
3	Fuelling facilities/capacity:	
4	De-icing facilities:	By arrangement with handling agent.
5	Hangar space for visiting aircraft:	By arrangement with Metro Business Aviation, BMA, BA.
6	Repair facilities for visiting aircraft:	Maintenance and repair (by arrangement).
7	Remarks:	

EGLL AD 2.5 – PASSENGER FACILITIES	
1 Hotels:	Hotels in vicinity.
2 Restaurants:	Restaurant, buffet and bar.
3 Transportation:	Train and Metro to Central London, Buses, Coaches, taxis and car hire.
4 Medical facilities:	Occupational Health Department. Tel: 0181-745 7211/7047/7048.
5 Bank and Post Office:	Post Office in terminals.
6 Tourist Office:	
7 Remarks:	

EGLL AD 2.6 – RESCUE AND FIRE FIGHTING SERVICES	
1 AD Category for fire fighting:	RFF Category 9.
2 Rescue equipment:	
3 Capability for removal of disabled aircraft:	By arrangement with nominated recovery company.
4 Remarks:	

EGLL AD 2.7 – SEASONAL AVAILABILITY – CLEARING	
1 Type of clearing equipment:	Mechanical, Chemical de-icing, Sanding/Gritting.
2 Clearance priorities:	Standard. See AD 1.2.2.
3 Remarks:	Braking action assessment by Grip Tester. Latest information from: SNOWTAM Officer 020-8745 3490. 020-8745 7920/1 or 020-8745 7373.

EGLL AD 2.8 – APRONS, TAXIWAYS AND CHECK LOCATIONS DATA

1	Apron surface:	Concrete except

Concrete except
November: Surface: Block Paving   Strength: 78
Stands C39-57:   Surface: Block paving   Strength:

2   Taxiway width, surface and strength:

Width: 23 m   Surface: Asphalt/Concrete   Strength: 78
Width: 30 m   Surface: Asphalt/Concrete   Strength:
Width: 37 m   Surface: Asphalt/Concrete   Strength:

3   Altimeter check location and elevation:

Central Area 77 ft amsl.   Southern helicopter pad 76 ft amsl.
Cargo Centre Southside 76 ft amsl.

4   VOR checkpoints:

   INS check points:

Stand No	Co-ordinates	Stand No	Co-ordinates	Stand No	Co-ordinates	Stand No	Co-ordinates
A1	*512829N 0002708W	E3	*512813N 0002657W	M24	*512822N 0002737W	X1	*512746N 0002731W
A3	*512829N 0002706W	E5	*512812N 0002655W	M26	*512823N 0002739W	X2	*512746N 0002734W
A5	*512829N 0002703W	E7	*512810N 0002652W	M28	*512825N 0002740W	X3	*512746N 0002739W
A9	*512829N 0002701W	E9	*512807N 0002647W	M30	*512827N 0002743W	X4	*512746N 0002742W
A76	*512829N 0002712W	E36	*512813N 0002652W	M31	*512825N 0002735W	X5	*512746N 0002746W
		E38	*512812N 0002649W	M32	*512829N 0002746W	X6	*512746N 0002750W
B2	*512824N 0002701W	E40	*512811N 0002648W	M33L	*512827N 0002737W	X7	*512744N 0002754W
B4	*512825N 0002659W	E42	*512810N 0002645W	M33	*512826N 0002737W	X8	*512744N 0002758W
B6	*512826N 0002657W			M33R	*512826N 0002737W	X9	*512744N 0002801W
B8	*512827N 0002654W	F2	*512809N 0002701W				
B10	*512828N 0002652W	F4	*512808N 0002701W	N74	*512829N 0002711W		**Royal Suite/VIP**
B12	*512829N 0002650W	F6	*512807N 0002658W	N76	*512829N 0002713W		
B17	*512823N 0002655W	F8	*512806N 0002657W	N78	*512830N 0002715W	RSA	*512746N 0002723W
B19	*512825N 0002654W	F10	*512805N 0002653W	N80	*512829N 0002716W	RSB	*512745N 0002721W
B21	*512825N 0002651W	F11	*512808N 0002704W	N82	*512830N 0002719W	RSC	*512745N 0002721W
B23	*512826N 0002649W	F13	*512807N 0002701W	N84	*512829N 0002722W	RSD	*512746N 0002722W
B25L	*512828N 0002645W	F15	*512805N 0002658W	N86	*512829N 0002724W	RSE	*512744N 0002722W
B25	*512828N 0002646W			N88	*512829N 0002725W	RSF	*512744N 0002722W
B25R	*512827N 0002646W	G12	*512804N 0002710W	N90	*512830N 0002728W		
B27L	*512830N 0002640W	G14	*512804N 0002706W	N92L	*512829N 0002730W		**Terminal 4**
B27	*512829N 0002641W			N92	*512829N 0002731W		
B27R	*512828N 0002643W	H5	*512809N 0002725W	N92R	*512829N 0002733W	S1	*512728N 0002654W
B29	*512830N 0002638W	H7	*512807N 0002724W			S2	*512727N 0002657W
		H9	*512804N 0002718W	R34L	*512829N 0002749W	S3	*512726N 0002700W
C14	*512817N 0002654W	H30	*512810N 0002721W	R34	*512829N 0002751W		
C16	*512818N 0002652W	H32	*512808N 0002719W	R34R	*512829N 0002751W	T4	*512730N 0002708W
C18	*512819N 0002650W	H34	*512807N 0002718W	R36L	*512829N 0002754W	T5	*512732N 0002706W
C20	*512820N 0002648W	H36	*512805N 0002716W	R36	*512829N 0002755W	T6	*512734N 0002703W
C22	*512820N 0002645W			R36R	*512829N 0002755W	T7	*512736N 0002700W
C24	*512822N 0002643W	J2	*512809N 0002736W	R38L	*512829N 0002758W	T8	*512738N 0002658W
C26	*512823N 0002640W	J4	*512808N 0002734W	R38	*512828N 0002801W	T9	*512740N 0002655W
C28	*512824N 0002638W	J6	*512807N 0002733W	R38R	*512829N 0002800W	T10	*512741N 0002652W
C30	*512825N 0002636W	J6A	*512807N 0002734W	R40	*512825N 0002801W	T11	*512743N 0002649W
C32	*512826N 0002634W	J8	*512805N 0002731W			T12	*512745N 0002647W
C34	*512828N 0002634W	J10	*512804N 0002730W		**Western Apron**	T31	*512728N 0002710W
C39	*512818N 0002645W	J13	*512807N 0002740W			T33	*512703N 0002730W
C41	*512819N 0002642W	J15	*512805N 0002737W	147	*512805N 0002819W		
C43	*512820N 0002640W			148	*512805N 0002816W	V14	*512737N 0002640W
C45	*512822N 0002636W	K14	*512804N 0002748W	149	*512805N 0002812W	V15	*512739N 0002639W
C57	*512823N 0002633W	K16	*512806N 0002751W	150	*512809N 0002810W	V16	*512741N 0002638W
		K17	*512808N 0002752W	151	*512811N 0002810W	V17	*512743N 0002637W
D46	*512811N 0002640W	K19	*512810N 0002755W	152	*512814N 0002810W	V18	*512745N 0002635W
D47	*512812N 0002640W	K21	*512812N 0002758W	154	*512819N 0002809W	V19	*512737N 0002636W
D48	*512813N 0002639W	K23	*512814N 0002800W	155	*512821N 0002809W	V20	*512739N 0002634W
D49	*512814N 0002637W	K25	*512817N 0002800W	156	*512826N 0002810W	V21	*512741N 0002633W
D50	*512814N 0002636W			157	*512829N 0002811W	V22	*512743N 0002631W
D52	*512816N 0002634W	L18	*512812N 0002745W	158	*512831N 0002812W	V31	*512742N 0002621W
D53	*512816N 0002634W	L20	*512814N 0002746W			V32	*512742N 0002616W
D54	*512817N 0002633W	L22	*512816N 0002749W		**Cargo**		
D55	*512818N 0002631W	L23	*512813N 0002740W			W1	*512731N 0002715W
D56	*512818N 0002630W	L24L	*512818N 0002751W	301	*512744N 0002824W	W2	*512733N 0002715W
D58L	*512819N 0002629W	L24	*512818N 0002752W	302	*512741N 0002824W	W3	*512735N 0002714W
D58	*512820N 0002628W	L24R	*512819N 0002752W	303	*512738N 0002824W	W4	*512739N 0002715W
D58R	*512820N 0002628W	L25	*512815N 0002743W	306	*512737N 0002819W	W5	*512741N 0002717W
		L26	*512818N 0002757W	307	*512740N 0002819W	W6	*512743N 0002719W
		L27	*512817N 0002745W	308	*512742N 0002819W	W40	*512737N 0002704W
		L29	*512819N 0002746W			W41	*512739N 0002701W
		L31	*512821N 0002748W				
		L33	*512822N 0002751W				
		L35L	*512822N 0002756W				
		L35	*512822N 0002754W				
		L35R	*512822N 0002754W				
		L37	*512822N 0002758W				

EGLL AD 2.9 – SURFACE MOVEMENT GUIDANCE AND CONTROL SYSTEM MARKINGS

1	**Use of aircraft stand ID signs:** **Taxiway guide lines and visual docking/** **parking guidance system of aircraft stands:**	Nose-in parking is in operation throughout the terminal apron areas and in the cargo terminal. Stand centre-lines are marked by yellow paint lines. Azimuth and stopping guidance are provided by APIS, AGNIS/PAPA, AGNIS/MIRROR, or AGNIS/STOP ARROW (painted on the apron) except for the stands listed below. The type of stopping guidance is marked beside the APIS or AGNIS at the head of the stand. Left and right markings painted as alternately coloured yellow and white centre-lines are provided on many large stands to enable certain smaller aircraft to double park. AGNIS facilities are provided for certain left and right stands; where these do not exist, a marshalling service is provided. The following stands are not equipped with stand entry guidance. Marshalling is provided: B12, G14, J6, K25, L24, L26, L37, N92, R38, T33, 154, 155, 156, 157, X1, X2, X3, X4, X5, X6, 301-308, RSA and RSF. Stand H9: All arriving aircraft which fall within the Code D criteria may access the stand from any direction. All code D aircraft wishing to depart from Stand H9 may do so by pushing into the Hotel cul-de-sac. All aircraft larger than code D, regardless of live or towed movements, must access Stand H9 via the following routes: 70 (O) – 70 (I) – H9. 72 (I) – 70 (I) – H9 Access via 69 (I) – 70 (I) is **NOT** permitted. For larger than code D aircraft departures from H9, a non-standard push into the inner taxiway to face west is required. Mirror Parking Aid. A319 and B737-500 aircraft using stands B2, C14, C16, C18, C20, E3, F2, F4, F6, F8, F11, F13 and J10 must have the port engine fully shut down before entering the stand. Between the hours of 0700-2300 (Local) all 3 and 4 engined aircraft entering the V apron for Stands V18, V19, V20, V21 and V22 are to stop and shut down engines **EITHER** in Block 118 abeam the stop board in the Grass Area 15 **or** on the V apron centre-line abeam the centre-line of V22. The aircraft will then be met by a tug and towed onto stand. B777 aircraft are now cleared to taxi LIVE onto stands V18-V22 inclusive with no operational restrictions.
2	**Runway and taxiway markings and lighting:**	Runway: Runway designation, threshold, centre-line, touchdown zone and fixed distance markings available on Runways 09L, 09R, 27L, 27R. Runway 23 has distance to go markings. Taxiway: Blue edge lights on taxiways in maintenance areas and holding areas, green centre-line lights with selective switching on all other taxiways and routes. Rapid Exit Taxiways from Runways 09L/27R and 09R/27L have green lights which run parallel and to the side of the runway centre-line before joining the turn-off centre-line. Other turn-offs are marked by green centre-line lights leading directly away from the runway centre-line. **Note:** All turn-off lights are colour coded amber and green alternate lights to the edge of the instrument strip.
3	**Stop bars:**	Red stop bars are provided where appropriate.
4	**Remarks:**	**Pilots should only request push-back when they are actually ready to do so.** Illuminated wind direction indicators are located adjacent to the touchdown zones of Runways 27L, 27R and 09R **Pilots should not enter an aircraft stand unless the Stand Entry Guidance is illuminated or a marshaller has signalled clearance to proceed.**

EGLL AD 2.10 – AERODROME OBSTACLES					
In Approach/Take-off areas			In circling area and at aerodrome		
1			2		
Runway/Area affected	Obstacle type Elevation Markings/lighting	Co-ordinates	Obstacle type Elevation Markings/lighting		Co-ordinates
a	b	c	a		b
		ft amsl		ft amsl	
23/Approach	Light Column	93	(0.16 nm from THR 23)	Chimney (Lgtd)	296	(040°T, 2.48 nm from ARP)
	Aerial	99	(0.18 nm from THR 23)	Radio Mast	260	(345°T, 2.07 nm from ARP)
	Aerial	96	(0.20 nm from THR 23)	Radar Tower	229	(351°T, 0.13 nm from ARP)
	Light Column	105	(0.23 nm from THR 23)	Radar	236	(085°T, 1.10 nm from ARP)
	Obst. Light (Lgtd)	116	(0.32 nm from THR 23)	TV Aerials (flats)	251	(133°T, 2.11 nm from ARP)
	Chimneys (2)	113	(0.30 nm from THR 23)			
	Light Column	105	(0.27 nm from THR 23)			
	Tree	138	(0.65 nm from THR 23)			
	Tree	144	(0.67 nm from THR 23)			
	Tree	163	(0.86 nm from THR 23)			
	Gasholder (Lgtd)	388	(2.45 nm from THR 23)			
23/Take-off	Floodlights	136	(† 2374 Across RCL))			
09L/Approach 27R/Take-off	ILS array (Lgtd)	95	(0.35 nm from THR 09)			
27R/Approach 09L/Take-off	Aerial	96	(0.18 nm from THR 27R)			
	ILS array (Lgtd)	101	(0.22 nm from THR 27R)			
	Light Column	92	(0.25 nm from THR 27R)			
	Light Column	93	(0.25 nm from THR 27R)			
	Air Vent	108	(0.27 nm from THR 27R)			
	Chimney	109	(0.36 nm from THR 27R)			
	Tree	132	(0.66 nm from THR 27R)			
	Tree	134	(0.67 nm from THR 27R)			
	Aerials	163	(0.67 nm from THR 27R)			
09R/Approach 27L/Take-off	Aerial	92	(0.37 nm from THR 09R)			
	Light Column	89	(0.38 nm from THR 09R)			
	Trees	99	(0.39 nm from THR 09R)			
	Light Column	102	(0.45 nm from THR 09R)			
	Trees	155	(0.80 nm from THR 09R)			
27L/Approach 09R/Take-off	ILS array (Lgtd)	92	(0.15 nm from THR 27L)			
	TV Aerial	108	(0.25 nm from THR 27L)			
	Approach Light (Lgtd)	103	(0.38 nm from THR 27L)			

3 Remarks:	Electricity pylons running on a line NE/SW and 2.6 nm W from ARP at 182 ft aal/262 ft amsl.
	† Distance (meters) beyond start of TORA. ‡ Distance (meters) Left (L) or Right (R) of extended centre-line.

EGLL AD 2.11 – METEOROLOGICAL INFORMATION PROVIDED

1	Associated MET Office:	Bracknell NMC.
2	Hours of service: MET Office outside hours:	H24.
3	Office responsible for TAF preparation: Periods of validity:	Bracknell NMC. 9, 18 hours.
4	Type of landing forecast: Interval of issuance:	TREND. 30 Minutes.
5	Briefing/consultation provided:	Self-briefing/Telephone.
6	Flight documentation: Language(s) used:	Charts, abbreviated plain language text. TAFs/METARs. English.
7	Charts and other information available for briefing or consultation:	
8	Supplementary equipment available for providing information:	
9	ATS units provided with information:	London Heathrow.
10	Additional Information (limitation of service etc):	

EGLL AD 2.12 – RUNWAY PHYSICAL CHARACTERISTICS

Designations Runway Number	True and MAG bearing	Dimensions of Runway (m)	Strength (PCN) and surface of Runway and stopway	Threshold co-ordinates	Threshold elevation highest elevation of TDZ of precision APP runway
1	2	3	4	5	6
23	221.8° GEO 226° MAG	1966 x 45 † §	60/R/B/W/T Grooved Asphalt	512834.95N 0002554.53W	THR 75 ft
09L	089.7° GEO 094° MAG	3902 x 45 † ø	83/R/A/W/T Concrete/Asphalt with porous friction course	512839.00N 0002906.11W	THR 79 ft
27R	269.7° GEO 274° MAG	3902 x 45 † ø	83/R/A/W/T Concrete/Asphalt with porous friction course	512839.63N 0002559.75W	THR 78 ft
09R	089.7° GEO 094° MAG	3658 x 45 †	83/R/A/W/T Asphalt with porous friction course	512753.25N 0002856.50W	THR 75 ft
27L	269.7° GEO 274° MAG	3658 x 45 †	83/R/A/W/T Asphalt with porous friction course	512753.82N 0002602.81W	THR 77 ft

Remarks:

† There is a 22.5 m wide weight bearing shoulder each side except for the western extension of Runways 27L/09R and 27R/09L which has 8 m weight bearing shoulders on each side and Runway 23 which has a 7.5 m weight bearing shoulder on each side.
Compass base: *512824N 0002612W
ø Note: Blocks 112, 113 and 114 concrete, blocks 9 to 19 inclusive, asphalt.
§ Runway 23. Take-off is permitted only by propeller driven aircraft up to MTOW 24 tonnes and by prior permission from the Heathrow Operations Duty Manager.

Slope of RWY/SWY	Stopway Dimensions (m)	Clearway Dimensions (m)	Strip Dimensions (m)	OFZ
7	8	9	10	11

12   **Remarks:**

EGLL AD 2.13 – DECLARED DISTANCES					
Runway Designator	TORA (m)	TODA (m)	ASDA (m)	LDA (m)	Remarks
1	2	3	4	5	6
23	1183	1243	1966	1966	‡ 09L and 09R landing thresholds are displaced by 305 m.
09L	3902	3902	3902	3597 ‡	
27R	3902	3902	3902	3902	
09R	3658	3658	3658	3353 ‡	
27L	3658	3658	3658	3658	
27R	3492	3492	3492		Take-off from intersection with Block 18.
27L	3218	3218	3218		Take-off from intersection with Block 86.
09R	2919	2919	2919		Take-off from intersection with Block 79.

EGLL AD 2.14 – APPROACH AND RUNWAY LIGHTING								
Runway	Approach lighting Type Length Intensity	Threshold lighting Colour Wingbars	PAPI VASIS Angle Dist from THR (MEHT)	TDZ lighting Length	Runway Centre-line lighting Length Spacing Colour Intensity	Runway edge lighting Length Spacing Colour Intensity	Runway End lighting Colour Wingbars	Stopway lighting Length (m) Colour
1	2	3	4	5	6	7	8	9
23	Coded centre-line with four crossbars 592 m HI	HI Green with HI wingbars	PAPI 3° 355 m (57 ft)			Bi-directional 46 m spacing		
09L	Coded centre-line with five crossbars. 853 m HI Supplementary lighting, inner 300 m.	HI Green with HI wingbars	PAPI 3° 425 m (67 ft)	914 m	HI Bi-directional colour coded 15 m spacing	Bi-directional 46 m spacing	Red	
27R	Coded centre-line with five crossbars. 905 m HI Supplementary lighting, inner 300 m.	HI and LI Green (semi-flush) with HI wingbars	PAPI 3° 470 m (74 ft)	914 m	HI Bi-directional colour coded 15 m spacing	Bi-directional 46 m spacing	Red	
09R	Coded centre-line with five crossbars. 939 m HI Supplementary lighting, inner 300 m.	HI Green with HI wingbars	PAPI 3° 425 m (67 ft)	914 m	HI Bi-directional colour coded 15 m spacing	Bi-directional 46 m spacing	Red	
27L	Coded centre-line with five crossbars. 922 m HI Supplementary lighting, inner 300 m.	HI and LI Green with HI wingbars	PAPI 3° 411 m (64 ft)	914 m	HI Bi-directional colour coded 15 m spacing	Bi-directional 46 m spacing	Red	
10 Remarks:	Runway 09L: Edge lights first 290 m red on 09L and 09R. Runway 09R: Edge lights first 290 m red on 09L and 09R. Runway 23: 7 sequenced flashing centre-line lights between 424 m crossbar and threshold.							

EGLL AD 2.15 – OTHER LIGHTING, SECONDARY POWER SUPPLY

1	ABN/IBN location, characteristics and hours of operation:	
2	LDI location and lighting: Anemometer location and lighting:	
3	Taxiway edge and centre-line lighting:	
4	Secondary power supply/switch-over time:	Yes.
5	Remarks:	Apron floodlighting. Obstacle lighting.

EGLL AD 2.16 – HELICOPTER LANDING AREA

1	Co-ordinates TLOF or THR of FATO:	
2	TLOF and/or FATO elevation (ft):	
3	TLOF and FATO area dimensions: Surface, Strength, Markings:	
4	True and MAG Bearing of FATO:	
5	Declared distance available:	
6	Approach and FATO lighting:	
7	Remarks:	Refer to AD 2.20 item 5.

EGLL AD 2.17 – ATS AIRSPACE

Designation and lateral limits:	Vertical limits	Airspace Classification
1	2	3
London Control Zone (CTR) 513611N 0004133W - 513611N 0001253W thence clockwise by an arc of a circle radius 12 nm centred on 512812N 0002713W to 512013N 0001255W - 512013N 0003800W - 512104N 0004242W thence clockwise by an arc of a circle radius 12 nm centred on 512812N 0002713W to 513611N 0004133W.	2500 ft ALT SFC	A
London Heathrow Aerodrome Traffic Zone (ATZ) Circle radius 2.5 nm centred on longest notified runway (09L/27R) 512839N 0002741W	2000 ft aal SFC	A

4	ATS unit callsign: Languages:	Heathrow Director English.
5	Transition Altitude:	6000 ft.
6	Remarks:	ATZ hours: H24. See EGLL AD 2.23, Flight Procedures, paragraph 14 for details of Local Flying Areas.

EGLL AD 2.18 – ATS COMMUNICATION FACILITIES					
Service Designation	**Callsign**	**Frequency MHz**	**Hours of Operation**		**Remarks**
			Winter	**Summer**	
1	2	3	4		5
APP	Heathrow Director	119.725 120.400 (4) 127.525 (4) 134.975 (4) 121.500 †	H24	H24	† Emergency Channel O/R. (1)Ground Movement Planning Departing aircraft are to make initial call to 'Heathrow Delivery' on this frequency during hours of operation. At other times call 'Heathrow Ground'. (2)Ground Movement Control. (3)Special VFR and Helicopter flights in the London CTR. These flights will additionally receive a service from the following: Thames Radar (132.700 MHz): 0630-0700 and 2030-2200. Heathrow Director (119.725 MHz): 2200-0630. (One hour earlier in summer). (4)When instructed by ATC.
TWR	Heathrow Tower	118.500	H24	H24	
		118.700	0700-2200 or as directed	0600-2200 or as directed	
		124.475 (4) 121.500 †	H24	H24	
	Heathrow Delivery	121.975 (1)	0630-2200 or as directed	0530-2200 or as directed	
	Heathrow Ground	121.900 (2) 121.700 (2)	As directed by ATC	As directed by ATC	
RAD	Heathrow Radar	119.900 (3)	0700-2030	0600-1930	
		125.625 (4) 121.500 †	H24	H24	
Arrival ATIS	Heathrow Information	123.900 113.750 115.100 115.300 115.600	0500-2300	0400-2200	Broadcast on Bovingdon VOR. Broadcast on Biggin VOR. Broadcast on Ockham VOR. Broadcast on Lambourne VOR.
Departure ATIS		121.850			
FIRE	Heathrow Fire	121.600	Available when Fire vehicle attending aircraft on the ground in an emergency.		Non-ATS frequency.

EGLL AD 2.19 – RADIO NAVIGATION AND LANDING AIDS						
Type Category (Variation)	**IDENT**	**Frequency**	**Hours of Operation**		**Antenna site co-ordinates**	**Elevation of DME transmitting antenna**	**Remarks**
			Winter	Summer			
			# and by arrangement				
1	2	3	4		5	6	7
LLZ 09L ILS CAT III	I AA	110.30 MHz	HO	HO	512839.70N 0002538.62W		See Note.
GP	I AA	335.00 MHz	HO	HO	512843.59N 0002850.58W		3° ILS Ref Datum Hgt 51 ft.
LLZ 09R ILS CAT III	I BB	109.50 MHz	HO	HO	512753.88N 0002543.12W		See Note.
GP	I BB	332.60 MHz	HO	HO	512748.36N 0002840.47W		3° ILS Ref Datum Hgt 52 ft.
LLZ 27L ILS CAT III	I LL	109.50 MHz	HO	HO	512753.13N 0002930.82W		See Note.
GP	I LL	332.60 MHz	HO	HO	512749.08N 0002619.56W		3° ILS Ref Datum Hgt 56 ft.
MLS 27R	M HER	Ch 522	H24	H24			
Azimuth					512838.89N 0002935.79W		
Elevation					512843.06N 0002614.96W		
LLZ 27R ILS CAT III	I RR	110.30 MHz	HO	HO	512838.88N 0002939.69W		See Note.
GP	I RR	335.00 MHz	HO	HO	512843.93N 0002617.62W		3° ILS Ref Datum Hgt 60 ft. Glide Path flag indication may be noticed when below Glide Path in the region of 8° left of the centre-line.
DME	I AA (RWY 09L) I RR (RWY 27R)	Ch 40X	HO	HO	512843.78N 0002732.86W		On AD. DME freq paired with ILS I AA and I RR. Zero range is indicated at THR of Runway 09L and Runway 27R.
DME	I BB (RWY 09R) I LL (RWY 27L)	Ch 32X	HO	HO	512749.56N 0002730.76W		On AD. DME freq paired with ILS I BB and I LL. Zero range is indicated at THR of Runway 09R and Runway 27L.
DME	HTT	Ch 44X	HO	HO	512825.76N 0002601.11W		On AD. DME related freq 110.700 MHz. Zero range is indicated at THR of Runway 23. Range 25 nm at an altitude of 20000 ft in an area ± 40° either side of the extended Runway 23 centre-line. Any ILS indications should be ignored.
L	HRW	424.0 kHz	H24	H24	512843.99N 0002733.96W		Range 20 nm.
							Note: Particular care should be exercised in selecting the appropriate ILS facility as more than one ILS will normally be radiating.

EGLL AD 2.20 – LOCAL TRAFFIC REGULATIONS

1   **Airport Regulations**

a   Use governed by regulations applicable to London CTR – IFR procedures apply in all weather conditions in London CTR.

b   The following conditions and procedures apply to single-engined and light twin-engined aircraft not fully equipped with radio apparatus (including ILS receiver) as specified at GEN 1.5 but carrying at least the VHF RTF frequencies to permit communication with London (Heathrow) Airport Approach/Director, Tower and Ground Movement Control:

    i   the flight must be made on a special VFR clearance under the weather conditions and along the routes specified in the EGLL AD 2.22, paragraph 13.

    ii   The first VHF RTF communication with Approach Control must include the words 'Customs required' if the flight is an international one.

c   An operator which has not operated a scheduled service or a series charter service from Heathrow prior to 1 November 1992 shall only be permitted to commence a scheduled service or a series charter service from Heathrow to a destination which was not served from the airport by any operator in the twelve months prior to 1 November 1992 if any jet aircraft to be used meets the requirements ICAO Annex 16, Chapter 3.

d   When applying for permission to commence a service falling within the terms of this Condition, documents attesting that jet aircraft comply with Chapter 3 Noise certification standards must be produced. If these documents are not produced the aircraft will be regarded as a non Chapter 3 aircraft.

e   All flights are at all times subject to PPR within the terms of the Heathrow Rule 1 Traffic Distribution Rules 1991. The filing of a flight plan with NATS or receipt of an ATC clearance does not constitute permission to use London Heathrow.

f   Availability: H24, subject to the approval of the Managing Director, Heathrow Airport Ltd, and the acceptance of the flight by the co-ordinator (Airport Co-ordination Ltd). The airport may be used also by aircraft other than those engaged on scheduled flights provided that prior permission and a clearance number for each flight is obtained from the Managing Director Heathrow. Use is also subject to limitations imposed by Night Noise Restrictions (See current Supplement).

g   Subject to paragraph (iv), Operators of General and Business aviation aircraft may only operate in the peak during any operating season if they obtain permission to do so from the airport operator as well as a slot in advance of each movement also from the airport operator. In practice, permission to operate in the peak will be deemed to have been granted under the terms of the Traffic Distribution Rule (copies available from Manager Heathrow Operations Centre, Heathrow Airport Limited), if a slot for each movement in the peak is granted. Operators of General and Business Aviation aircraft, who arrive and depart at any time, may operate the movement provided that the aircraft departs or arrives to the approved slot time. Those who fail to operate to the approved slot time or who operate in the peak without first obtaining permission and a slot from the airport operator, are liable to be prohibited from operating in the peak thereafter, unless the airport operator is satisfied that the movement amounted to an emergency or other circumstances beyond the control of the operator or the commander of the aircraft. This permission will not be given for flights for recreational, charity and record breaking purposes. Light single and twin engined private aircraft will not be permitted to use the airport. Applications for prior permission must be made not more than 10 days and not less than 24 hours before the proposed flight and should be addressed to the Manager, Airport Co-ordination Ltd, by telephone 020-8759 4871, or 020-8759 2995, availability office hours; or by telex 934892 LHR Ltd. These applications must include the following information:

    i   Aircraft owner/operator;

    ii   Aircraft type and registration;

    iii   Origin and/or destination;

    iv   ETA and ETD;

    v   Number of passengers;

    vi   A handling agent (Airline Operator or Handling Agent based at Heathrow) is a requirement for all flights including general aviation and helicopter movements.

h   Diversion Procedure – Airline and other operators are advised that before filing London Heathrow as an alternate, they are required to have made arrangements for ground handling with an airline from the appropriate terminal. Nothing in this procedure shall, however, prevent an aircraft that has declared an emergency from landing.

i   The operation of the Antonov An124 and An225 is subject to prior approval of the Airfield Operations Manager, Tel: 020-8745 5923.

j   Fixed-wing and rotary aircraft using London Heathrow Airport do so in accordance with the Heathrow Conditions of Use document. A copy of the document will be made available on request by Tel: 020-8745 4076

EGLL AD 2.20 – LOCAL TRAFFIC REGULATIONS

2 **Ground Movement**

a **General**

   i Ground Movement Control (GMC) is in continuous operation and all surface movement of aircraft, vehicles and personnel on the Manoeuvring Area is subject to ATC authority.

   ii Directions issued by ATC should be followed specifically. RTF transmissions must be brief, concise and kept to the minimum number.

   iii Procedures – within the Movement Area, pilots will be cleared to and from the aircraft stands under general direction from GMC, they are reminded of the extreme importance of maintaining a careful lookout at all times.

   iv **It is the aircraft Commander's responsibility not to accept an ATC clearance into a block not approved for the type of aircraft.**

   v Runway Holding Areas for aircraft departing on Runways 27L/09R and 27R/09L are marked by signs displaying 'Holding Area'. The areas are illustrated on pages AD 2-EGLL-2-3/4. Within these areas, revised Air Traffic Control procedures are as follows:

     1 At all times in good visibility an ATIS message will remind pilots that they remain responsible for wing tip clearance. In the hours of darkness, selectable reds and greens are used in the 27L and 27R holding areas. The holding areas for 09L and 09R have blue edge lights.

     2 In promulgated holding areas, flight crew will be expected to follow conditional line-up clearances to maximise runway utilisation, which may entail overtaking and passing other aircraft in the holding areas. It is stressed that during these manoeuvres, avoidance of other aircraft is the responsibility of the flight crew involved. If doubt exists as to whether other aircraft can be overtaken then ATC must be informed that the conditional clearance that has been received cannot be complied with.

   vi ATC will clear aircraft to the holding point of the departure runway in use. Until a line-up clearance of sequence instruction is issued, commanders are to position their aircraft in such a way that the entrances to the runways are not obstructed.

b **Manoeuvring Area**

   i The Manoeuvring Area is divided into blocks as illustrated on page AD 2-EGLL-2-1. Boards are placed at the boundary of each block, bearing two groups of numbers, the left group representing the block in which the aircraft is standing and the right group the block immediately ahead. The Central Terminal Area Outer Taxiway blocks are indicated by an 'O' after the numbers, and the Inner Taxiway blocks by an 'I' after the numbers. The numbers on these boards should be used when reporting positions.

   ii The Manoeuvring Area is equipped with the following forms of taxiway guidance:

     1 Yellow painted taxiway centre-lines;

     2 Yellow painted holding position lines at the approach to runways;

     3 Yellow painted special holding position lines related to runway protection in CAT II/III;

     4 Green taxiway centre-line lights and red stopbars controlled from the Tower;

     5 An illuminated red stopbar means STOP. Aircraft must not proceed until the stopbar is extinguished or ATC permission is received;

     6 Runway Guard Lights: Pairs of alternately flashing ground mounted yellow lights at each side of the taxiways, where they connect with a runway, operate H24.

     7 During the winter period, reflective yellow 'snow edge markers' are placed at the edge of the taxiways.

   iii Ground movement of large aircraft

   The following restrictions apply to aircraft with wingspans exceeding 52 m.

     1 Inner taxiway Blocks 72(I) and 61(I). All large aircraft must be towed through this route.

     2 All B747 aircraft must be towed on any route through Block 25 (I). This route is available to all other aircraft without restriction.

     3 Southern taxiway Block 108. Pilots of B747/400 aircraft must exercise caution when using the block as wing-tip clearances to the south are minimal.

     4 Stand cul-de-Sacs. The Charlie, Echo, Foxtrot and Hotel cul-de-sacs are not available for large aircraft.

     5 The Bravo cul-de-sac is available to the B747/400 only to access stand B25. Access to stand B27 must be into the inner taxiway. All other large aircraft may use this cul-de-sac without restriction.

     6 All B747/400 aircraft in the Zulu cul-de-sac must be under tow. All other large aircraft may use this cul-de-sac without restriction.

     7 Manoeuvring of aircraft with a wingspan greater than 65 m must only be undertaken with the adjacent airside road closed.

   iv The taxiway route 95-118-86 is restricted in use to daylight and good visibility only and for aircraft up to 27.70 m wingspan (Gulfstream V).

c **Runway Crossing Procedure (Runway 09R/27L and 23)**

   i Aircraft and vehicles which are required to cross active runways will be issued instructions by the Ground Movement Controller, which will include a taxiway block as a clearance limit, in which the aircraft or vehicle will be required to hold short of the active runway.

   ii When reaching the clearance limit specified in the taxiing instructions, the aircraft or vehicle will be instructed to change frequency to that of the Air Controller of the appropriate runway.

   iii After crossing the runway and having reported 'runway vacated' with the Air Controller, the aircraft or vehicle will be instructed to revert to the GMC frequency for further clearance. In the absence of further clearance it is essential that the aircraft or vehicle does not proceed beyond the first Block clear of the runway.

EGLL AD 2.20 – LOCAL TRAFFIC REGULATIONS

d **Start-up Procedures**

i Pilots are to report their aircraft type, QNH and the identification letter of the received ATIS information on first contact with 'Heathrow Delivery'.

ii ATC are responsible for clearance delivery and start-up approval as a separate function from Ground Movement Control. Pilots are warned that start-up approval applies only to those engines which may be started-up on stand. If push-back is necessary to complete start-up, approval must be obtained from Ground Movement Control.

iii Following push-back from cul-de-sac stands, all aircraft must pull forward to a minimum of 100 m from the blast screen before disconnecting the tug. In the F, H and M cul-de-sacs, due to exhaust fume ingestion within the buildings at the end of the cul-de-sacs, engine start should be delayed until the aircraft has reached the 100 m mark, under no circumstances must tail mounted engines be started until the aircraft has been pulled forward to the minimum 100 m clear of the blast screen.

iv When requesting start-up clearance pilots should give the full callsign, type and stand number. All jet aircraft are to advise ATC, if for any reason they are unable to accelerate after noise abatement procedures to 250 kt. Note that poor departure speed performance may result in a short delay at peak periods. Aircraft must be ready to start before calling on frequency.

v Sequencing of aircraft ground movement for take-off in low visibility. When the RVR is below 400 meters pilots are not to request start-up clearance until the reported RVR is equal to or greater than the appropriate value in the following table:

AIRCRAFT TAKE-OFF MINIMA	MINIMUM RVR FOR START-UP CLEARANCE
350 meters RVR	300 meters RVR
300 " "	250 " "
250 " "	200 " "
200 " "	150 " "
150 " "	150 " "
100 " "	100 " "
75 " "	75 " "

It is emphasised that these measures will apply only when the reported RVR is below 400 meters and the co-operation of all pilots is sought in maintaining the safety level in low minima operations.

vi If within 30 minutes of a previously issued ADT the flight is unable to comply with that CTOT, the Pilot should advise ATC as soon as possible.

vii Pilots are advised that delays in excess of 10 minutes can be expected at the Holding Point. Sufficient time should be allowed for start, push-back and taxi to take account of such a delay especially if required to comply with an CTOT.

e **Departures – Minimum Runway Occupancy Time**

i On receipt of line-up clearance pilots should ensure, commensurate with safety and standard operating procedures, that they are able to taxi into the correct position at the hold and line up on the runway as soon as the preceding aircraft has commenced its take-off roll.

ii Pilots who require to back-track the runway (including line up from Block 75 onto Runway 27L) must notify ATC prior to arrival at the holding point.

iii Whenever possible, cockpit checks should be completed prior to line up and any checks requiring completion whilst on the runway should be kept to the minimum required. Pilots should ensure that they are able to commence the take-off roll immediately take-off clearance is issued.

iv Pilots not able to comply with these requirements should notify ATC as soon as possible once transferred to Heathrow Tower Departures Frequency.

EGLL AD 2.20 – LOCAL TRAFFIC REGULATIONS

3   CAT II/III Operations

a Runways 09R, 09L, 27R and 27L, subject to serviceability of the required facilities, are suitable for Category II and III operations by operators whose minima have been accepted by the Civil Aviation Authority.

b During Category II and III operations, special ATC procedures (ATC Low Visibility Procedures) will be applied. Pilots will be informed when these procedures are in operation by Arrival and Departure ATIS or by RTF.

c Departing Aircraft: ATC will require departing aircraft to use the Category III holding points at the following blocks. However, other departure points may be used at ATC discretion in which case due allowance will be made by ATC for the necessary ILS protection.

Runway 27L: Blocks 75/87, 94/88, 95/88, 118/86 and 130/86.

Runway 27R: Blocks 40/19, 101/8, 134/19, 137/18, 40/8 and 31(O)/17.

Runway 09L: Block 115/112.

Runway 09R: Blocks 98/100 and 106/107.

d Arriving Aircraft: Ground Movement Radar (GMR) is normally available and all runway exits will then be illuminated. Pilots should select the first convenient exit.

In the event that GMR is not available only the exits at the following designated blocks will be illuminated.

Runway 27L: Blocks 79/89 or 79/106.

Runway 27R: Block 9/36.

Runway 09L: Block 17/31(O).

Runway 09R: Blocks 87/75 or 87/88.

e Pilots are to delay the call 'runway vacated' until the aircraft has completely passed the end of the green/yellow colour coded taxiway centre-line lights.

f When Low Visibility Procedures are in force a much reduced landing rate can be expected due to the requirement for increased spacing between arriving aircraft. In addition to the prevailing weather conditions, such factors as equipment serviceability may also have an effect on actual landing rates. For information and planning purposes, the approximate landing rates that can be expected are:

IRVR (m)	Expected Landing Rate
Between 1000 and 600	34
Between 600 and 150	24
Less than 150	less than 20

4   Warnings

a The letters 'LH' and an arrow pointing to Runway 23 at Heathrow are painted on the gasholder at Southall on its north-east side. Letters and arrows are in white and 30 ft high.

b Pilots are warned, when landing on Runway 27R in strong southerly/south westerly winds, of the possibility of building-induced turbulence and large windshear effects.

c Similarly, Runway 27L arrivals may be affected by winds with a strong Northerly component. Building-induced turbulence may be experienced at the mid sections of each runway from winds with a strong Southerly, or strong Northerly component.

5   Helicopter Operations

a Helicopter aiming points are located as follows:

North eastern end of Block 97 (marked with a 18 m sided triangle with a conventional 'H'). This aiming point is lit and available for use throughout operational hours. The take-off and climb surface has been protected to 8% to the East and West of the FATO.

b Helicopters are at all times subject to **PPR.**

c Helicopters alighting at the aiming point will ground or air-taxi to parking areas as directed by ATC.

EGLL AD 2.20 – LOCAL TRAFFIC REGULATIONS

6   Use of Runways

Special runway utilisation procedures are detailed at GEN 3.3.3.

Preferential Runway System

a   In weather conditions when the tail wind component is no greater than 5 knots on the main Runway 27R and 27L, these runways will normally be used in preference to Runways 09R and 09L, provided the runway (s) surface is dry.

b   When the associated crosswind component on these main runways exceeds 25 knots, Runway 23 will normally be made available if there is a lesser crosswind component affecting it.

c   Pilots who ask for permission to use the runway into wind when, in accordance with these procedures, Runway 27R or 27L are in use, should understand that their arrival or departure may be delayed.

d   Runway 23 has painted DISTANCE TO GO markings at 1000 m and 500 m from the end of the runway.

e   Departures – Wake Vortex separation.
Wake vortex separations are in accordance with the 5 Group Scheme and are as detailed in UK AIC 17/1999. On departure, when in receipt of line up clearance, the pilot must inform ATC if greater wake vortex separation will be required behind the preceeding aircraft than that laid down in UK AIC 17/1999. Failure to do so may resut in additional delay.

f   The use of Runway 23 for take-off is restricted to propeller-driven aircraft of less than 24 tonnes (24 000 kgs/52 910 lbs).

g   Arrivals – Minimum Runway Occupancy Time
Pilots are reminded that rapid exit from the landing runway enables ATC to apply minimum spacing on final approach that will achieve maximum runway utilisation and will minimise the occurrence of 'go-arounds'.

7   Aircraft Separation

In certain weather conditions 2.5 nm radar separation may be applied on final approach. The conditions when this separation may be utilised are:

By day when the runway is dry;

Visibility and cloud ceiling equal to or better than 10 km and 1500 ft;

Only applied to those aircraft not requiring wake vortex separation;

Only applied within 15 nm from touchdown when on the localizer or on a closing heading.

EGLL AD 2.21 – NOISE ABATEMENT PROCEDURES

Notice under Section 78(1) of the Civil Aviation Act 1982
Whereas:

(1) By virtue of the Civil Aviation (Designation of Aerodromes) Order 1981(a) Heathrow Airport – London is a designated aerodrome for the purpose of Section 78 of the Civil Aviation Act 1982 (b);

(2) the requirements specified in this notice appear to the Secretary of State to be appropriate for the purpose of limiting, or of mitigating the effect of noise and vibration connected with the taking off or, as the case may be, landing of aircraft at Heathrow Airport – London;

Now, therefore, the Secretary of State in exercise of the powers conferred on him by Section 78 (1) and (12) of the Civil Aviation Act 1982, by this notice published in the manner prescribed by the Civil Aviation (Notices) Regulations 1978 (c), hereby provides as follows:

1   This notice may be cited as the Heathrow Airport – London (Interim Noise Abatement Requirements) Notice (2) 1999 and shall come into operation on 2 December 1999.

2   The Heathrow Airport – London (Interim Noise Abatement Requirements) Notice 1999 (d) is hereby revoked.

3   It shall be the duty of every person who is the operator of any aircraft which is to take off or land at Heathrow Airport – London to secure that, after the aircraft takes off or, as the case may be, before it lands at the aerodrome the following requirements are complied with:

  (1)   After take-off the aircraft shall be operated in such a way that it is at a height of not less than 1000 ft aal at the point at which the aircraft is closest to that site of a noise monitoring terminal which is nearest to the departure track of that aircraft.

  (2)   The sites of the noise monitoring terminals relating to Heathrow Airport - London are:

Description	OS Co-ordinates	Elevation above aerodrome	Latitude	Longitude
Site 6	TQ 0204 7510	-6 m	*512756N	0003157W
Site A	TQ 0263 7700	-4 m	*512857N	0003124W
Site C	TQ 0219 7566	-6 m	*512814N	0003148W
Site D	TQ 0197 7477	-7 m	*512745N	0003201W
Site E	TQ 0169 7409	-7 m	*512724N	0003216W
Site F	TQ 1151 7606	-3 m	*512821N	0002345W
Site G	TQ 1166 7560	-3 m	*512806N	0002338W
Site I	TQ 1164 7398	-4 m	*512714N	0002341W

  (3)   Subject to sub-paragraph (4), any aircraft other than Concorde shall, after take-off, be operated in such a way that it will not cause more than 97 dBA Lmax by day (from 0700 hours to 2300 hours local time) or 89 dBA Lmax by night (from 2300 hours to 0700 hours local time) as measured at any noise monitoring terminal at any of the sites referred to in sub-paragraph (2) above.

  (4)   The limits specified in sub-paragraph (3) above shall be adjusted in accordance with the following table in respect of any noise monitoring terminal at any of the sites referred to in the table in sub-paragraph (2) above to take account of the location of that terminal and its ground elevation relative to the aerodrome elevation.

EGLL AD 2.21 – NOISE ABATEMENT PROCEDURES

Description	Adjustment dBA
Site 6	minus 0.3
Site A	plus 0.6
Site C	minus 0.3
Site D	minus 0.7
Site E	minus 1.3
Site F	plus 0.2
Site G	minus 0.1
Site I	minus 0.3

(5) Where the aircraft is a jet aircraft, after passing the site referred to in sub-paragraph (1) above, it shall, maintain a gradient of climb of not less than 4% to an altitude of not less than 4000 ft. The aircraft shall be operated in such a way that progressively reducing noise levels at points on the ground under the flight path beyond that site are achieved.

(6) After the aircraft takes off from any runway specified in the first column of the following table, the aircraft shall follow the Noise Preferential Routeing Procedure specified in the third column of the table which relates to the ATC clearance previously given to the aircraft and specified in the second column of the table, whether flying in IMC or VMC:

Provided that nothing in this sub-paragraph (6) shall apply:

(a) to any propeller driven aircraft whose MTWA does not exceed 5700 kg; or

(b) during the period between 0600 hours and 2330 hours (local time): to any propeller driven aircraft whose MTWA does not exceed 17000 kg, or to any Dash 7 aircraft.

Take-off Runway	ATC Clearance	Procedure
27R	Via Woodley	Straight ahead to intercept LON VOR RDL 259 until LON DME 7 then turn right onto QDM 273 to WOD NDB (LON DME 16).
	Via Chiltern	Straight ahead to be established on BUR NDB QDM 301 by LON DME 4. At LON DME 6 turn right onto QDM 057 to CHT NDB.
	Via Burnham/ WOBUN	Straight ahead to be established on BUR NDB QDM 301 by LON DME 4. At LON DME 7 turn right to follow BUR NDB QDR 360 to abeam BNN VOR (LON DME 16).
	Via Midhurst	Straight ahead to intercept LON VOR RDL 259 until LON DME 5 then turn left onto BUR NDB QDR 165. At LON DME 12 turn right onto MID VOR RDL 015 to MID VOR.
	Via Epsom/ Detling	Straight ahead until LON DME 2 then turn left onto QDM 141 to EPM NDB then left onto DET VOR RDL 274 to abeam Biggin (DET DME 21).

Take-off Runway	ATC Clearance	Procedure
27L	Via Woodley	Straight ahead to intercept LON VOR RDL 259 until LON DME 7 then turn right onto QDM 273 to WOD NDB (LON DME 16).
	Via Chiltern	Straight ahead to be established on BUR NDB QDM 301 by LON DME 3. At LON DME 6 turn right onto QDM 057 to CHT NDB.
	Via Burnham/ WOBUN	Straight ahead to to be established on BUR NDB QDM 301 by LON DME 3. At LON DME 7 turn right to follow BUR NDB QDR 360 to abeam BNN VOR (LON DME 16).
	Via Midhurst	Straight ahead to intercept LON VOR RDL 243 until LON DME 5.5 then turn left onto BUR NDB QDR 165. At LON DME 12 turn right onto MID VOR RDL 015 to MID VOR
	Via Epsom/ Detling	Straight ahead until I-LL DME 1.0 (LON DME 2) then turn left onto QDM 141 to EPM NDB then left onto DET VOR RDL 274 to abeam Biggin (DET DME 21).

EGLL AD 2.21 – NOISE ABATEMENT PROCEDURES

Take-off Runway	ATC Clearance	Procedure	Take-off Runway	ATC Clearance	Procedure
09R	Via Woodley	Straight ahead until LON DME 2 then turn right onto QDM 285 to WOD NDB (LON DME 16).	09L	Via Woodley	Straight ahead until LON DME 1.5 then turn right onto QDM 285 to WOD NDB (LON DME 16).
	Via Ockham/ Southampton	Straight ahead until LON DME 2 then turn right onto LON VOR RDL 128 until LON DME 5 then right onto OCK VOR RDL 045. At OCK DME 2 turn right onto OCK VOR RDL 257 by OCK DME 3.		Via Ockham/ Southampton	Straight ahead until LON DME 1.5 then turn right onto LON VOR RDL 128 until LON DME 5 then right onto OCK VOR RDL 045. At OCK DME 2 turn right onto OCK VOR RDL 257 by OCK DME 3.
	Via Midhurst	Straight ahead until LON DME 2 then turn right onto LON VOR RDL 128 until LON DME 3.5 then turn right onto MID VOR RDL 030 to MID VOR.		Via Midhurst	Straight ahead until LON DME 1.5 then turn right onto LON VOR RDL 128 until LON DME 3.5 then turn right onto MID VOR RDL 030 to MID VOR.
	Via Detling	Straight ahead until LON DME 2 then turn right onto track 125° MAG. At LON DME 4 turn left to establish on DET VOR RDL 286 by DET DME 34 to DET DME 20.		Via Detling	Straight ahead until LON DME 1.5 then turn right onto track 125° MAG. At LON DME 4 turn left to establish on DET VOR RDL 286 by DET DME 34 to DET DME 20.
	Via BUZAD	Straight ahead until LON DME 2 then turn left onto track 054°MAG to intercept LON VOR RDL 074. At LON DME 10 turn left onto BIG VOR RDL 333 to BUZAD.		Via BUZAD	Straight ahead until LON DME 1.5 then turn left onto track 054°MAG to intercept LON VOR RDL 074. At LON DME 10 turn left onto BIG VOR RDL 333 to BUZAD.
	Via Brookmans Park	Straight ahead until LON DME 2 then turn left onto track 054°MAG to intercept LON RDL 074. At LON DME 10 turn left onto BPK VOR RDL 200 to BPK VOR.		Via Brookmans Park	Straight ahead until LON DME 1.5 then turn left onto track 054°MAG to intercept LON VOR RDL 074. At LON DME 10 turn left onto BPK VOR RDL 200 to BPK VOR.

(7)   Where the aircraft is approaching the aerodrome to land it shall commensurate with its ATC clearance minimise noise disturbance by the use of continuous descent and low power, low drag operating procedures (referred to in AD 2-EGLL-1-21 of the UK AIP). Where the use of these procedures is not practicable, the aircraft shall maintain as high an altitude as possible.

(8)   Subject to sub-paragraph (9) below, where the aircraft is using the ILS in IMC or VMC it shall not descend on the glidepath below an altitude of 2500 ft (Heathrow QNH) before being established on the localizer, nor thereafter fly below the glidepath. An aircraft approaching without assistance from the ILS shall follow a descent path which will not result in its being at any time lower than the approach path that would be followed by an aircraft using the ILS glidepath, and shall follow a track to intercept the extended runway centre-line at or above 2500 ft.

(9)   Nothing in sub-paragraph (8) above shall apply to any propeller driven aircraft whose MTWA does not exceed 5,700 kg.

(10)   Without prejudice to the provisions of sub-paragraphs (1) - (9) above, the aircraft shall at all times be operated in a manner which is calculated to cause the least disturbance practicable in areas surrounding the aerodrome.

(11)   The requirements set out in sub-paragraphs (1) - (10) above may at any time be departed from to the extent necessary for avoiding immediate danger or for complying with the instructions of an Air Traffic Control Unit.

4   In this notice, except where the context otherwise requires:

'local time' means, during any period of summer time, the time fixed by or under the Summer Time Act 1972 (e), and outside that period, Universal Co-ordinated Time;

'dBA' means a decibel unit of sound level measured on the A-weighted scale, which incorporates a frequency dependent weighting approximating the characteristics of human hearing;

'Lmax' means the highest instantaneous sound level recorded (with the noise monitoring terminal set at the slow meter setting);

other abbreviations used are defined in GEN 2-2 of the United Kingdom Aeronautical Information Publication (Air Pilot).

E J Duthie
Divisional Manager
Aviation Environmental Division
15 September 1999   **Department of the Environment, Transport and the Regions**

(a) S.I.1981/651.
(b) 1982 c.16.
(c) S.I.1978/1303.
(d) The Heathrow Airport – London (Interim Noise Abatement Requirements) Notice 1999 signed by E J Duthie on 8 January 1999.
(e) 1972 c.6.

EGLL AD 2.21 – NOISE ABATEMENT PROCEDURES

Notes
(These notes are not part of the notice)

(1) The Noise Preferential Routeing Procedures specified in the above notice are compatible with normal ATC requirements. The use of the routeings specified above is supplementary to noise abatement take-off techniques as used by piston-engined, turbo-prop, turbo-jet and turbo-fan aircraft.

(2) The attention of operators is drawn to the provisions of Section 78(2) of the Civil Aviation Act 1982, under which if it appears to the Secretary of State that any of the requirements in this notice have not been complied with as respects any aircraft, he may direct the manager of the aerodrome to withhold facilities for using the aerodrome from the operator of the aircraft.

(3) Noise from ground running of aircraft engines is controlled in accordance with instructions issued by Heathrow Airport Limited.

(4) In the interests of noise abatement, certain restrictions are imposed on the operation of training flights at this aerodrome. Operators concerned are advised to obtain details from Heathrow Airport Limited.

(5) To minimise disturbance in areas adjacent to the aerodrome, commanders of aircraft are requested to avoid the use of reverse thrust after landing, consistent with the safe operation of the aircraft, between 2330 hours and 0600 hours (local time).

(6) Full details concerning the maximum number of occasions and the types of aircraft which are permitted to take off or land at night during specified periods at this aerodrome are promulgated by Supplement.

EGLL AD 2.22 – FLIGHT PROCEDURES

1　　Radio Communication Failure Procedures

a　In the event of complete radio communication failure in an aircraft, the pilot is to adopt the appropriate procedures described at ENR 1.6, with the exceptions described below.

b　**Aircraft inbound to London Heathrow**

When complete communication failure occurs in the aircraft before ETA, or, before EAT, when this has been received and acknowledged, the aircraft will:

　i　Fly to the appropriate holding point, Ockham (Epsom NDB when applicable), Biggin (WEALD when applicable), Lambourne (TAWNY when applicable) or Bovingdon (BOVVA when applicable);

　ii　hold at the last assigned level until the last acknowledged ETA plus 10 minutes, or EAT when this has been given;

　iii　then commence descent for landing in accordance with paragraph 12, b (when applicable) and effect a landing within 30 minutes (or later if able to approach and land visually).

c　If complete radio communication failure occurs after an aircraft has reported to ATC on reaching the holding point, the aircraft will:

　i　Maintain the last assigned holding level at Ockham (Epsom NDB when applicable), Biggin (WEALD when applicable), Lambourne (TAWNY when applicable) or Bovingdon (BOVVA when applicable) until:

　　1　ATA over the holding point plus 10 minutes or 10 minutes after the last acknowledged communication with ATC, whichever is the later; or
　　2　EAT when this has been received and acknowledged;

　ii　then commence descent for landing in accordance with paragraph 12, b (when applicable) and effect a landing within 30 minutes (or later if able to approach and land visually).

d　When complete radio communication failure occurs during final approach under radar control the procedures to be followed are detailed at AD 2-EGLL-5-1.

e　When complete radio communication failure occurs in the aircraft following a missed approach the aircraft will:

　i　Fly to the appropriate missed approach holding point;

　ii　complete at least one holding pattern at the level associated with the missed approach holding point;

　iii　then commence descent for landing in accordance with paragraph 13, (when applicable) and effect a landing within 30 minutes (or later if able to approach and land visually).

f　The routes and levels to be used when leaving the London CTR or Holding Area in accordance with the procedures given at ENR 1.6, are shown in the table below; the route to be followed is dependent on the position of the aircraft at the time the decision to leave the Zone is made.

Position at time of decision	Route
Ockham VOR (Epsom when applicable)	Track 270°T at last assigned level.
Bovingdon VOR (BOVVA when applicable)	Track 030°T at last assigned level.
Lambourne VOR (TAWNY when applicable)	Track 020°T at last assigned level.
Biggin VOR (WEALD when applicable)	Track 089°T at last assigned level.
Epsom NDB (after Missed Approach Procedure)	Track 270°T at 3000 ft.
Chiltern NDB (after Missed Approach Procedure)	Track 270°T at 4000 ft.

g　**Aircraft outbound from London Heathrow**

　i　Aircraft departing under radar control from London Heathrow may be instructed by the radar controller, via Aerodrome Control, to maintain specific headings immediately after take-off.

　ii　If, after having been instructed to maintain a specific heading immediately after take-off, a pilot experiences radio failure, he shall climb on the assigned heading to the first specific altitude detailed in the clearance, maintain this heading and altitude for two minutes, and then proceed in accordance with the published radio failure procedures.

2   **Inbound other than on Airways**

a   Aircraft wishing to enter the London CTR direct from the London FIR will be required to use the procedure for flights joining Airways. Aircraft departing from aerodromes outside the geographical limits of the London TMA will normally be required to route via one of the following reporting points:

Bovingdon VOR   Lambourne VOR   Ockham VOR

b   Pilots of aircraft departing from an aerodrome less than 10 minutes flying time from the CTR boundary are to contact the Heathrow ATC Watch Supervisor on 0181-745 3328 before take-off.

c   It is in the interests of all pilots to contact the Heathrow Supervisor before take-off to establish the extent of any delays and to be allocated an entry time should it be required.

Note: Aircraft proceeding between Gatwick and Heathrow must adhere to the appropriate Standard Instrument Departure.

3   **Standard Terminal Arrival Routes (STAR)**

To London Heathrow see AD 2-EGLL-7-1 to AD 2-EGLL-7-5.

4   **Holding**

a   Main Stacks

Unless otherwise instructed by ATC, inbound aircraft using the Airways System will, after initial Airways Routeing, follow the appropriate arrival route to the holding fixes at Biggin, Bovingdon, Lambourne or Ockham, illustrated at AD 2-EGLL-7-1 to AD 2-EGLL-7-5. The holding procedures at WEALD, BOVVA, TAWNY or Epsom NDB will be used when the respective VORs are out of service.

Note: The Epsom hold on an axis of 274° (3000 ft) is only to be used, when instructed by ATC, following a missed approach.

b   From the holding points, aircraft will normally be directed by the Radar Controller to a position from which a straight-in approach can be made. Exceptionally, when traffic permits or circumstances necessitate, pilots may be permitted or instructed by ATC to carry out the appropriate approach procedure without radar control in paragraph 13.

5   **Approach Procedures with Radar Control**

When inbound traffic is being sequenced by Surveillance Radar, that part of the approach between the holding fix and the Final Approach Track (FAT) will be flown under directions from the Radar Controller. Once the aircraft is under Approach Radar Control, changes of heading or flight level/altitude will be made only on instructions from the Radar Controller except in the case of radio communication failure in the aircraft or at the radar unit.

6   **Detailed Procedures**

a   Headings and flight levels at which to leave the holding facility will be passed by ATC. Radar vectors will be given, and descent clearance will include an estimate of track distance to touchdown. Further distance information will be given between initial descent clearance and intercept heading to the ILS. On receipt of descent clearance the pilot will descend at the rate he judges will be best suited to the achievement of continuous descent, the object being to join the glidepath at the appropriate height for the distance without recourse to level flight.

b   Procedures require that aircraft should fly at 210 kt during the approach phase. ATC will request speed reductions to within the band 160 kt to 180 kt on, or shortly before, the closing heading to the ILS, and 160 kt when established on the ILS to 4 DME; all speeds to be flown as accurately as possible. Aircraft unable to conform to these speeds should inform ATC and state what speeds will be used. In the interests of accurate spacing, pilots are requested to comply with speed adjustments as promptly as is feasible within their own operational constraints, advising ATC if circumstances necessitate a change of speed for aircraft performance reasons.

c   The system is designed to maximize arrival capacity at London Heathrow and to minimize noise disturbance in the areas overflown during the approach and aircraft commanders are requested to conform to low-power, low-drag procedures.

d   The Final Approach may be carried out by means of ILS or other available instrument approach system at the discretion of the pilot.

e   The spacing provided between aircraft will be designed to achieve maximum runway utilization within the parameters of safe separation minima (including vortex effect) and runway occupancy. It is important to the validity of the separation provided, and to the achievement of optimum runway capacity, that runway occupancy time is kept to a minimum consistent with the prevailing conditions.

f   Missed Approach Procedures are contained on the Instrument Approach charts, but see also paragraph 11.

EGLL AD 2.22 – FLIGHT PROCEDURES

7 Pressure Settings

When below the Transition Altitude, pilots are to fly on the aerodrome QNH until established on final approach, at which point QFE or any other desired setting may be used.

8 Radar Failure

In the event of radar failure, fresh instructions will be issued to each aircraft under radar control and the procedures in paragraph 4.5 will be brought into use.

9 Radio Communication Failure at the Radar Unit

If radio communication completely fails at the radar unit when under Radar Control, pilots are to contact Aerodrome Control for fresh instructions.

10 Approaches without Radar

a When traffic is not being sequenced by Surveillance Radar, aircraft will be cleared from the holding facility to carry out the approach procedure appropriate to the runway-in-use, as outlined below.

b **Approach Procedures** without Radar

Holding Point	Runway	Leaving Altitude	Procedure	Final Approach Aid	Missed Approach Procedure
Ockham VOR	09L 09R	*7000 ft	Leave on OCK VOR RDL 306° descending to 2500 ft. At 14 DME OCK turn right to intercept ILS localizer course. Continue approach as detailed on the Instrument Approach Chart.	ILS	As detailed on the Instrument Approach Charts
	27L 27R	*7000 ft	Leave on OCK VOR RDL 067° descending to 2500 ft. At 13 DME OCK turn left to intercept ILS localizer course. Continue approach as detailed at on the Instrument Approach Chart.	ILS	
Epsom NDB (for use when Ockham VOR is out of service)	09L 09R	*7000 ft	Leave EPM on track 297° MAG maintaining *7000 ft. On passing LON VOR RDL 184° commence descent to 2500 ft. At LON VOR/DME 254°/12 nm turn right to intercept ILS localizer course. Continue approach as detailed at on the Instrument Approach Chart.	ILS	
	27L 27R	*7000 ft	Leave EPM on track 068° MAG descending to 2500 ft. At LON VOR/DME 114°/14 nm turn left to intercept ILS localizer course. Continue approach as detailed at on the Instrument Approach Chart.	ILS	
Bovingdon VOR	09L 09R	*7000 ft	Leave on BNN VOR RDL 222° maintaining *7000 ft. At 4 DME BNN commence descent to 2500 ft. At 15 DME BNN turn left to intercept ILS localizer course. Continue approach as detailed on the Instrument Approach Chart.	ILS	
	27L 27R	*7000 ft	Leave on BNN VOR RDL 137° maintaining *7000 ft. At 7 DME BNN commence descent to 2500 ft. At 19 DME BNN turn right to intercept ILS localizer course. Continue approach as detailed on the Instrument Approach Chart.	ILS	
Biggin VOR	09L 09R	*7000 ft	Leave on BIG VOR RDL 287° maintaining *7000 ft. On passing LON VOR RDL 176° commence descent to 2500 ft. At 12 DME LON turn right to intercept ILS localizer course. Continue approach as detailed on the Instrument Approach Chart.	ILS	
	27L 27R	*7000 ft	Leave on BIG VOR RDL 335° descending to 2500 ft. At LON VOR RDL 105° turn left to intercept ILS localizer course. Continue approach as detailed on the Instrument Approach Chart.	ILS	
Lambourne VOR	09L 09R	*7000 ft	Leave on LAM VOR RDL 265° maintaining *7000 ft. On passing LON VOR RDL 007° commence descent to 2500 ft. At 11 DME LON turn left to intercept ILS localizer course. Continue approach as detailed on the Instrument Approach Chart.	ILS	
	27L 27R	*7000 ft	Leave on LAM VOR RDL 234° descending to 2500 ft. On passing LON VOR RDL 096° turn right to intercept ILS localizer course. Continue approach as detailed on the Instrument Approach Chart.	ILS	

Note 1: These procedures are designed for manoeuvring speeds of up to 220 kt IAS and assume aircraft can maintain a descent gradient of approximately 300 ft per nm.

Note 2: Change to altimeter setting for landing when cleared to 2000 ft or below.
*As these altitudes are above the Transition Altitude aircraft will normally be instructed by ATC to fly at the appropriate equivalent Flight Level.

EGLL AD 2.22 – FLIGHT PROCEDURES

11   Modified procedures for use in the event of Missed Approach and Radio Failure - London Heathrow

Holding Point	Runway	Leaving Altitude	Procedure	Final Approach Aid	Missed Approach Procedure
Epsom NDB	09R	3000 ft	Leave EPM on track 297° MAG maintaining 3000 ft. On passing LON VOR RDL 230° descend to 2500 ft. At LON VOR/DME 254°/12 nm turn right to intercept the ILS localizer course. Continue approach as detailed on the Instrument Approach Chart.	ILS	As detailed on the Instrument Approach Chart
	23	3000 ft	Provided that Runway 27L has been notified as available and broadcast as so on the ATIS, carry out procedure to Runway 27L as detailed below. If unable, carry out at least one hold at EPM, then proceed to Alternate aerodrome, leaving London CTR/EPM Holding Area on track 270°T at 3000 ft.	SRA	
	27L	3000 ft	Leave EPM on track 068° MAG maintaining 3000 ft. On passing LON VOR RDL 135° descend to 2500 ft. At LON VOR/DME 114°/14 nm turn left to intercept the ILS localizer course. Continue approach as detailed on the Instrument Approach Chart.	ILS	
Chiltern NDB	09L	3000 ft	Leave CHT on track 247° MAG maintaining 3000 ft. On passing LON VOR RDL 320° descend to 2500 ft. At LON VOR/DME 286°/12.5 nm turn left to intercept the ILS localizer course. Continue approach as detailed on the Instrument Approach Chart.	ILS	
	27R	3000 ft	Leave CHT on track 121° MAG maintaining 3000 ft. On passing LON VOR RDL 050° descend to 2500 ft. At LON VOR/DME 086°/11.5 nm turn right to intercept the ILS localizer course. Continue approach as detailed on the Instrument Approach Chart.	ILS	

Note 1: These procedures are designed for manoeuvring speeds of up to 220 kt IAS and assume aircraft can maintain a descent gradient of approximately 300 ft per nm.

Note 2: These procedures are only to be used in the event of complete radio failure following a missed approach.

Note 3: These procedures are only to be used in the event of complete radio failure after being directed to the CHT hold following a missed approach.

12   **Speed Limitation**

a   In order to improve the departure flow and at the same time maintain separation between aircraft, particularly those following the same routeing over the first 20 to 30 miles of flight, a speed limit of 250 kt. IAS applies to all aircraft departing from London Heathrow.

b   This speed limit applies to all departures from London Heathrow whilst flying below FL 100 unless previously removed by ATC. ATC will endeavour to remove the speed limit as soon as possible and will use the phrase 'No ATC Speed Restriction'. This phrase must not be interpreted as relieving the pilot of his responsibility for the observance of any noise abatement procedures which may include a speed/power limitation.

c   In certain weather conditions and perhaps for other reasons of safety pilots may not be able to comply with the speed limit of 250 kt IAS. When such circumstances are anticipated, a pilot should inform ATC when requesting start-up clearance stating the minimum speed acceptable. In this event, pilots will be informed before take-off of any higher speed limitation. Similarly should such circumstances arise during flight the pilot should immediately advise ATC again stating the minimum speed acceptable.

d   Should weather conditions or other factors necessitate any modification of this procedure, this will be promulgated on the Departure ATIS.

EGLL AD 2.22 – FLIGHT PROCEDURES

13　**Special VFR Clearance in the London Control Zone**

　　a　Special VFR clearances for flights within the London CTR may be requested and will be given whenever traffic conditions permit. These flights are subject to the general conditions laid down for Special VFR flights and will normally be given only to aircraft which carry RTF including the appropriate frequencies listed in EGLL AD 2.18.

　　　　Note: Pilots holding a Private Pilots Licence (Aeroplanes) are reminded of the visibility requirements for Special VFR flights laid down in Schedule 8 of the Air Navigation (No 2) Order 1995 and the related notification at ENR 1-4-1 paragraph 2.1.1.1 (vi) (a).

　　b　The use of Special VFR clearances is intended to be confined to the following types of flight:

　　　　i　Light aircraft which cannot comply with full IFR requirements and which wish to proceed to or from an aerodrome within the Zone or to transit the Zone at the lower levels; except that aircraft using the local flying areas and the access lanes notified for Brooklands, Denham, Fairoaks and White Waltham and complying with the published procedures will be considered as complying with a Special VFR clearance;

　　　　ii　aircraft carrying out special flights, such as photographic survey flights, which may require penetration of the Zone in VMC.

　　c　Clearance to make Special VFR flights below 1500 ft (London Heathrow QNH) other than for helicopters flying on the approved helicopter routes, will not be given in the sector of the London CTR enclosed by the bearings 020°T and 140°T from London Heathrow except where the flight has been subject to a Non-Standard Flight approval as specified at ENR 1-1-13, paragraph 9.

　　d　Requests for clearances for Special VFR flights in the Zone should be made to the Supervisor London Heathrow Special VFR Unit (call sign 'Heathrow Radar') - telephone 0181-745 3328.

　　e　Pilots who wish to leave Heathrow on a Special VFR clearance should pass brief details of their flight to the Flight Briefing Unit, either in person or by telephone 0181-745 3163, and not to ATC by RTF.

　　f　Non-scheduled flights by single-engined and light twin-engined aircraft will be cleared to London Heathrow only on a Special VFR clearance, not above 1000 ft (London Heathrow QNH) and subject to the following:

　　　　The weather conditions must be such as to permit the pilot to navigate by visual means, with a visibility of at least 10 km and a cloud base of not less than 1200 ft. (If the weather observations at London Heathrow are below either of these minima, clearance to enter the Zone will not be granted).

　　g　Aircraft may be given radar service whilst within the Zone if, due to the traffic situation, ATC considers it advisable. It will remain the responsibility of the pilot to remain at all times in flight conditions which will enable him to determine his flight path and to keep clear of obstacles, and to ensure that he is able to comply with the relevant low flying restrictions of Rule 5 of the Rules of the Air Regulations 1996, with particular regard to Rule 5 (1) (a) (i). Pilots must inform the Radar Controller if compliance with the above entails a change of heading or height.

　　h　Special VFR flights may be subject to delay when parts of the route are outside radar cover or when they cannot be fitted readily into gaps in the main traffic flow. Pilots should therefore always ensure that they have adequate fuel reserves and are able to divert to another aerodrome if necessary.

14　**Local Flying arrangements and Special Access Lanes for Brooklands, Denham, Fairoaks, Northolt, White Waltham and Wycombe Air Park/Booker Aerodromes.**

　　a　Flights within the local flying areas of aerodromes within, or adjacent to, the London CTR, may be made subject to certain conditions. Details of those for Denham, Fairoaks, Northolt, White Waltham and Wycombe Air Park/Booker appear in the relevant AD Sections. An additional local flying area is established for the unlicensed aerodrome at Brooklands and is detailed below.

　　b　Brooklands

　　　　i　Within a local flying area of 1 nm radius, centred on the mid-point of the runway (512103N 0002812W), but excluding that part to the east of the B374 road and a line bearing 180°T from the A245/B374 road junction and excluding the area south of the southern boundary of the London CTR, flights without compliance with IFR requirements may take place subject to the following conditions:

　　　　　1　Aircraft to remain below cloud and in sight of the ground;

　　　　　2　Maximum altitude: 800 ft QNH when London Heathrow runway 23 is in use, otherwise 1000 ft QNH;

　　　　　3　Minimum flight visibility: 3 km;

　　　　　4　Prior permission **must** be obtained from Brooklands Museum Trust Ltd.

　　Note 1: Pilots of aircraft flying in the local flying area are responsible for providing their own separation from other aircraft operating in the relevant airspace.

　　Note 2: Brooklands will advise ATC London Heathrow prior to commencement of flying and ascertain the runway-in-use.

EGLL AD 2.22 – FLIGHT PROCEDURES

15   **Non-IFR Helicopter Flights in the London CTR**

a   General Arrangements

  i   Non-IFR helicopter flying in the London CTR is normally restricted to flight at or below specified altitudes along defined routes. These routes have been selected to provide maximum safety by avoiding built up areas as much as possible. Details of the major landmarks on these routes, the altitudes and reporting points are listed at paragraph 27 and are illustrated at AD 2-EGLL-3-2. The precise routes are overprinted on the 1: 50 000 Map entitled - Helicopter Routes in the London Control Zone. This map is available from accredited chart agents. The illustration also shows the Specified Area of central London over which flight by single-engined helicopters is virtually prohibited except along the River Thames because of the requirement to be able to land clear of the area in the event of engine failure.

  ii   All non-IFR helicopter flying in the London CTR is subject to Special VFR clearance. In addition, permission in writing from the Civil Aviation Authority is required for flight within the Specified Area by single-engined helicopters.

  iii   The following routes are not available to single-engined helicopters at night: H7, H9 (Hayes to Gutteridge) and H10 (Gutteridge to Kew Bridge).

b   Procedures for flight along Helicopter Routes

  i   Non-IFR flights in the London Control Zone are not to be operated unless helicopters can remain in a flight visibility of at least 1 km, except when crossing over, taking-off from or landing at London Heathrow, when the reported visibility at Heathrow must be at least 2 km. Non-IFR helicopters must remain clear of cloud and in sight of the surface.

  ii   Altimeter setting will be London Heathrow QNH.

  iii   Maximum altitudes are shown in column 3 at paragraph 17. Special VFR clearances will be issued in the form 'not above...... feet' except for flights along route H10 between Perivale Golf Course and Chiswick Bridge. Between these points, pilots are required to fly at the altitudes (known as Standard Operating Altitudes) specified in column 3 at paragraph 17. These altitudes must be maintained accurately in order to ensure adequate terrain clearance and separation from fixed-wing traffic approaching Runway 27R at London Heathrow.

  iv   Pilots should fly the precise routes as depicted on the 1: 50 000 Map entitled Helicopter Routes in the London Control Zone. 'Corner cutting' is to be avoided. In order to obtain sufficient lateral separation from opposite direction traffic, pilots may temporarily deviate to the right of the route.

  v   When flying along the River Thames within the Specified Area, pilots should normally fly over that part of the river bed lying between high water marks, but not so near the banks as to become a nuisance on account of noise. When deviating from the river, in accordance with paragraph iv above, single-engined helicopters must at all times be able to return to the river in the event of engine failure, in order to alight clear of the Specified Area.

c   Noise

  i   On all routes, in order to minimize noise nuisance, pilots should maintain the maximum altitude compatible with their ATC clearance and with the prevailing cloud conditions. They are reminded that there is no relaxation from the need to comply with Rule 5(1)(e) of the Rules of the Air Regulations 1996, which precludes flight closer than 500 ft to any person, vessel, vehicle or structure.

  ii   Pilots are requested wherever possible to avoid overflying hospitals and schools.

  iii   Within the London CTR, civil helicopters will not be permitted to fly in formation. Military pilots should not normally fly in formations of more than two helicopters.

d   Air Traffic Control Clearance

  i   During the hours of operation of Thames Radar/Heathrow Radar, pilots must obtain a Special VFR clearance. Thames Radar provides the control service for all Special VFR flights within the London CTR east of London Westland Heliport; Heathrow Radar provides a similar service to the west of the Heliport. Pilots are requested to contact ATC three minutes before reaching the Zone Boundary, giving details of call sign, route, ETA at Zone entry point and destination.

  ii   Outside the hours of operation of Thames Radar/Heathrow Radar clearance for Special VFR flights within the London CTR must be obtained from Heathrow Director (LATCC(TC)).

  iii   When the destination is London Heathrow, the Special VFR clearance, for entry into the Zone, will include routeing and other instructions.

e   Holding

    Non-IFR helicopters, particularly those using London Heathrow or the routes close to it, may be required to hold at any of the locations on the route, shown in column 1 at paragraph 17 and on the illustration at AD 2-EGLL-3-2.

f   Communications

  i   Helicopters using Westland Heliport or the route east of it must be able to communicate with the Heliport (Battersea Tower).

  ii   Helicopters flying along the routes west of Westland Heliport must be able to communicate with Heathrow Radar, and in the case of H9 and H10, also with Approach Control at Northolt (126.450 MHz). Compulsory and On-Request reporting points are shown in column 1 at paragraph 17.

  iii   Helicopters using London Heathrow must also be able to communicate with Heathrow Tower.

EGLL AD 2.22 – FLIGHT PROCEDURES

g   Loss of Communications Procedures

   i   In the event of communications failure in a helicopter operating in accordance with these procedures, the pilot is to adopt the procedure detailed at ENR 1.2 except as described below.

   ii   If a Special VFR clearance has been received to transit the CTR along a Helicopter Route continue the flight in accordance with the clearance.

   iii   Where an intermediate clearance limit has been given (or clearance issued for only a part of the requested transit), proceed to the specified clearance limit and hold for 3 minutes. Then proceed via the requested Helicopter Route at the published maximum altitude for the Route.

   iv   For helicopters overflying or landing at London Heathrow Airport. (See paragraph l).

   v   If no onward clearance has been received before reaching, or when holding at, Sipson or Bedfont, **reverse track and leave the CTR** via H10/Cookham if approaching Sipson or H9 if approaching Bedfont. **Do not attempt to cross London Heathrow Airport.**

      1   Between Sipson and Bedfont:

         aa   if the landing runway has been crossed, cross the departure runway downwind of the threshold, exercising extreme caution with regard to possible landing traffic; and **leave the CTR via H10/Cookham or H9** as appropriate;

         bb   if the departure runway has been crossed, with instructions given to hold at the dual taxiways or fuel farm, **reverse track** to cross the departure runway downwind of the threshold, exercising extreme caution with regard to the possibility of landing traffic; and **leave the CTR via H10/Cookham or H9** as appropriate.

      2   If landing at London Heathrow Airport, and having crossed the runways if necessary as detailed above, proceed to land at the helicopter aiming point by day or at 27L threshold by night (then vacate 27L and await vehicle escort).

h   Separation between Non-IFR helicopters

   i   Separation may be decided between helicopters on the Helicopter Routes, on the basis that pilots of helicopters will be asked by ATC to maintain visual separation from other helicopter traffic, provided that:

      1   the visibility at London Heathrow is 6 km or more and the helicopters can operate clear of cloud and in sight of the ground or water and remain in a flight visibility of at least 6 km;

      2   there is agreement between the helicopter pilots concerned;

      3   the current route structure, the altitudes applicable and communication procedures are adhered to;

      4   appropriate traffic information is passed to the helicopter pilots. (Normally for this purpose it will only be necessary for ATC to pass general traffic information eg..... 'Two helicopters westbound along H10 at 1000 ft in the vicinity of Perivale - acknowledge.').

   ii   If a pilot refuses or considers that the conditions are such that he is unable to maintain visual separation, he will be provided with the separations currently in force.

i   Procedures for Helicopter Operations at London Heathrow Airport

   i   General Requirements

      1   In order to facilitate the expeditious transit of London Heathrow by helicopters operating under SVFR clearance, procedures have been devised based on ATC's ability in certain circumstances, to reduce standard separations by visual reference to traffic operating in the immediate vicinity of the airport.

      2   Procedures based on visual separation will only be applied when:

         aa   The London Heathrow reported weather conditions are equal to or better than 6 km visibility and 1000 ft lowest reported cloud base;

         bb   it is during daylight hours;

      3   fixed-wing operations are aligned on westerly Runways 27L/27R or on easterly runways when 09R is in use for departure and/or 09L is in use for landing.

      Note: At this stage, visual procedures have not been devised for the few occasions when Runway 23 is in use, when Runway 09R is in use for landing, or when Runway 09L is in use for departures. In all other circumstances, standard separation will be applied.

      4   Helicopter operations are to commence at Feltham, Bedfont or Sipson as appropriate. Helicopters holding at Feltham or Sipson are separated for both ATC and wake vortex purposes from fixed-wing aircraft landing on, departing from or carrying out missed approaches to, Runways 27L/27R, 09L/09R, 23.

      5   Helicopters may be held at Bedfont only when the conditions detailed in paragraph 2 exist and are separated for wake vortex purposes and visual separation only. Pilots must remain in visual contact with aircraft on approach to Runway 27L and are warned that the missed approach procedure to Runway 27L requires a left turn at 1000 ft aal.

      6   Helicopters are to transit the airport at not less than 800 ft (Heathrow QNH).

      7   SVFR Helicopter operations are not permitted at London Heathrow when the reported MET visibility is less than 2 km.

      8   Loss of communications procedures for helicopters overflying or landing at London Heathrow Airport are detailed at paragraph 20, d.

EGLL AD 2.22 – FLIGHT PROCEDURES

j   Crossing Runway 27L/27R

   i   Unless otherwise instructed, helicopter pilots are to ensure that the departure runway, as advised by ATC, is crossed downwind of the threshold.

   ii   Clearance to commence crossing the landing runway will not be issued until such time as a suitable gap in the landing stream exists. ATC will then pass traffic information pertaining to the fixed-wing landing aircraft after which the crossing is to be effected, and having issued the clearance to cross behind that aircraft, will expect the helicopter to execute the manoeuvre as expeditiously as possible.

   iii   Where the departure runway is crossed first, holding prior to crossing the landing runway will be permitted between the two main runways. This operation is to be contained between the departure runway threshold and a line drawn east-west through the Dual Taxiways (see diagram at AD 2-EGLL-4-1); under no circumstances is the helicopter to transgress this line until such time as a clearance to cross the landing stream has been received.

   Note: It is essential that helicopters execute the holding manoeuvre in an orbit and not in the hover.

k   Crossing Runway 09L/09R

   i   Crossing of these runways will only be permitted when Runway 09L is in use for landing and/or Runway 09R is in use for departure.

   ii   In order to provide both ATC and wake vortex separation from departing fixed-wing traffic, the crossing route is as follows:

   Feltham - Bedfont - Duke of Northumberland River - West of Runway 09R threshold - Fuel Farm - (direct, or as instructed by ATC) - Sipson: and vice versa (AD 2-EGLL-4-1).

   However, whilst the helicopter is in transit between Bedfont and Runway 09R threshold, ATC will pass traffic information, in respect of this movement to fixed-wing departures from 09R.

   iii   Clearance to commence crossing the landing runway will not be issued until such times as a suitable gap in the landing stream exists. ATC will then pass traffic information pertaining to the fixed-wing landing aircraft after which the crossing is to be effected, and having issued the clearance to cross behind that aircraft, will expect the helicopter to execute the manoeuvre as expeditiously as possible.

   iv   Where Runway 09R is crossed first, holding prior to crossing Runway 09L will be permitted between Runway 09R threshold and a line drawn east-west through the southern edge of the Fuel Farm (AD 2-EGLL-4-1). Under no circumstances is the helicopter to transgress this line until such time as clearance to cross the landing stream has been received.

   Note: It is essential that helicopters execute the holding manoeuvre in an orbit and not in the hover.

l   Landing and Departure Procedures

   i   During the hours of daylight a helicopter landing at/departing from London Heathrow will normally be effected from/to either holding area at Sipson or that at Bedfont or Feltham if appropriate from/to the helicopter aiming point located as follows:

   Eastern end of Block 111 (marked with a conventional 'H');

   Caution must be exercised when using this aiming point which is on a live taxiway.

   ii   During the hours of darkness a helicopter landing at/departing from London Heathrow will normally be effected from/to either the holding area at Sipson or that at Bedfont to use Runway 09R/27L as directed by ATC.

   iii   In circumstances as detailed in paragraph 23 visual separation will be applied whilst the helicopter is in transit between the holding area and the helicopter aiming point. In all other circumstances, standard separation will be applied.

   iv   Vortex wake separation will be applied in respect of landing/departing helicopters and fixed-wing operations on the adjacent runway.

16   **Aerodrome Operating Minima - Non-Public Transport Flights:** Refer to AD 1.1.2 before application.

Approach Lighting Category	Runway	Approach Aid	OCH (ft) Acft CAT A See IAC for other CATs	Minima	
				DH/MDH (ft) **Caution** See AD 1.1.2	RVR (m)
1	2	3	4	5	6
Runway 09L Full	09L	ILS/DME I AA	148	250	600
	09L	LLZ/DME I AA	400	400	900
	09L	SRA (2 nm termination range)	650	650	1200
Runway 09R Ful	09R	ILS/DME I BB	154	250	600
	09R	LLZ/DME I BB	400	400	900
	09R	SRA (2 nm termination range)	650	650	1200
Runway 23 Intermediate	23	SRA (2 nm termination range)	650	650	1500
Runway 27L Full	27L	ILS/DME I LL	151	250	600
	27L	LLZ/DME I LL	410	410	900
	27L	SRA (2 nm termination range)	650	650	1200
Runway 27R Full	27R	ILS/DME I RR	145	250	600
	27R	LLZ/DME I RR	400	400	900
	27R	SRA (2 nm termination range)	650	650	1200

EGLL AD 2.22 – FLIGHT PROCEDURES

17   Helicopter Routes in the London Control Zone and London/City Control Zone

   a   Abbreviations:

      H — Holding Point   ▲ — Compulsory Reporting Point   Δ — On Request Reporting Point

   Map references are to the 1: 50 000 Ordnance Survey Map of Great Britain.

   b   The precise routes which must be adhered to are portrayed on Edition 7 of the 1: 50 000 Map entitled Helicopter Routes in the London Control Zone — see paragraph 17,a,i. An indication of the routes network is shown on the illustration at AD 2-EGLL-3-2.

   c   Pilots are required to be at the lower altitudes on arrival at the point at which the lower altitude applies.

HELICOPTER ROUTES

Route Designator Significant Points and National Grid & Lat/Long	Description of Reference	Maximum Altitude (London Heathrow QNH)	Holding Point	Remarks
1	2	3	4	5
H2				
Δ Iver TQ 035 826 513157N 0003031W	Delaford Park		H	
Δ West Drayton TQ 052 784 512939N 0002909W	M4 Motorway Crossing of River Colne (1.25 nm west of Airport Spur)	1000 ft		
▲ Airport Spur TQ 075 786 512944N 0002715W	Junction of M4 Motorway and Motorway Spur to London Heathrow		H	Note.

Note: Unless otherwise cleared by ATC, pilots are not to fly south of the M4 between West Drayton and Airport Spur.

H3				
▲ Bagshot Mast SU 908 619 512057N 0004157W	Intersection of London Control Zone/M3 Motorway		H	Note 1.
		2000 ft		
	M3 Motorway Junction 3			
		1000 ft		
Δ Thorpe TQ 018 679 512402N 0003216W	M3 Motorway south of Thorpe Green (M25 intersection)		H	
		800 ft		
▲ Sunbury Lock TQ 112 683 512409N 0002416W	Midway between Sunbury Lock and the middle of Knight reservoir		H	Note 2.
Δ Teddington TQ 170 714 512547N 0001906W	Weir on River Thames		H	
	North edge of Richmond Park	1000 ft		
▲ Barnes TQ 234 765 512827N 0001325W	River Thames at Barn Elms Park			
▲ Westland Heliport TQ 266 762 512812N 0001046W	Westland Heliport		H	

Note 1: When London Heathrow Runway 23 is in use, the H3 route from Bagshot Mast to its intersection with H9 route (Sunbury Lock Reporting Point), will not be available to helicopters.

Note 2: Helicopter Route H3 will normally be closed whenever easterly operations are taking place at London Heathrow Airport. Helicopter pilots are recommended to obtain Heathrow runway information on the ATIS frequency 123.900 MHz before contacting Heathrow Radar on 119.900 MHz, or Westland Heliport on 122.900 MHz. Notwithstanding the above, special arrangements for the use of Route H3 in connection with certain events may be made as required. Such arrangements will be promulgated by NOTAM.

EGLL AD 2.22 – FLIGHT PROCEDURES

HELICOPTER ROUTES

Route Designator Significant Points and National Grid & Lat/Long	Description of Reference	Maximum Altitude (London Heathrow QNH)	Holding Point	Remarks
1	2	3	4	5
H4				
▲ Isle-of-Dogs TQ 381 781 512902N 0000042W	Specified Area Boundary crossing River Thames			Note 2.
△ London Bridge TQ 330 805 513027N 0000504W	London Bridge (road bridge)	2000 ft		Note 3. Note 4.
▲ Vauxhall Bridge TQ 302 782 512915N 0000736W	CTR Boundary crossing River Thames			
▲ Chelsea Bridge TQ 286 778 512902N 0000900W	Chelsea Road Bridge	1500 ft		
▲ Westland Heliport TQ 266 762 512812N 0001046W	Westland Heliport		H	Note 1.

Note 1: There are no Holding Points on H4 east of Westland Heliport. The nearest Holding Point is at Greenwich Marshes, outside the 'Specified Area'.

Note 2: The sector of Route H4, Isle-of-Dogs — Vauxhall Bridge, is established and notified for the purposes of Rule 5(2)(a) of the Rules of the Air Regulations 1996.

Note 3: Captive passenger carrying balloons up to 36 meters in diameter operate from sites adjacent to H4 at Tower Bridge and Vauxhall Bridge to the south and at Hyde Park to the north, by day only, to a maximum height of 400 ft agl.

Note 4: Ferris Wheel: The London Eye Ferris Wheel (464 ft amsl) lies within the boundary of H4 at Jubilee Gardens (513012N 0000711W) between London Bridge and Vauxhall Bridge. Pilots are reminded of the application of Rule 5(1)(e).

H5				
▲ Northwood TQ 071 906 513612N 0002719W	Zone Boundary midway between Harefield and Northwood	2000 ft (Note 1)	H	Note 2.
△ Uxbridge Common TQ 062 855 513332N 0002813W	Roundabout on A40 road north of Uxbridge Common		H	

Note 1: When London Heathrow Runway 23 is in use, the maximum altitude between Uxbridge Common and Northwood will be 1000 ft London Heathrow QNH.

Note 2: On H5 between Northwood and Airport Spur pilots may be required to communicate with Northolt Approach (126.450 MHz).

H7				
▲ Banstead TQ 243 614 512014N 0001301W	Golf course northwest of town	2000 ft	H	
	Sutton/Epsom railway	1500 ft		
△ Morden TQ 229 672 512327N 0001407W	Cemetery northeast of Gas Works			
△ Caesar's Camp TQ 220 711 512532N 0001446W	Golf course southwest corner of Wimbledon Common	1000 ft	H	
▲ Barnes TQ 234 765 512827N 0001325W	River Thames at Barn Elms Park			
▲ Westland Heliport TQ 266 762 512812N 0001046W	Westland Heliport		H	

EGLL AD 2.22 – FLIGHT PROCEDURES

HELICOPTER ROUTES

Route Designator Significant Points and National Grid & Lat/Long	Description of Reference	Maximum Altitude (London Heathrow QNH)	Holding Point	Remarks
1	2	3	4	5
H9				
▲ Oxshott West TQ 101 609 512014N 0002512W	Intersection of London Control Zone/A3 Trunk Road		H	
△ Esher Common TQ 136 621 512051N 0002210W	A3 Trunk Road West of A3/A244 intersection	2000 ft		
OR				
▲ Oxshott East TQ 160 611 512014N 0002006W	Prince's Coverts		H	
△ Arbrook TQ 149 624 512058N 0002100W	Intersection of A3 Trunk Road / Railway Line	2000 ft		
△ Esher Common TQ 136 621 512051N 0002210W	A3 Trunk Road West of A3/A244 intersection	1500 ft		
THEN				
	London/Woking railway			
▲ Sunbury Lock TQ 112 683 512409N 0002416W	Midway between Sunbury Lock and the middle of Knight reservoir		H	
△ Feltham TQ 095 726 512630N 0002533W	Open space south of Railway Line	800 ft	H	Note 4.
△ Bedfont TQ 088 745 512732N 0002610W	East of Terminal Four, South of the A30		H	Note 4.
△ Sipson TQ 076 772 512900N 0002706W	Open space northeast of the junction Motorway Spur and Main Road A4 at north perimeter of London Heathrow		H	
▲ Airport Spur TQ 075 786 512944N 0002715W	Junction of M4 Motorway and Motorway Spur to London Heathrow	1000 ft	H	
△ Hayes TQ 082 804 513042N 0002636W	Gravel pits at Goulds Green	1500 ft (Note 1)	H	Note 2.
△ Gutteridge TQ 097 845 513254N 0002507W	A40, south of Northolt Aerodrome Runway Intersection	2000 ft (Note 1)	H	Note 3.
▲ Northwood TQ 071 906 513612N 0002719W	Zone Boundary midway between Harefield and Northwood		H	

Note 1: When London Heathrow Airport Runway 23 is in use the maximum altitude between Hayes and Northwood will be 1000 ft London Heathrow QNH.

Note 2: Between Northwood and Airport Spur, pilots may be required to communicate with Northolt Approach (126.450 MHz).

Note 3: The holding manoeuvre is to be carried out to the south of the Northolt Aerodrome Boundary.

Note 4: Helicopters will be held at Bedfont during daylight hours when the reported weather conditions are equal to or better than 6 km visibility and 1000 ft lowest reported cloud and will be held at Feltham at all other times.

Warning: Runway 27L missed approach procedure requires a left turn at 1000 ft aal. Pilots holding at Bedfont must remain in visual contact with aircraft on final approach to Runway 27L.

EGLL AD 2.22 – FLIGHT PROCEDURES

HELICOPTER ROUTES

Route Designator Significant Points and National Grid & Lat/Long	Description of Reference	Maximum Altitude (London Heathrow QNH)	Holding Point	Remarks
1	2	3	4	5
H10				
	CTR Boundary	2000 ft		
▲ Cookham SU 898 857 513345N 0004222W	Bridge over River Thames north of Cookham		H	
∆ Iver TQ 035 826 513157N 0003031W	Delaford Park	1500 ft (Note 1)	H	
∆ Uxbridge Common TQ 062 855 513332N 0002813W	Roundabout on A40 road north of Uxbridge Common		H	Note 4.
∆ Gutteridge TQ 097 845 513254N 0002507W	A40, south of Northolt Aerodrome Runway Intersection		H	Note 5.
∆ Perivale TQ 170 828 513156N 0001852W	On A40, north of Golf course divided by River Brent	1200 ft (Note 2)	H	Note 3.
∆ Brentford TQ 186 791 512951N 0001731W	Gunnersbury Park north of Chiswick Fly-over		H	
	North edge of Gunnesbury Park			
∆ Kew Bridge TQ 190 778 512914N 0001716W	Bridge across River Thames at northeast corner of Gardens and Common	750 ft (Note 2)	H	
	Chiswick Bridge			
▲ Barnes TQ 234 765 512827N 0001325W	River Thames at Barn Elms Park	1000 ft		
▲ Westland Heliport TQ 266 762 512812N 0001046W	Westland Heliport		H	

Note 1: When London Heathrow Runway 23 is in use the maximum altitude between Iver and Gutteridge will be 1000 ft London Heathrow QNH, and the route between Gutteridge and Kew Bridge will not normally be available.

Note 2: This is the Standard Operating Altitude for this segment of the route (Perivale to Chiswick Bridge).

Note 3: The holding manoeuvre is to be contained to the west of Perivale.

Note 4: Between Iver and Perivale, pilots may be required to communicate with Northolt Approach (126.450 MHz).

Note 5: The holding manoeuvre is to be carried out to the south of the Northolt Aerodrome Boundary.

Note 6: Due to conflict with departure procedures severe delays can be expected by helicopters requesting Helicopter route H10 when London Heathrow Runways 09 are in use.

EGLL AD 2.22 – FLIGHT PROCEDURES

18   **The Specified Area**

Straight lines joining:

Kew Bridge (512912N 0001716W) —

Eastern extremity of Brent Reservoir (513420N 0001407W) —

Gospel Oak Station (513318N 0000904W) —

Southeast corner of Springfield Park (513409N 0000318W) —

Bromley-by-Bow Station (513130N 0000045W) —

Southwest corner of Hither Green (512645N 0000044W) —

Herne Hill Station (512712N 0000610W) —

Wimbledon Station (512515N 0001222W) —

Northwest corner of Castlenau Reservoir (512854N 0001407W) —

Kew Bridge (512912N 0001716W),

excluding so much of the bed of the River Thames as lies within that area between the ordinary high water marks on each of its banks.

EGLL AD 2.23 – ADDITIONAL INFORMATION

EGLL AD 2.24 – CHARTS RELATED TO THE AERODROME

Page

For take-off obstacles see Aerodrome Obstacle Chart ICAO Type A

AERODROME CHART – ICAO

LONDON HEATHROW
EGLL

512839N 0002741W

HORIZ DATUM WGS 84
(CO-ORD IN DEG, MIN, SEC)

ELEV 80FT

BEARINGS ARE MAGNETIC
ELEVATIONS IN FEET AMSL 231
HEIGHTS IN FEET ABOVE AD .....(141)

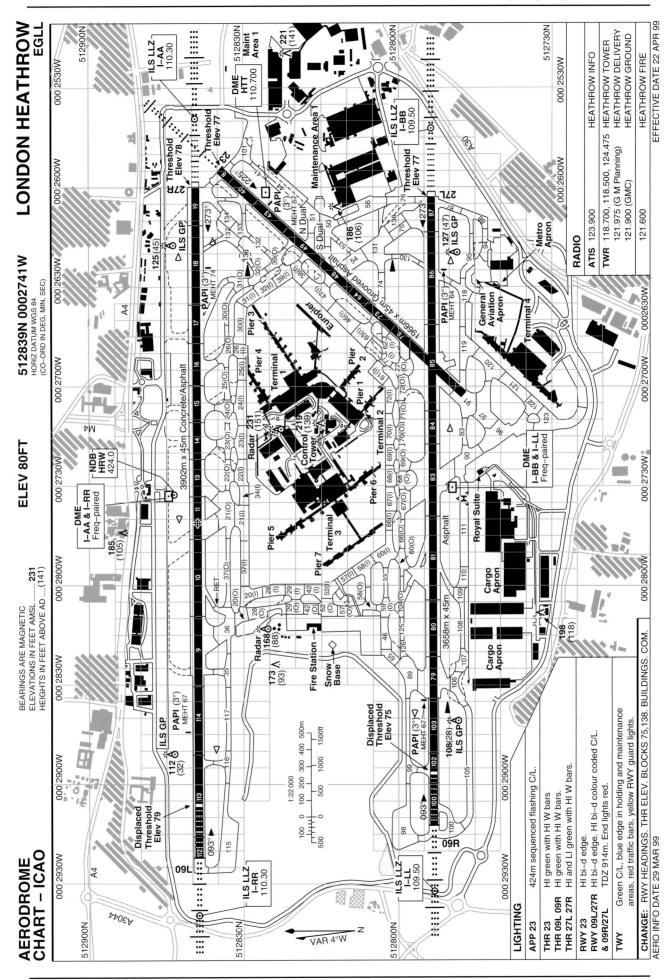

RADIO

HEATHROW INFO	
HEATHROW TOWER	
HEATHROW DELIVERY	
HEATHROW GROUND	
HEATHROW FIRE	

ATIS 123.900

TWR 118.700, 118.500, 124.475
121.975 (G M Planning)
121.900 (GMC)
121.600

EFFECTIVE DATE 22 APR 99

LIGHTING

APP 23	424m sequenced flashing C/L.
THR 23	HI green with HI W bars
THR 09L 09R	HI green with HI W bars.
THR 27L 27R	HI and LI green with HI W bars.
RWY 23	HI bi-d edge.
RWY 09L/27R	HI bi-d edge. HI bi-d colour coded C/L.
& 09R/27L	HI bi-d edge. HI bi-d colour coded C/L. TDZ 914m. End lights red.
TWY	Green C/L, blue edge in holding and maintenance areas, red traffic bars, yellow RWY guard lights.

CHANGE: RWY HEADINGS. THR ELEV. BLOCKS 75,138. BUILDINGS. COM.

AERO INFO DATE 29 MAR 99

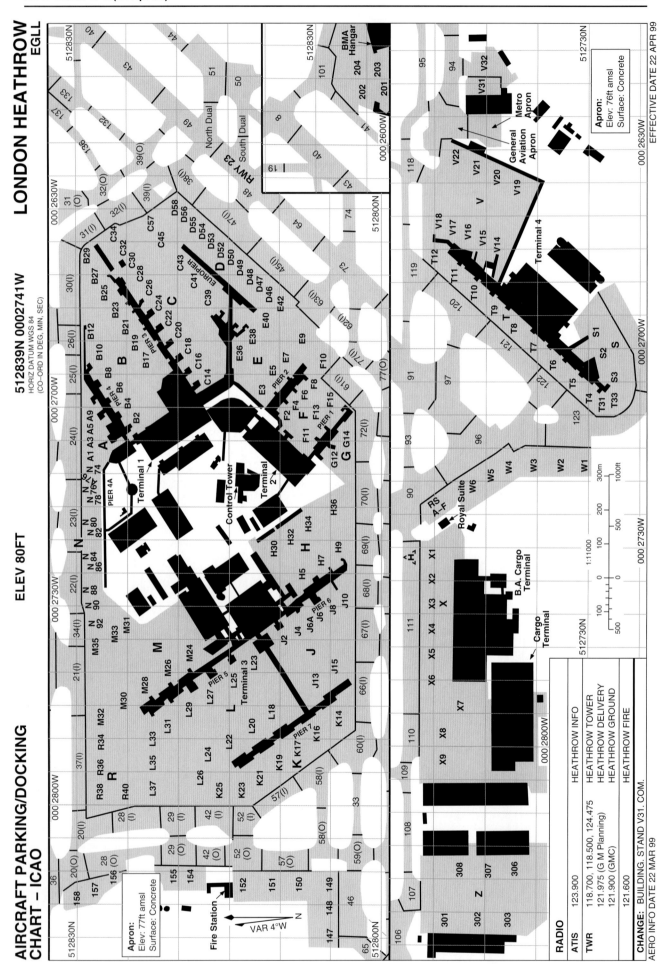

AIRCRAFT PARKING/DOCKING
CHART – ICAO

ELEV 80FT

512839N 0002741W
HORIZ DATUM WGS 84
(CO–ORD IN DEG, MIN, SEC)

LONDON HEATHROW
EGLL

Apron:
Elev: 77ft amsl
Surface: Concrete

Apron:
Elev: 76ft amsl
Surface: Concrete

VAR 4°W

RADIO		
ATIS	123.900	HEATHROW INFO
TWR	118.700, 118.500, 124.475	HEATHROW TOWER
	121.975 (G M Planning)	HEATHROW DELIVERY
	121.900 (GMC)	HEATHROW GROUND
	121.600	HEATHROW FIRE

CHANGE: BUILDING. STAND V31. COM.

AERO INFO DATE 22 MAR 99

EFFECTIVE DATE 22 APR 99

HOLDING AREAS FOR RUNWAY 27R/27L

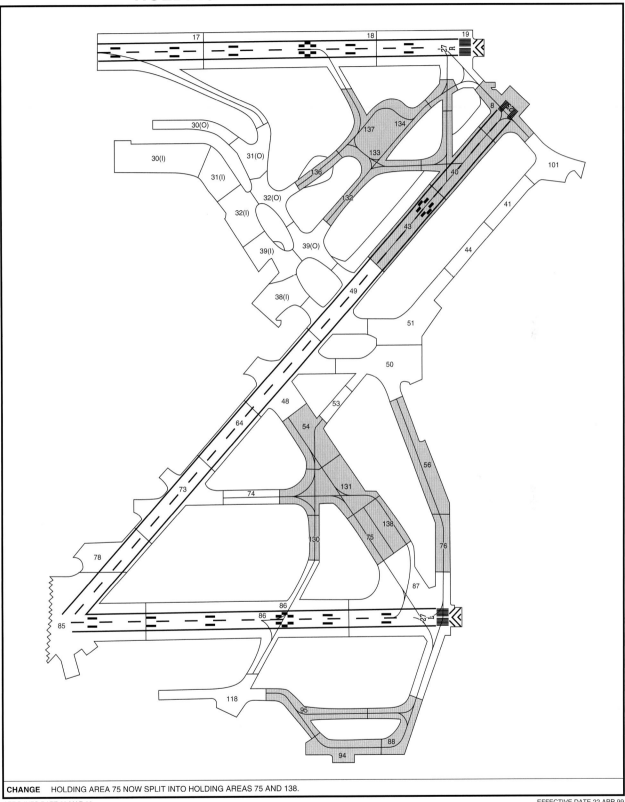

CHANGE   HOLDING AREA 75 NOW SPLIT INTO HOLDING AREAS 75 AND 138.

AERO INFO DATE 29 MAR 99   EFFECTIVE DATE 22 APR 99

HOLDING AREAS FOR RUNWAY 09R/09L

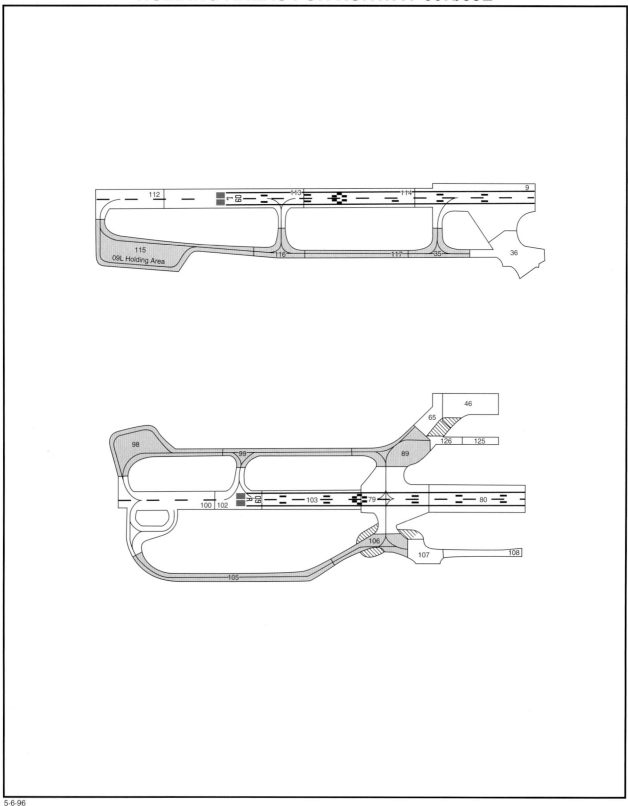

5·6·96

LONDON CTR LOCAL FLYING AND ENTRY/EXIT PROCEDURES

LEGEND

Local Flying Area

MAX ALT 800' when
RWY 23 is in use at
LONDON HEATHROW

STANSTED CTA D
1500'-2500' ALT

NORTH WEAL

STAPLEFORD
LAM

659
(325)

5130N
453
(443)
660
(650)
711
(690)
443
(433)

DAMYNS
HALL

LONDON/CITY CTR D
SFC-2500' ALT

R107/2

LONDON
/City
LCY

397
(371)

806
(788)

375
(350)

672
(622)

522
(502)

338
(329)

374
(361)

464
(451)

425
(400)

562
(328)

956
(581)

425
(300)

R151/1.5

BIG
BIGGIN HILL

G KENLEY
cables
Intense
Gliding

TMA A
Lower Limit 2500' ALT

BPK

728
(328)

920
(500)

369
(328)

627
(328)

538
(473)

458
(374)

739
(656)

362
(337)

542
(328)

492
(394)

480
(433)

404
(351)

343
(330)

LONDON
/Westland
H

387
(312)

LONDON CTR A
SFC-2500' ALT

EPM

Eastern end closed when
RWY 23 is in use at
LONDON HEATHROW

MAX ALT 2000'

NORTHOLT RMA

414
(318)

528
(430)

331
(312)

415
(300)

367
(300)

BROOKLANDS

OCK

LUTON
CTA D
Lower Limit
2500' ALT

LONDON TMA A Lower Limit 2500' ALT

781
(427)

725
(371)

M PLAISTOWS

790
(325)

RUSSELLS FARM
H

BNN

ELSTREE

508

NORTHOLT

H HAYES

HRW

LON

LONDON
HEATHROW

All local and lane flying
to be below cloud and
in sight of the ground.

MAX ALT 1000'÷
MIN VIS 3KM

FAIROAKS
FOS

876

HEN

TMA A
Lower Limit
3500' ALT

811

MAX ALT 2000'

CHT

DENHAM

MAX ALT 1000'
MIN VIS 3KM

470
(365)

BUR

MAX ALT 1500'
MIN VIS 3KM

ASCOT
H

MAX ALT 1500'÷
MIN VIS 3KM

759
(341)

D133/1-2
OCNL/2-4

D133A/1-2

D132/2-5

FARNBOROUGH

AYLESBURY/
Thame

PRINCES
RISBOROUGH

TMA A
Lower Limit
4500' ALT

1121
(327)
837

WYCOMBE AIR
PARK/Booker

Intense
Gliding

WYCOMBE ATZ

WHITE WALTHAM

473

WOD

5130N

TMA A Lower Limit
3500' ALT

BLACKBUSHE
BLK

HELICOPTER ROUTES IN THE LONDON CTR AND LONDON/City CTR

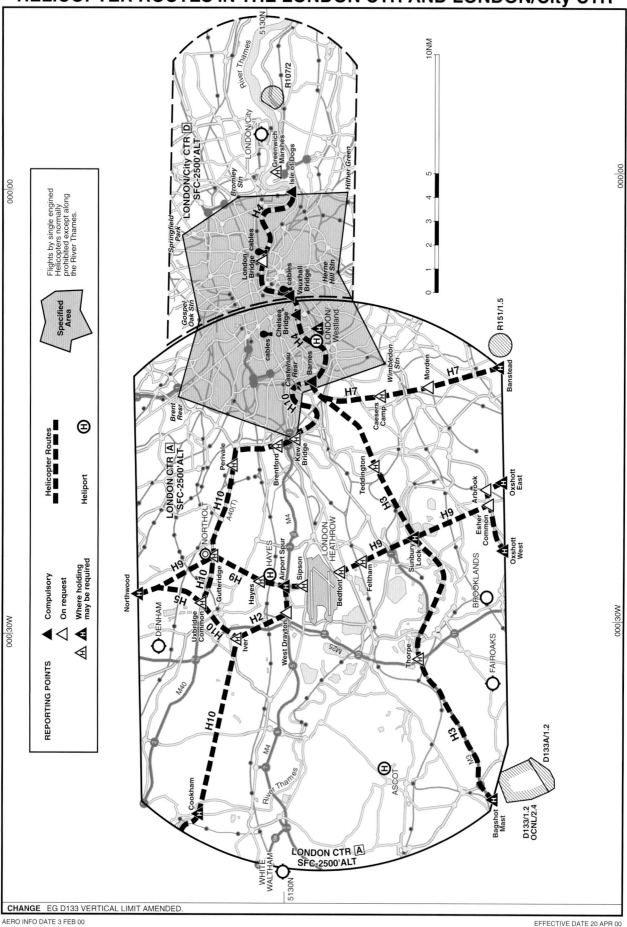

CHANGE EG D133 VERTICAL LIMIT AMENDED.

LONDON HEATHROW - HELICOPTER CROSSING OPERATIONS

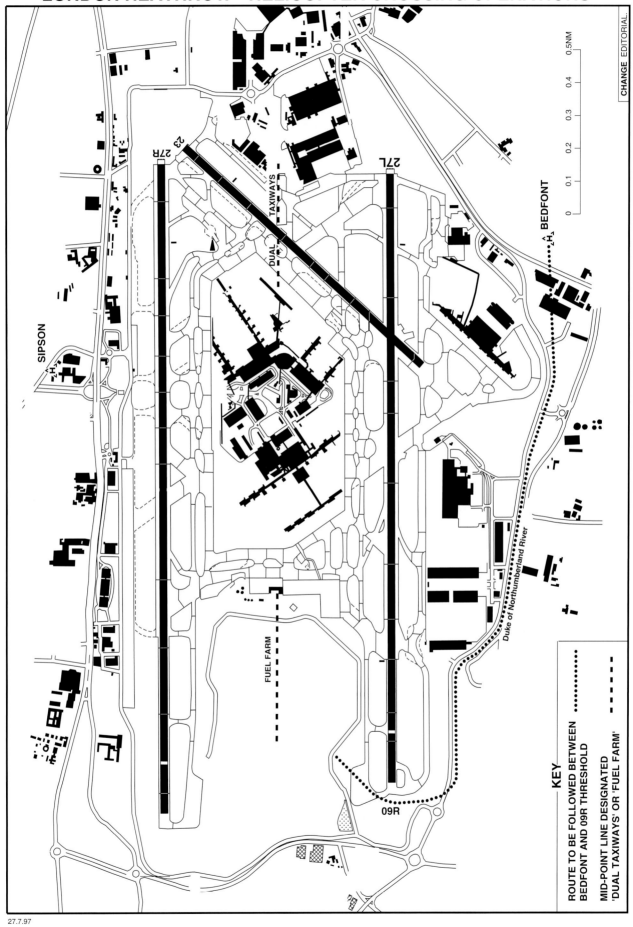

CHANGE EDITORIAL

0 0.1 0.2 0.3 0.4 0.5NM

SIPSON

DUAL TAXIWAYS

BEDFONT

27R 23

27L

Duke of Northumberland River

FUEL FARM

09R

KEY

· · · · · · · ROUTE TO BE FOLLOWED BETWEEN
BEDFONT AND 09R THRESHOLD

– – – – – MID-POINT LINE DESIGNATED
'DUAL TAXIWAYS' OR 'FUEL FARM'

27.7.97

INTENTIONALLY   BLANK

RADAR VECTORING AREA

LONDON HEATHROW

GENERAL INFORMATION
1. All bearings are magnetic.
2. Levels shown are based on QNH.
3. Only significant obstacles and dominant spot heights are shown.
4. The minimum levels shown within the Radar Vectoring Area ensure terrain clearance in conformity with Rule 29 of the Rules of the Air Regulations in respect of obstacles within the RVA.
5. Minimum Sector Altitudes are based on obstacles and spot heights within 25NM of the 2 Aerodrome Reference Points.

NORTHOLT : Aircraft being vectored for RWY 07 at Northolt may be given descent clearance to 1500FT when on 40° leg north of Bagshot Mast.

Within the Radar Vectoring Area the minimum initial altitude to be allocated by the radar controller is:

a) 2000FT in the sector south of the line HOOK-OCK VOR-BIG VOR-BIG VOR R075 except within 3NM of Wrotham Mast where the minimum altitude is 2300FT.

b) 1800FT in the sector north of the line HOOK-OCK VOR-BIG VOR-BIG VOR R075 except within the 3-5NM circle enclosing the Crystal Palace Masts where the minimum altitude is 2100FT.

LOSS OF COMMUNICATIONS PROCEDURES

Initial Approach
Continue visually or by means of an appropriate approved final approach aid. If not possible proceed to EPM NDB† (RWY 09R, 23, 27L) or CHT NDB†(RWY 09L, 27R) at 3000FT or last assigned level if higher.

Intermediate and Final Approach
Continue visually or by means of an appropriate approved final approach aid. If not possible follow the Missed Approach Procedure to EPM NDB† (RWY 09R, 23, 27L) or CHT NDB (RWY 09L, 27R).

†In all cases where the aircraft returns to the holding facility the procedure to be adopted is the Basic Radio Failure Procedure detailed at ENR 1.1.3 or the Modified Procedure for the Missed Approach Radio Failure detailed at AD 2-EGLL-1.23.

Elevation 80FT

Transition ALT 6000FT

RADAR VECTORING AREA

CHANGE EG D133 VERTICAL LIMIT AMENDED. OBSTACLES.

AERO INFO DATE 11 APR 00

EFFECTIVE DATE 20 APR 00

INTENTIONALLY   BLANK

COMPTON SIDs

LONDON HEATHROW

GENERAL INFORMATION

1. SIDs reflect Noise Preferential Routeings. See EGLL AD 2.21 for Noise Abatement Procedures.
2. Initial climb straight ahead to 580' QNH (500' QFE).
3. Cross Noise Monitoring Points not below 1080' QNH (1000' QFE) thereafter maintain minimum 4% climb gradient to 4000' (Note climb gradients greater than 4% may be required for ATC and airspace purposes) to comply with Noise Abatement requirements.
4. Callsign for RTF frequency used **when instructed** after take-off 'London Control'. Report callsign, SID designator, current altitude and cleared altitude on first contact with 'London Control'.
5. Callsign for frequency marked * will be 'Heathrow Director'.
6. En-route cruising level will be issued after take-off by 'London Control'. **Do not climb above SID levels until instructed by ATC.**
7. Maximum IAS 250KT below FL100 unless otherwise authorised.

NOT TO SCALE

TRANSITION ALT 6000'

WARNING
Due to interaction with other routes do NOT climb above 6000FT until cleared by ATC.

	AVERAGE TRACK MILEAGE TO **WOD NDB**
CPT 3F	15
CPT 3G	15
CPT 3H	16
CPT 5J	21
CPT 4K	22

SID	RWY	ROUTEING (incl. Noise Preferential Routeing)	ALTITUDES	AIRWAY ROUTE
CPT 3F 134·125	27R	Straight ahead to intercept **LON VOR** R259 until **LON** D7, then turn right onto QDM 273° to **WOD NDB** (**CPT** D13), then to **CPT VOR**.	Cross **LON** D11 (**CPT** D17) above 3000' **WOD NDB** (**CPT** D13) above 4000' **CPT** D8 at 6000' 27L/R 5%, 09L/R 3·5%.	Via **CPT** G1-Westbound
CPT 3G 134·125	27L	Straight ahead to intercept **LON VOR** R259 until **LON** D7, then turn right onto QDM 273° to **WOD NDB** (**CPT** D13), then to **CPT VOR**.		
CPT 3H 134·125	23	Straight ahead to **LON** D2, then turn right onto QDM 278° to **WOD NDB** (**CPT** D13), then to **CPT VOR**.		
CPT 5J *134·975	09R	Straight ahead to **LON** D2, then turn right onto QDM 285° to **WOD NDB** (**CPT** D13), then to **CPT VOR**.		
CPT 4K *134·975	09L	Straight ahead to **LON** D1·5, then turn right onto QDM 285° to **WOD NDB** (**CPT** D13), then to **CPT VOR**.		

CHANGE RADIALS UPDATED.

AERO INFO DATE 4 OCT 99

EFFECTIVE DATE 2 DEC 99

MIDHURST SIDs   # LONDON HEATHROW

GENERAL INFORMATION
1 SIDs reflect Noise Preferential Routeings. See EGLL AD 2.21 for Noise Abatement Procedures.
2 Initial climb straight ahead to 580' QNH (500' QFE).
3 Cross Noise Monitoring Points not below 1080' QNH (1000' QFE) thereafter maintain minimum 4% climb gradient to 4000' (Note climb gradients greater than 4% may be required for ATC and airspace purposes) to comply with Noise Abatement requirements.
4 Callsign for RTF frequency used **when instructed** after take-off 'London Control'. Report callsign, SID designator, current altitude and cleared altitude on first contact with 'London Control'.
5 En-route cruising level will be issued after take-off by 'London Control'. **Do not climb above SID levels until instructed by ATC.**
6 Maximum IAS 250KT below FL100 unless otherwise authorised.

NOT TO SCALE

TRANSITION ALT 6000'

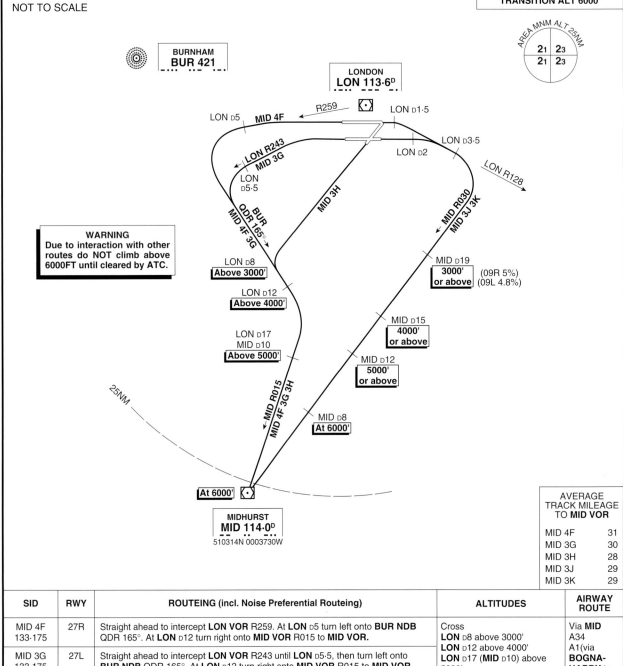

WARNING
Due to interaction with other routes do NOT climb above 6000FT until cleared by ATC.

AVERAGE TRACK MILEAGE TO **MID VOR**

MID 4F	31
MID 3G	30
MID 3H	28
MID 3J	29
MID 3K	29

SID	RWY	ROUTEING (incl. Noise Preferential Routeing)	ALTITUDES	AIRWAY ROUTE
MID 4F 133·175	27R	Straight ahead to intercept **LON VOR** R259. At **LON** D5 turn left onto **BUR NDB** QDR 165°. At **LON** D12 turn right onto **MID VOR** R015 to **MID VOR**.	Cross **LON** D8 above 3000' **LON** D12 above 4000' **LON** D17 (**MID** D10) above 5000' **MID VOR** at 6000'.	Via **MID** A34 A1(via **BOGNA-HARDY** to join M605) Southbound.
MID 3G 133·175	27L	Straight ahead to intercept **LON VOR** R243 until **LON** D5·5, then turn left onto **BUR NDB** QDR 165°. At **LON** D12 turn right onto **MID VOR** R015 to **MID VOR**.		
MID 3H 133·175	23	Straight ahead to intercept **BUR NDB** QDR 165°. At **LON** D12 turn right onto **MID VOR** R015 to **MID VOR**.		
MID 3J 133·175	09R	Straight ahead to **LON** D2, then turn right onto **LON VOR** R128 until **LON** D3·5, then turn right onto **MID VOR** R030 to **MID VOR**.	Cross **MID** D19 at 3000' or above (09R 5%) (09L 4.8%) **MID** D15 at 4000' or above **MID** D12 at 5000' or above **MID** D8 at 6000'.	
MID 3K 133·175	09L	Straight ahead to **LON** D1·5, then turn right onto **LON VOR** R128 until **LON** D3·5, then turn right onto **MID VOR** R030 to **MID VOR**.		

CHANGE AIRWAY A47 REDESIGNATED M605. RADIALS UPDATED.

BROOKMANS PARK SIDs

LONDON HEATHROW

GENERAL INFORMATION
1. SIDs reflect Noise Preferential Routeings. See EGLL AD 2.21 for Noise Abatement Procedures.
2. Initial climb straight ahead to 580' QNH (500' QFE).
3. Cross Noise Monitoring Points not below 1080' QNH (1000' QFE) thereafter maintain minimum 4% climb gradient to 4000' (Note climb gradients greater than 4% may be required for ATC and airspace purposes) to comply with Noise Abatement requirements.
4. Callsign for RTF frequency used **when instructed** after take-off 'London Control'. Report callsign, SID designator, current altitude and cleared altitude on first contact with 'London Control'.
5. En-route cruising level will be issued after take-off by 'London Control'. **Do not climb above SID levels until instructed by ATC.**
6. Maximum IAS 250KT below FL100 unless otherwise authorised.

NOT TO SCALE

TRANSITION ALT 6000'

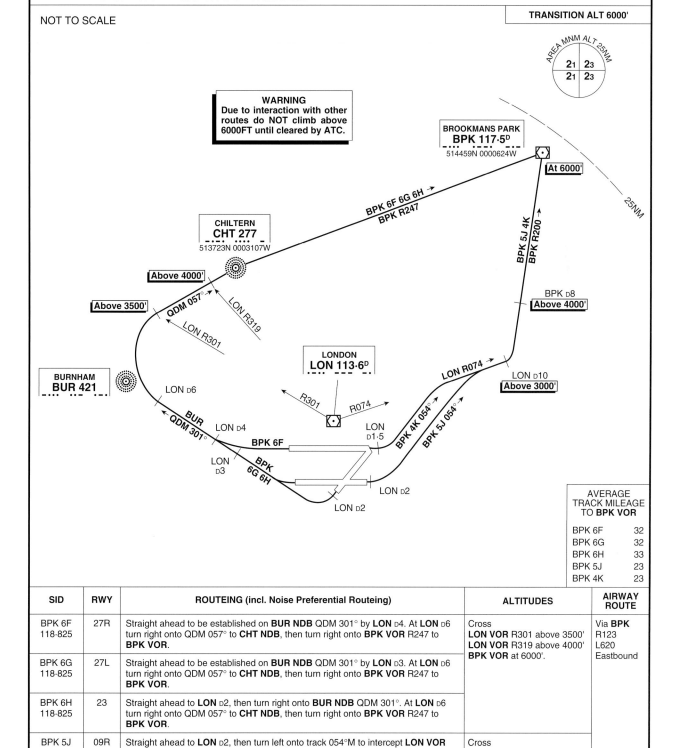

		AVERAGE TRACK MILEAGE TO **BPK VOR**
BPK 6F		32
BPK 6G		32
BPK 6H		33
BPK 5J		23
BPK 4K		23

SID	RWY	ROUTEING (incl. Noise Preferential Routeing)	ALTITUDES	AIRWAY ROUTE
BPK 6F 118·825	27R	Straight ahead to be established on **BUR NDB** QDM 301° by **LON** D4. At **LON** D6 turn right onto QDM 057° to **CHT NDB**, then turn right onto **BPK VOR** R247 to **BPK VOR**.	Cross **LON VOR** R301 above 3500' **LON VOR** R319 above 4000' **BPK VOR** at 6000'.	Via **BPK** R123 L620 Eastbound
BPK 6G 118·825	27L	Straight ahead to be established on **BUR NDB** QDM 301° by **LON** D3. At **LON** D6 turn right onto QDM 057° to **CHT NDB**, then turn right onto **BPK VOR** R247 to **BPK VOR**.		
BPK 6H 118·825	23	Straight ahead to **LON** D2, then turn right onto **BUR NDB** QDM 301°. At **LON** D6 turn right onto QDM 057° to **CHT NDB**, then turn right onto **BPK VOR** R247 to **BPK VOR**.		
BPK 5J 118·825	09R	Straight ahead to **LON** D2, then turn left onto track 054°M to intercept **LON VOR** R074. At **LON** D10 turn left onto **BPK VOR** R200 to **BPK VOR**.	Cross **LON** D10 above 3000' **BPK** D8 above 4000' **BPK VOR** at 6000'.	
BPK 4K 118·825	09L	Straight ahead to **LON** D1·5, then turn left onto track 054°M to intercept **LON VOR** R074. At **LON** D10 turn left onto **BPK VOR** R200 to **BPK VOR**.		

CHANGE AIRWAY R12 REDESIGNATED L620. RADIALS UPDATED.

WOBUN SIDs/BUZAD SIDs

LONDON HEATHROW

GENERAL INFORMATION

1. SIDs reflect Noise Preferential Routeings. See EGLL AD 2.21 for Noise Abatement Procedures.
2. Initial climb straight ahead to 580' QNH (500' QFE).
3. Cross Noise Monitoring Points not below 1080' QNH (1000' QFE) thereafter maintain minimum 4% climb gradient to 4000' (Note climb gradients greater than 4% may be required for ATC and airspace purposes) to comply with Noise Abatement requirements.
4. Callsign for RTF frequency used **when instructed** after take-off 'London Control'. Report callsign, SID designator, current altitude and cleared altitude on first contact with 'London Control'.
5. En-route cruising level will be issued after take-off by 'London Control'. **Do not climb above SID levels until instructed by ATC.**
6. Maximum IAS 250KT below FL100 unless otherwise authorised.

TRANSITION ALT 6000'

NOT TO SCALE

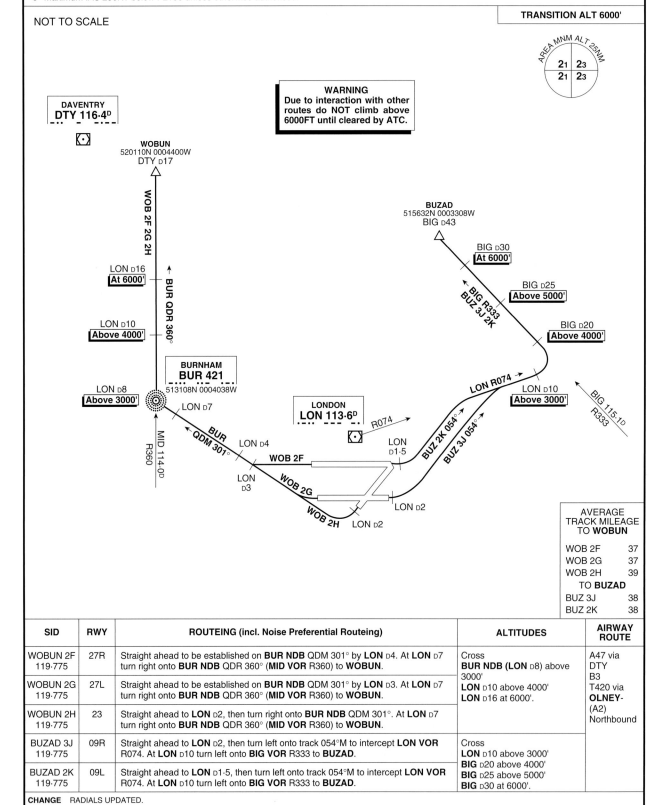

AVERAGE TRACK MILEAGE TO **WOBUN**	
WOB 2F	37
WOB 2G	37
WOB 2H	39
TO BUZAD	
BUZ 3J	38
BUZ 2K	38

SID	RWY	ROUTEING (incl. Noise Preferential Routeing)	ALTITUDES	AIRWAY ROUTE
WOBUN 2F 119·775	27R	Straight ahead to be established on **BUR NDB** QDM 301° by **LON** D4. At **LON** D7 turn right onto **BUR NDB** QDR 360° (**MID VOR** R360) to **WOBUN**.	Cross **BUR NDB (LON** D8) above 3000' **LON** D10 above 4000' **LON** D16 at 6000'.	A47 via DTY B3 T420 via **OLNEY-** (A2) Northbound
WOBUN 2G 119·775	27L	Straight ahead to be established on **BUR NDB** QDM 301° by **LON** D3. At **LON** D7 turn right onto **BUR NDB** QDR 360° (**MID VOR** R360) to **WOBUN**.		
WOBUN 2H 119·775	23	Straight ahead to **LON** D2, then turn right onto **BUR NDB** QDM 301°. At **LON** D7 turn right onto **BUR NDB** QDR 360° (**MID VOR** R360) to **WOBUN**.		
BUZAD 3J 119·775	09R	Straight ahead to **LON** D2, then turn left onto track 054°M to intercept **LON VOR** R074. At **LON** D10 turn left onto **BIG VOR** R333 to **BUZAD**.	Cross **LON** D10 above 3000' **BIG** D20 above 4000' **BIG** D25 above 5000' **BIG** D30 at 6000'.	
BUZAD 2K 119·775	09L	Straight ahead to **LON** D1·5, then turn left onto track 054°M to intercept **LON VOR** R074. At **LON** D10 turn left onto **BIG VOR** R333 to **BUZAD**.		

CHANGE RADIALS UPDATED.

DOVER SIDs/DETLING SIDs

LONDON HEATHROW

GENERAL INFORMATION

1. SIDs reflect Noise Preferential Routeings. See EGLL AD 2.21 for Noise Abatement Procedures.
2. Initial climb straight ahead to 580' QNH (500' QFE).
3. Cross Noise Monitoring Points not below 1080' QNH (1000' QFE) thereafter maintain minimum 4% climb gradient to 4000' (Note climb gradients greater than 4% may be required for ATC and airspace purposes) to comply with Noise Abatement requirements.
4. Callsign for RTF frequency used **when instructed** after take-off 'London Control'. Report callsign, SID designator, current altitude and cleared altitude on first contact with 'London Control'.
5. En-route cruising level will be issued after take-off by 'London Control'. **Do not climb above SID levels until instructed by ATC.**
6. Maximum IAS 250KT below FL100 unless otherwise authorised.
7. DVR 4G/DET 2G first turn point: when I-LL DME is out of service use LON d2.

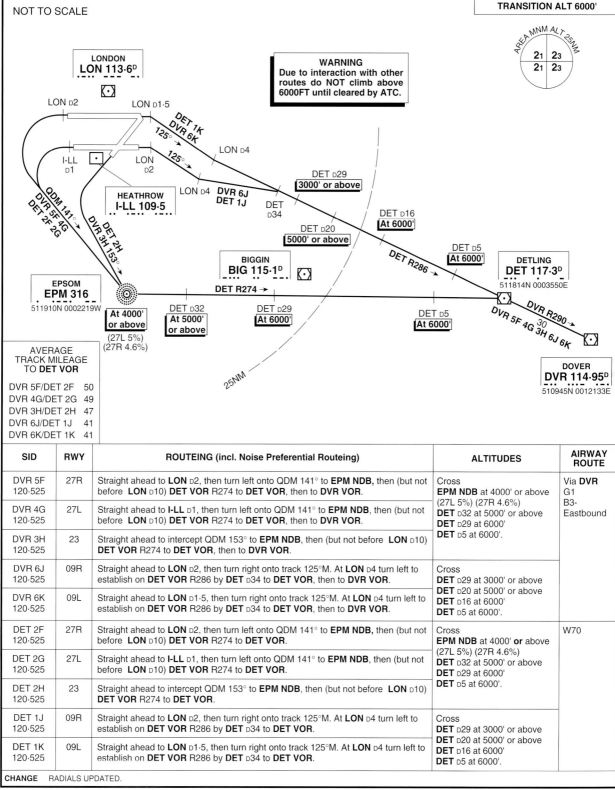

SID	RWY	ROUTEING (incl. Noise Preferential Routeing)	ALTITUDES	AIRWAY ROUTE
DVR 5F 120·525	27R	Straight ahead to **LON** d2, then turn left onto QDM 141° to **EPM NDB,** then (but not before **LON** d10) **DET VOR** R274 to **DET VOR**, then to **DVR VOR**.	Cross **EPM NDB** at 4000' or above (27L 5%) (27R 4.6%) **DET** d32 at 5000' or above **DET** d29 at 6000' **DET** d5 at 6000'.	Via **DVR** G1 B3- Eastbound
DVR 4G 120·525	27L	Straight ahead to **I-LL** d1, then turn left onto QDM 141° to **EPM NDB**, then (but not before **LON** d10) **DET VOR** R274 to **DET VOR**, then to **DVR VOR**.		
DVR 3H 120·525	23	Straight ahead to intercept QDM 153° to **EPM NDB**, then (but not before **LON** d10) **DET VOR** R274 to **DET VOR**, then to **DVR VOR**.		
DVR 6J 120·525	09R	Straight ahead to **LON** d2, then turn right onto track 125°M. At **LON** d4 turn left to establish on **DET VOR** R286 by **DET** d34 to **DET VOR**, then to **DVR VOR**.	Cross **DET** d29 at 3000' or above **DET** d20 at 5000' or above **DET** d16 at 6000' **DET** d5 at 6000'.	
DVR 6K 120·525	09L	Straight ahead to **LON** d1·5, then turn right onto track 125°M. At **LON** d4 turn left to establish on **DET VOR** R286 by **DET** d34 to **DET VOR**, then to **DVR VOR**.		
DET 2F 120·525	27R	Straight ahead to **LON** d2, then turn left onto QDM 141° to **EPM NDB,** then (but not before **LON** d10) **DET VOR** R274 to **DET VOR**.	Cross **EPM NDB** at 4000' **or** above (27L 5%) (27R 4.6%) **DET** d32 at 5000' or above **DET** d29 at 6000' **DET** d5 at 6000'.	W70
DET 2G 120·525	27L	Straight ahead to **I-LL** d1, then turn left onto QDM 141° to **EPM NDB**, then (but not before **LON** d10) **DET VOR** R274 to **DET VOR**.		
DET 2H 120·525	23	Straight ahead to intercept QDM 153° to **EPM NDB**, then (but not before **LON** d10) **DET VOR** R274 to **DET VOR**.		
DET 1J 120·525	09R	Straight ahead to **LON** d2, then turn right onto track 125°M. At **LON** d4 turn left to establish on **DET VOR** R286 by **DET** d34 to **DET VOR**.	Cross **DET** d29 at 3000' or above **DET** d20 at 5000' or above **DET** d16 at 6000' **DET** d5 at 6000'.	
DET 1K 120·525	09L	Straight ahead to **LON** d1·5, then turn right onto track 125°M. At **LON** d4 turn left to establish on **DET VOR** R286 by **DET** d34 to **DET VOR**.		

CHANGE RADIALS UPDATED.

SOUTHAMPTON SIDs

LONDON HEATHROW

GENERAL INFORMATION
1 SIDs reflect Noise Preferential Routeings. See EGLL AD 2.21 for Noise Abatement Procedures.
2 Initial climb straight ahead to 580' QNH (500' QFE).
3 Cross Noise Monitoring Points not below 1080' QNH (1000' QFE) thereafter maintain minimum 4% climb gradient to 4000' (Note climb gradients greater than 4% may be required for ATC and airspace purposes) to comply with Noise Abatement requirements.
4 Callsign for RTF frequency used **when instructed** after take-off 'London Control'. Report callsign, SID designator, current altitude and cleared altitude on first contact with 'London Control'.
5 En-route cruising level will be issued after take-off by 'London Control'. **Do not climb above SID levels until instructed by ATC.**
6 Maximum IAS 250KT below FL100 unless otherwise authorised.

TRANSITION ALT 6000'

NOT TO SCALE

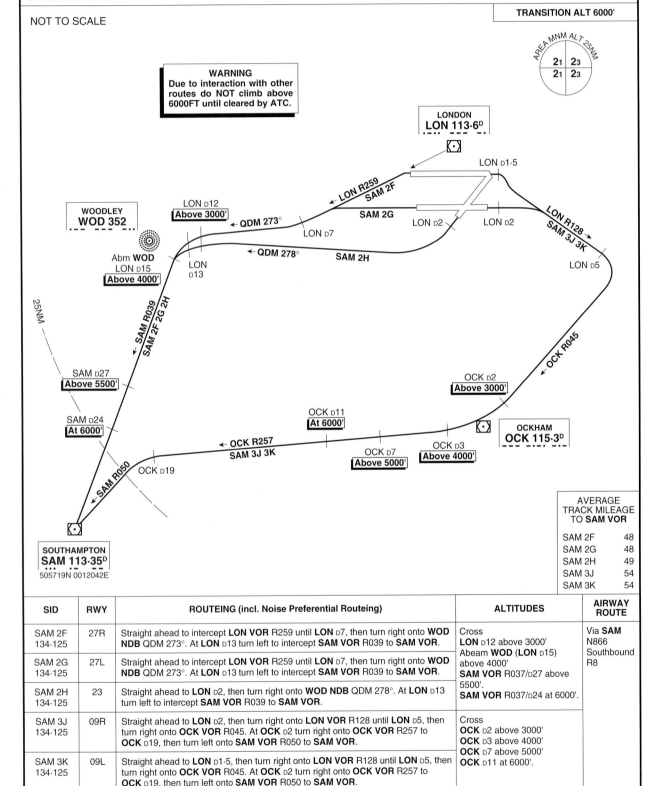

WARNING
Due to interaction with other routes do NOT climb above 6000FT until cleared by ATC.

AVERAGE TRACK MILEAGE TO **SAM VOR**	
SAM 2F	48
SAM 2G	48
SAM 2H	49
SAM 3J	54
SAM 3K	54

SID	RWY	ROUTEING (incl. Noise Preferential Routeing)	ALTITUDES	AIRWAY ROUTE
SAM 2F 134·125	27R	Straight ahead to intercept **LON VOR** R259 until **LON** D7, then turn right onto **WOD NDB** QDM 273°. At **LON** D13 turn left to intercept **SAM VOR** R039 to **SAM VOR**.	Cross **LON** D12 above 3000' Abeam **WOD** (**LON** D15) above 4000' **SAM VOR** R037/D27 above 5500'. **SAM VOR** R037/D24 at 6000'.	Via **SAM** N866 Southbound R8
SAM 2G 134·125	27L	Straight ahead to intercept **LON VOR** R259 until **LON** D7, then turn right onto **WOD NDB** QDM 273°. At **LON** D13 turn left to intercept **SAM VOR** R039 to **SAM VOR**.		
SAM 2H 134·125	23	Straight ahead to **LON** D2, then turn right onto **WOD NDB** QDM 278°. At **LON** D13 turn left to intercept **SAM VOR** R039 to **SAM VOR**.		
SAM 3J 134·125	09R	Straight ahead to **LON** D2, then turn right onto **LON VOR** R128 until **LON** D5, then turn right onto **OCK VOR** R045. At **OCK** D2 turn right onto **OCK VOR** R257 to **OCK** D19, then turn left onto **SAM VOR** R050 to **SAM VOR**.	Cross **OCK** D2 above 3000' **OCK** D3 above 4000' **OCK** D7 above 5000' **OCK** D11 at 6000'.	
SAM 3K 134·125	09L	Straight ahead to **LON** D1·5, then turn right onto **LON VOR** R128 until **LON** D5, then turn right onto **OCK VOR** R045. At **OCK** D2 turn right onto **OCK VOR** R257 to **OCK** D19, then turn left onto **SAM VOR** R050 to **SAM VOR**.		

CHANGE   RADIALS UPDATED.

MAYFIELD SIDs   # LONDON HEATHROW

GENERAL INFORMATION
1 SIDs reflect Noise Preferential Routeings. See EGLL AD 2.21 for Noise Abatement Procedures.
2 Initial climb straight ahead to 580' QNH (500' QFE).
3 Cross Noise Monitoring Points not below 1080' QNH (1000' QFE) thereafter maintain minimum 4% climb gradient to 4000' (Note climb gradients greater than 4% may be required for ATC and airspace purposes) to comply with Noise Abatement requirements.
4 Callsign for RTF frequency used **when instructed** after take-off 'London Control'. Report callsign, SID designator, current altitude and cleared altitude on first contact with 'London Control'.
5 Maximum IAS 250KT en-route. MAY VOR at IAS 220KT or less.
6 Aircraft VOR or DME failure advise ATC and comply with ATC instructions.
7 MAY 2G first turn point: when I-LL DME is out of Service use LON D2.

TRANSITION ALT 6000'

NOT TO SCALE

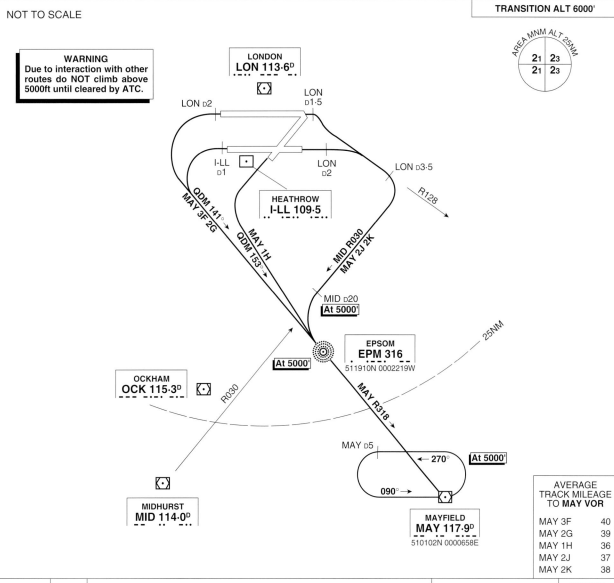

WARNING
Due to interaction with other routes do NOT climb above 5000ft until cleared by ATC.

AVERAGE TRACK MILEAGE TO **MAY VOR**	
MAY 3F	40
MAY 2G	39
MAY 1H	36
MAY 2J	37
MAY 2K	38

SID	RWY	ROUTEING (incl. Noise Preferential Routeing)	ALTITUDES	AIRWAY ROUTE
MAY 3F 126·825	27R	Straight ahead to **LON** D2, then turn left onto QDM 141° to **EPM NDB**, then (but not before **LON** D10) **MAY VOR** R318 to **MAY VOR**.	Cross **EPM NDB** at 5000' **MAY VOR** at 5000'.	For landing at LONDON GATWICK Airport only.
MAY 2G 126·825	27L	Straight ahead to **I-LL** D1, then turn left onto QDM 141° to **EPM NDB**, then (but not before **LON** D10) **MAY VOR** R318 to **MAY VOR**.		
MAY 1H 126·825	23	Straight ahead to intercept QDM 153° to **EPM NDB**, then (but not before **LON** D10) turn left onto **MAY VOR** R318 to **MAY VOR**.		
MAY 2J 126·825	09R	Straight ahead to **LON** D2, then turn right onto **LON VOR** R128 until **LON** D3·5, then turn right onto **MID VOR** R030 until **MID** D20, then turn left onto **MAY VOR** R318 to **MAY VOR**.	Cross **MID** D20 at 5000' **MAY VOR** at 5000'.	
MAY 2K 126·825	09L	Straight ahead to **LON** D1·5, then turn right onto **LON VOR** R128 until **LON** D3·5 then turn right onto **MID VOR** R030 until **MID** D20, then turn left onto **MAY VOR** R318 to **MAY VOR**.		

CHANGE RADIALS UPDATED.

INTENTIONALLY BLANK

STARs via BIGGIN

LONDON HEATHROW

GENERAL INFORMATION
1 Standard Routes may be varied at the discretion of ATC.
2 Cross SLPs or 3 min before holding facility at 250KT IAS or less.
3 When **BIG VOR** is out of service the route will be to **WEALD**. (For aircraft on AWY A2 the routeing will be **SANDY-DET -WEALD**). Designators will become WEALD 3A (WLD 3A), 3B, 3C, 3D, 1E and 1F.
4 As lowest level in **BIG/WEALD** holding stack (7000') is above transition altitude, aircraft will be instructed by ATC to fly at the appropriate flight level.
5 The routes shown also apply to aircraft inbound to **Northolt**.
6 **In order to provide airspace management flexibility during periods of congestion in the London TMA, STARs BIG 1E and BIG 3D are to facilitate the transfer of traffic to BIG VOR which would normally route via LAM VOR. These STARs are for use only as directed by ATC and must not be used for flight planning purposes.**
7 **During periods of congestion in the London TMA, traffic may be routed to OCKHAM hold via OCK 1G STAR as directed by ATC. Not to be used for flight planning purposes.**

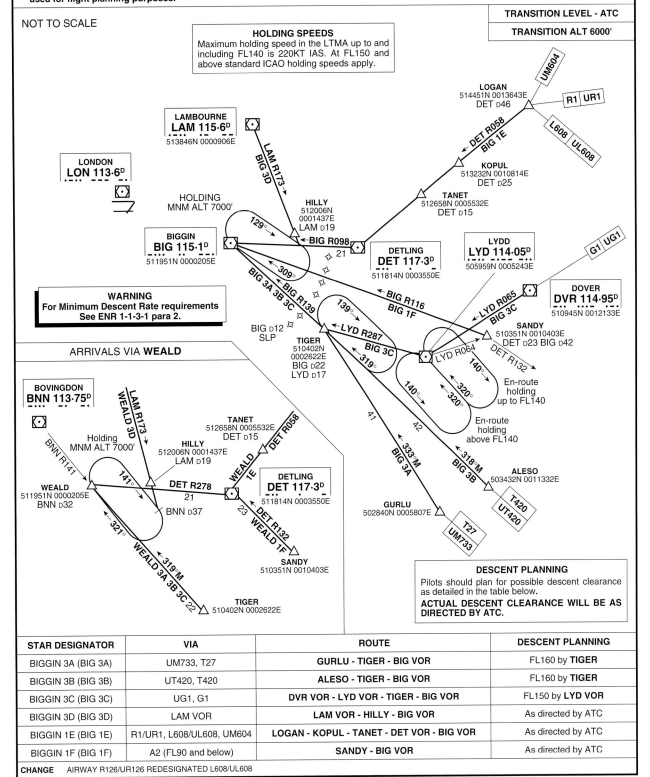

STAR DESIGNATOR	VIA	ROUTE	DESCENT PLANNING
BIGGIN 3A (BIG 3A)	UM733, T27	**GURLU - TIGER - BIG VOR**	FL160 by **TIGER**
BIGGIN 3B (BIG 3B)	UT420, T420	**ALESO - TIGER - BIG VOR**	FL160 by **TIGER**
BIGGIN 3C (BIG 3C)	UG1, G1	**DVR VOR - LYD VOR - TIGER - BIG VOR**	FL150 by **LYD VOR**
BIGGIN 3D (BIG 3D)	LAM VOR	**LAM VOR - HILLY - BIG VOR**	As directed by ATC
BIGGIN 1E (BIG 1E)	R1/UR1, L608/UL608, UM604	**LOGAN - KOPUL - TANET - DET VOR - BIG VOR**	As directed by ATC
BIGGIN 1F (BIG 1F)	A2 (FL90 and below)	**SANDY - BIG VOR**	As directed by ATC

CHANGE   AIRWAY R126/UR126 REDESIGNATED L608/UL608

STARs via BOVINGDON

LONDON HEATHROW

GENERAL INFORMATION
1 Standard Routes may be varied at the discretion of ATC.
2 Cross SLPs or 3 min before holding facility at 250KT IAS or less.
3 When **BNN VOR** is out of service route to **BOVVA**. Designators become BOVVA 1A (BVA 1A), 1B, 1C, 1D & 1E.
4 As lowest level in **BNN/BOVVA** holding stack (7000') is above transition altitude, aircraft will be instructed by ATC to fly at the appropriate flight level.
5 The routes shown also apply to aircraft inbound to **Northolt**.
6 In order to provide airspace management flexibility during periods of congestion in the London TMA, STARs **BNN 1D/BNN 1E** are to facilitate the transfer of traffic between terminal holding facilities. These STARs are for use only as directed by ATC and must not be used for flight planning purposes.

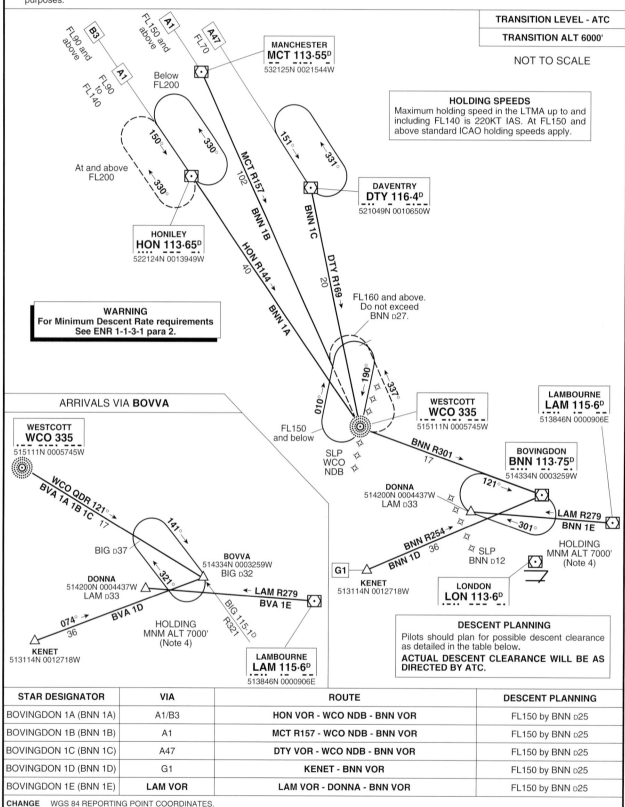

TRANSITION LEVEL - ATC

TRANSITION ALT 6000'

NOT TO SCALE

HOLDING SPEEDS
Maximum holding speed in the LTMA up to and including FL140 is 220KT IAS. At FL150 and above standard ICAO holding speeds apply.

WARNING
For Minimum Descent Rate requirements
See ENR 1-1-3-1 para 2.

ARRIVALS VIA **BOVVA**

DESCENT PLANNING
Pilots should plan for possible descent clearance as detailed in the table below.
ACTUAL DESCENT CLEARANCE WILL BE AS DIRECTED BY ATC.

STAR DESIGNATOR	VIA	ROUTE	DESCENT PLANNING
BOVINGDON 1A (BNN 1A)	A1/B3	**HON VOR - WCO NDB - BNN VOR**	FL150 by BNN D25
BOVINGDON 1B (BNN 1B)	A1	**MCT R157 - WCO NDB - BNN VOR**	FL150 by BNN D25
BOVINGDON 1C (BNN 1C)	A47	**DTY VOR - WCO NDB - BNN VOR**	FL150 by BNN D25
BOVINGDON 1D (BNN 1D)	G1	**KENET - BNN VOR**	FL150 by BNN D25
BOVINGDON 1E (BNN 1E)	**LAM VOR**	**LAM VOR - DONNA - BNN VOR**	FL150 by BNN D25

CHANGE   WGS 84 REPORTING POINT COORDINATES.

STARs via LAMBOURNE

LONDON HEATHROW

GENERAL INFORMATION
1 Standard Routes may be varied at the discretion of ATC.
2 Cross SLPs or 3 min before holding facility at 250KT IAS or less.
3 Due to proximity of Danger Area EG D138 do not fly south of track Abeam **CLN VOR** until **BRASO**.
4 When **LAM VOR** is out of service inbound aircraft approaching from East will proceed to **TAWNY**. Designators become TAWNY 3A (TNY 3A).
5 As lowest level in **LAM/TAWNY** holding stacks (7000') is above transition altitude, aircraft will be instructed by ATC to fly at the appropriate flight level.
6 The routes shown also apply to aircraft inbound to **Northolt**.
7 During periods of congestion in the London TMA, traffic may be routed to **BIGGIN**, **BOVINGDON** or **OCKHAM** holds via BIG 3D, BIG 1E, BNN 1E,
   OCK 1H STARs as directed by ATC. Not to be used for flight planning purposes.

NOT TO SCALE	**TRANSITION LEVEL - ATC**
	TRANSITION ALT 6000'

HOLDING SPEEDS
Maximum holding speed in the LTMA up to and including FL140 is 220KT IAS. At FL150 and above standard ICAO holding speeds apply.

WARNING
For Minimum Descent Rate requirements
See ENR 1-1-3-1 para 2.

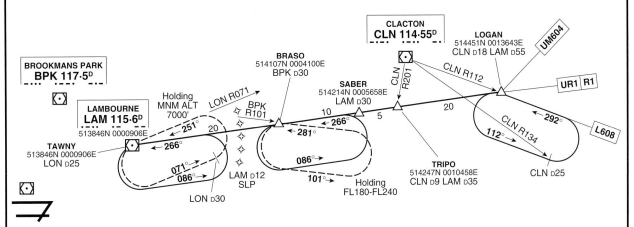

STAR DESIGNATOR	VIA	ROUTE	DESCENT PLANNING
LAMBOURNE 3A (LAM 3A)	UM604, UR1/R1 L608	**LOGAN - LAM VOR**	FL250 by **LOGAN** (aircraft Flight Planned at FL280 and above and all aircraft via UM604) FL150 by **SABER** FL200 by **LOGAN** (aircraft Flight Planned at FL270 and below) FL150 by **SABER**

DESCENT PLANNING
Pilots should plan for possible descent clearance as detailed in the table below.
ACTUAL DESCENT CLEARANCE WILL BE AS DIRECTED BY ATC.

CHANGE   BRASO HOLD RADIAL CHANGE. R126 REDESIGNATED L608.

STARs via OCKHAM (north and west)

LONDON HEATHROW

GENERAL INFORMATION
1 Standard Routes may be varied at the discretion of ATC.
2 Cross SLPs or 3 min before holding facility at 250KT IAS or less.
3 When **OCK VOR** is out of service route to **TOMMO** hold. Designators will become TOMMO 1A (TOM 1A), 1D & 1F.
4 As lowest level in **OCK/TOMMO** holding stacks (7000') is above transition altitude, aircraft will be instructed by ATC to fly at the appropriate flight level.
5 The routes shown also apply to aircraft inbound to **Northolt**.
6 STAR **OCK 1D/TOM 1D** is an overload procedure to be used only when instructed by ATC. NOT to be used for flight planning purposes.

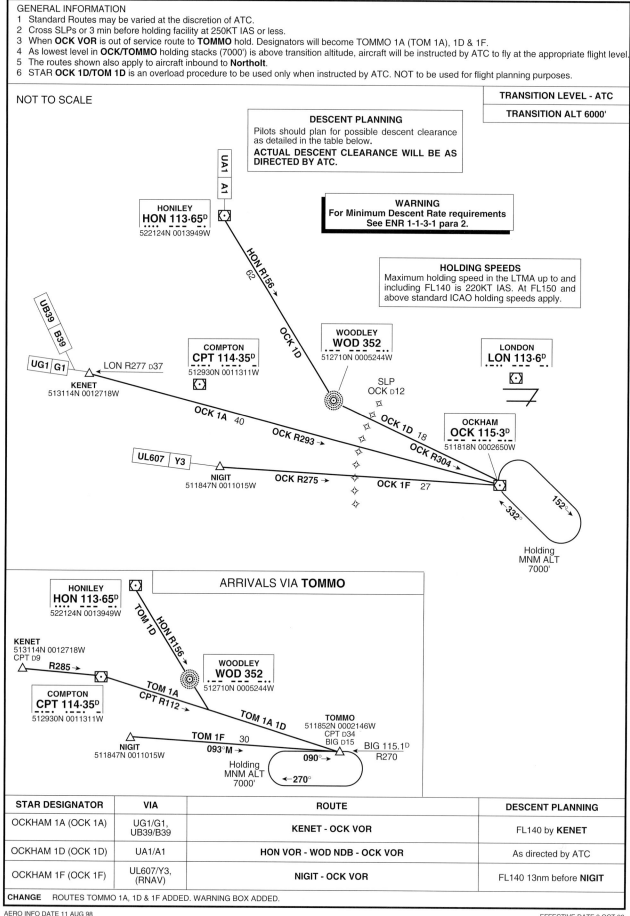

STAR DESIGNATOR	VIA	ROUTE	DESCENT PLANNING
OCKHAM 1A (OCK 1A)	UG1/G1, UB39/B39	**KENET - OCK VOR**	FL140 by **KENET**
OCKHAM 1D (OCK 1D)	UA1/A1	**HON VOR - WOD NDB - OCK VOR**	As directed by ATC
OCKHAM 1F (OCK 1F)	UL607/Y3, (RNAV)	**NIGIT - OCK VOR**	FL140 13nm before **NIGIT**

CHANGE   ROUTES TOMMO 1A, 1D & 1F ADDED. WARNING BOX ADDED.

STARs via OCKHAM (southwest)

LONDON HEATHROW

GENERAL INFORMATION
1. Standard Routes may be varied at the discretion of ATC.
2. Cross SLPs or 3 min before holding facility at 250KT IAS or less.
3. When **OCK VOR** is out of service route via **FIMLI** to **TOMMO** hold. Designators will become TOMMO 1B, 1C & 1E.
4. As lowest level in **OCK/TOMMO** holding stacks (7000') is above transition altitude, aircraft will be instructed by ATC to fly at the appropriate flight level.
5. The routes shown also apply to aircraft inbound to **Northolt**.
6. En-route holding at **ELDER** (FL110-FL150) or **BEWLI** (via **OLGUD - TARAN**) as directed by ATC.

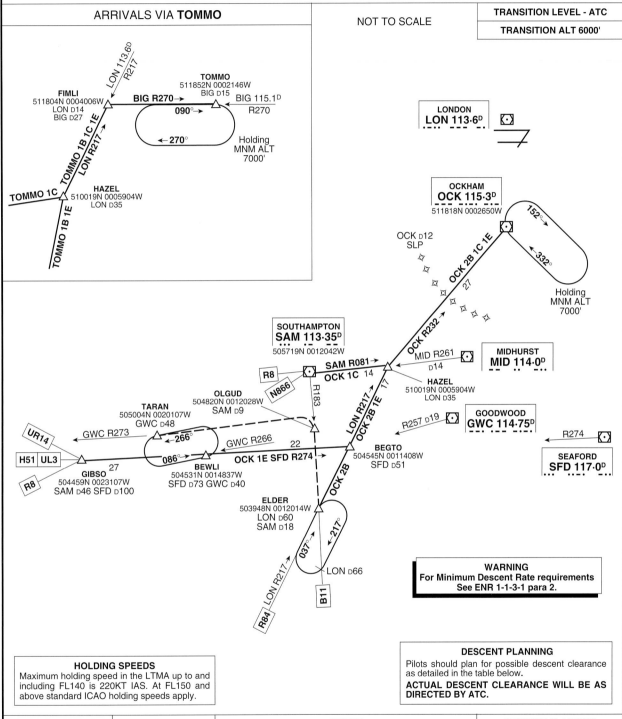

HOLDING SPEEDS
Maximum holding speed in the LTMA up to and including FL140 is 220KT IAS. At FL150 and above standard ICAO holding speeds apply.

DESCENT PLANNING
Pilots should plan for possible descent clearance as detailed in the table below.
ACTUAL DESCENT CLEARANCE WILL BE AS DIRECTED BY ATC.

STAR DESIGNATOR	VIA	ROUTE	DESCENT PLANNING
OCKHAM 2B (OCK 2B)	R84, B11	**ELDER - HAZEL - OCK VOR** (Note 6)	FL130 by **HAZEL**
OCKHAM 1C (OCK 1C)	N866 (FL100 & below), R8 (FL150 & below),	**SAM VOR - HAZEL - OCK VOR**	FL130 by **HAZEL**
OCKHAM 1E (OCK 1E)	R8 (FL160 & above), UL3/H51, UR14	**GIBSO - BEWLI - BEGTO - HAZEL - OCK VOR**	FL130 by **HAZEL**

CHANGE WGS 84 REPORTING POINT COORDINATES.

STARs via OCKHAM (east)   # LONDON HEATHROW

GENERAL INFORMATION
1. In order to provide airspace management flexibility during periods of congestion in the London TMA, STARs OCK 1G/OCK 1H, and TOM 1G/TOM 1H are to facilitate the transfer of traffic between terminal holding facilities. These STARs are for use only as directed by ATC and must not be used for flight planning purposes.
2. Standard Routes may be varied at the discretion of ATC.
3. Cross SLPs or 3 MIN before holding facility at 250KT IAS or less. (See BIG & LAM STARs).
4. When **OCK VOR** is out of service route to **TOMMO** hold. Designators will become TOMMO 1G & 1H.
5. As lowest level in **OCK/TOMMO** holding stacks (7000') is above transition altitude, aircraft will be instructed by ATC to fly at the appropriate flight level.
6. The routes shown also apply to aircraft inbound to **Northolt**.

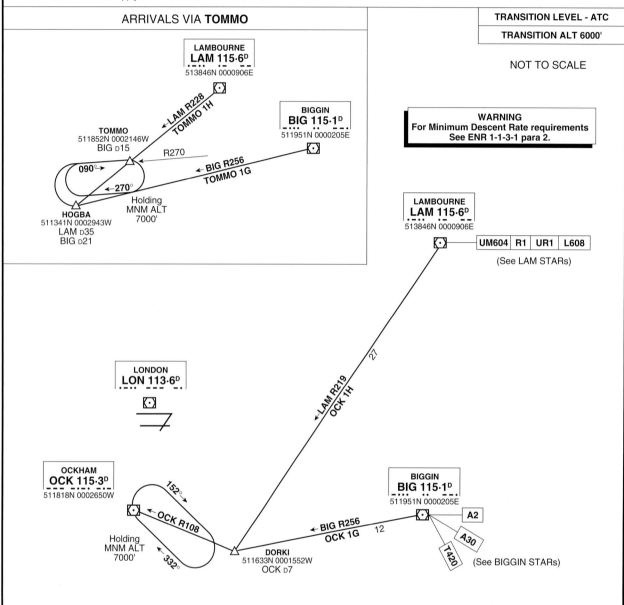

| TRANSITION LEVEL - ATC |
| TRANSITION ALT 6000' |

NOT TO SCALE

WARNING
For Minimum Descent Rate requirements
See ENR 1-1-3-1 para 2.

HOLDING SPEEDS
Maximum holding speed in the LTMA up to and including FL140 is 220KT IAS. At FL150 and above standard ICAO holding speeds apply.

DESCENT PLANNING
Pilots should plan for possible descent clearance as detailed in the table below.
ACTUAL DESCENT CLEARANCE WILL BE AS DIRECTED BY ATC.

STAR DESIGNATOR	VIA	ROUTE	DESCENT PLANNING	
OCKHAM 1G (OCK 1G)	T420/A2/A30	**BIG VOR - DORKI - OCK VOR**	FL150 by **TIGER** FL150 by **BIG** D25 FL140 by **LYD VOR**	as appropriate to AWY Route
OCKHAM 1H (OCK 1H)	UM604, R1/UR1, L608	**LAM VOR - DORKI - OCK VOR**	FL150 by **SABER**	

CHANGE   AIRWAY R126 REDESIGNATED L608.

LONDON HEATHROW
ILS DME RWY 09L
I-AA 110.30

EGLL CAT A,B,C,D			
AD ELEV **80FT**	THR ELEV **79FT**	TRANSITION ALT **6000**	VAR **4°W**

MSA 25NM ARP

ATIS	APPROACH	RADAR	TOWER (GMC)
123.900,113.750,115.100	119.725,120.400,134.975	125.625,119.900	118.700,118.500,124.475,(121.900)

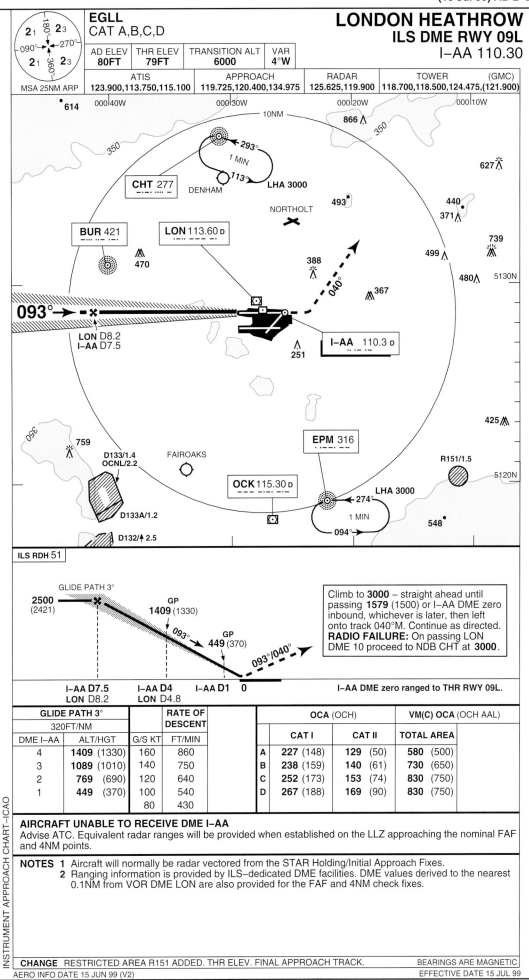

ILS RDH 51

Climb to **3000** – straight ahead until passing **1579** (1500) or I-AA DME zero inbound, whichever is later, then left onto track 040°M. Continue as directed.
RADIO FAILURE: On passing LON DME 10 proceed to NDB CHT at **3000**.

I-AA D7.5 LON D8.2	I-AA D4 LON D4.8	I-AA D1	0

I-AA DME zero ranged to THR RWY 09L.

GLIDE PATH 3°						OCA (OCH)		VM(C) OCA (OCH AAL)
320FT/NM			RATE OF DESCENT					
DME I-AA	ALT/HGT		G/S KT	FT/MIN		CAT I	CAT II	TOTAL AREA
4	**1409**	(1330)	160	860	**A**	**227** (148)	**129** (50)	**580** (500)
3	**1089**	(1010)	140	750	**B**	**238** (159)	**140** (61)	**730** (650)
2	**769**	(690)	120	640	**C**	**252** (173)	**153** (74)	**830** (750)
1	**449**	(370)	100	540	**D**	**267** (188)	**169** (90)	**830** (750)
			80	430				

AIRCRAFT UNABLE TO RECEIVE DME I-AA
Advise ATC. Equivalent radar ranges will be provided when established on the LLZ approaching the nominal FAF and 4NM points.

NOTES
1 Aircraft will normally be radar vectored from the STAR Holding/Initial Approach Fixes.
2 Ranging information is provided by ILS-dedicated DME facilities. DME values derived to the nearest 0.1NM from VOR DME LON are also provided for the FAF and 4NM check fixes.

INSTRUMENT APPROACH CHART-ICAO

CHANGE RESTRICTED AREA R151 ADDED. THR ELEV. FINAL APPROACH TRACK.

BEARINGS ARE MAGNETIC

AERO INFO DATE 15 JUN 99 (V2)

EFFECTIVE DATE 15 JUL 99

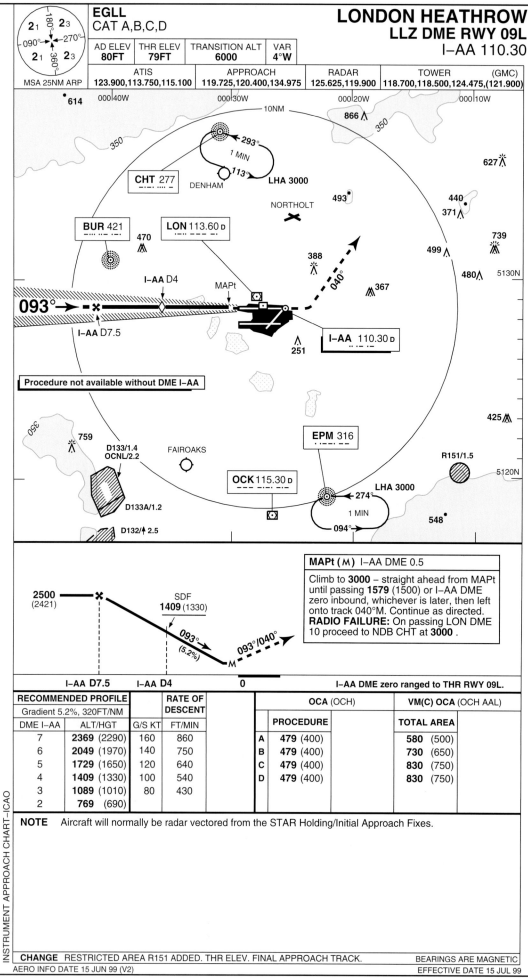

EGLL
CAT A,B,C,D

LONDON HEATHROW
LLZ DME RWY 09L
I–AA 110.30

AD ELEV	THR ELEV	TRANSITION ALT	VAR
80FT	**79FT**	**6000**	**4°W**

MSA 25NM ARP

ATIS	APPROACH	RADAR	TOWER	(GMC)
123.900,113.750,115.100	119.725,120.400,134.975	125.625,119.900	118.700,118.500,124.475,(121.900)	

Procedure not available without DME I–AA

MAPt (M) I–AA DME 0.5

Climb to **3000** – straight ahead from MAPt until passing **1579** (1500) or I–AA DME zero inbound, whichever is later, then left onto track 040°M. Continue as directed.
RADIO FAILURE: On passing LON DME 10 proceed to NDB CHT at **3000** .

2500 (2421)

SDF **1409** (1330)

093° (5.2%)

093°/040°

I–AA DME zero ranged to THR RWY 09L.

I–AA D7.5	I–AA D4	0	

RECOMMENDED PROFILE		RATE OF DESCENT		OCA (OCH)		VM(C) OCA (OCH AAL)	
Gradient 5.2%, 320FT/NM							
DME I–AA	ALT/HGT	G/S KT	FT/MIN		PROCEDURE		TOTAL AREA
7	**2369** (2290)	160	860	A	**479** (400)		**580** (500)
6	**2049** (1970)	140	750	B	**479** (400)		**730** (650)
5	**1729** (1650)	120	640	C	**479** (400)		**830** (750)
4	**1409** (1330)	100	540	D	**479** (400)		**830** (750)
3	**1089** (1010)	80	430				
2	**769** (690)						

NOTE Aircraft will normally be radar vectored from the STAR Holding/Initial Approach Fixes.

INSTRUMENT APPROACH CHART–ICAO

CHANGE RESTRICTED AREA R151 ADDED. THR ELEV. FINAL APPROACH TRACK.

BEARINGS ARE MAGNETIC

AERO INFO DATE 15 JUN 99 (V2)

EFFECTIVE DATE 15 JUL 99

LONDON HEATHROW
SRA RTR 2NM RWY 09L
RADAR 125.625

EGLL
CAT A,B,C,D

MSA 25NM ARP

AD ELEV **80FT**	THR ELEV **79FT**	TRANSITION ALT **6000**	VAR **4°W**

ATIS **123.900,113.750,115.100**	APPROACH **119.725,120.400,134.975**	RADAR **125.625,119.900**	TOWER **118.700,118.500,124.475,(121.900)** (GMC)

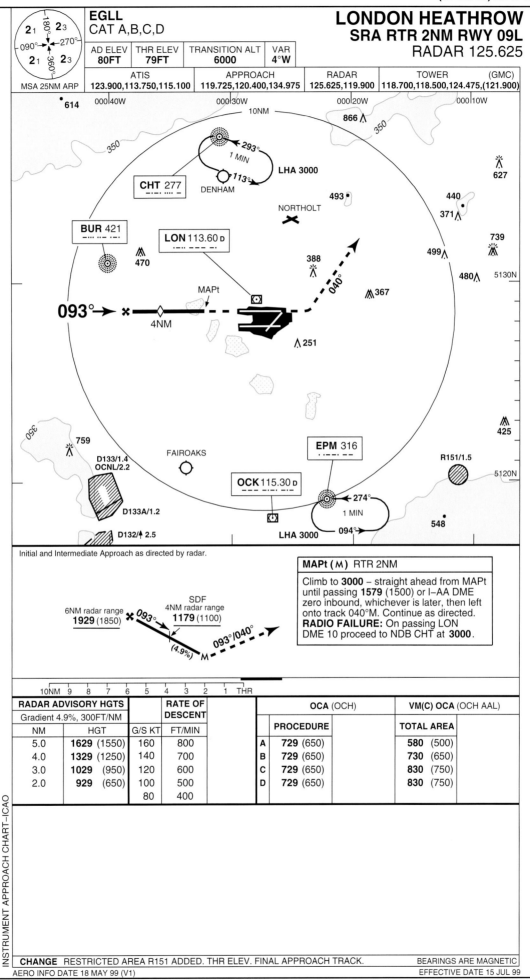

Initial and Intermediate Approach as directed by radar.

MAPt (M) RTR 2NM

Climb to **3000** – straight ahead from MAPt until passing **1579** (1500) or I–AA DME zero inbound, whichever is later, then left onto track 040°M. Continue as directed.
RADIO FAILURE: On passing LON DME 10 proceed to NDB CHT at **3000**.

RADAR ADVISORY HGTS		RATE OF DESCENT			OCA (OCH)			VM(C) OCA (OCH AAL)	
Gradient 4.9%, 300FT/NM									
NM	HGT	G/S KT	FT/MIN		PROCEDURE			TOTAL AREA	
5.0	**1629** (1550)	160	800	A	**729** (650)			**580** (500)	
4.0	**1329** (1250)	140	700	B	**729** (650)			**730** (650)	
3.0	**1029** (950)	120	600	C	**729** (650)			**830** (750)	
2.0	**929** (650)	100	500	D	**729** (650)			**830** (750)	
		80	400						

INSTRUMENT APPROACH CHART–ICAO

CHANGE RESTRICTED AREA R151 ADDED. THR ELEV. FINAL APPROACH TRACK.

AERO INFO DATE 18 MAY 99 (V1)

BEARINGS ARE MAGNETIC

EFFECTIVE DATE 15 JUL 99

LONDON HEATHROW
ILS DME RWY 09R
I–BB 109.50

EGLL
CAT A,B,C,D

MSA 25NM ARP

AD ELEV	THR ELEV	TRANSITION ALT	VAR
80FT	**75FT**	**6000**	**4°W**

ATIS	APPROACH	RADAR	TOWER (GMC)
123.900,113.750,115.100	119.725,120.400,134.975	125.625,119.900	118.700,118.500,124.475,(121.900)

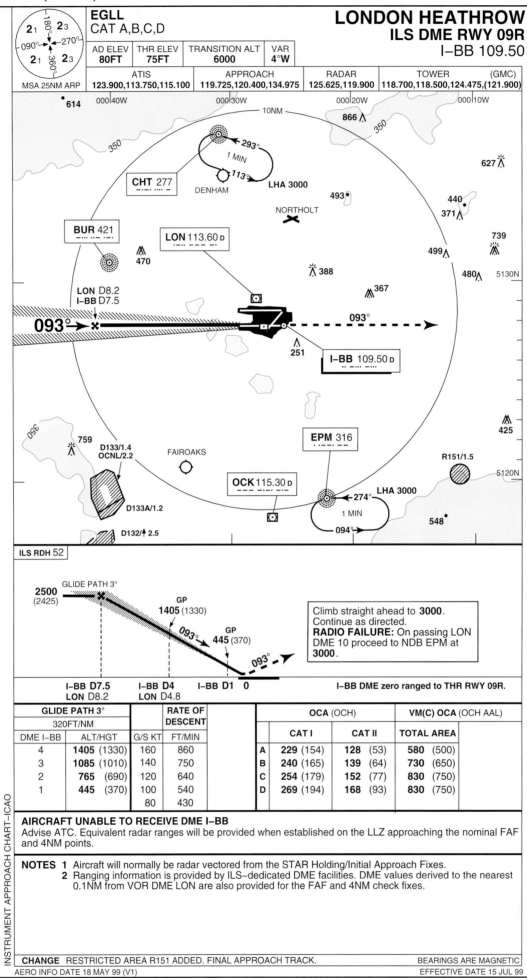

Climb straight ahead to **3000**.
Continue as directed.
RADIO FAILURE: On passing LON
DME 10 proceed to NDB EPM at
3000.

I–BB DME zero ranged to THR RWY 09R.

DME I–BB	ALT/HGT		G/S KT	FT/MIN
	GLIDE PATH 3°		RATE OF	DESCENT
	320FT/NM			
4	**1405** (1330)		160	860
3	**1085** (1010)		140	750
2	**765** (690)		120	640
1	**445** (370)		100	540
			80	430

	OCA (OCH)		VM(C) OCA (OCH AAL)
	CAT I	CAT II	TOTAL AREA
A	**229** (154)	**128** (53)	**580** (500)
B	**240** (165)	**139** (64)	**730** (650)
C	**254** (179)	**152** (77)	**830** (750)
D	**269** (194)	**168** (93)	**830** (750)

AIRCRAFT UNABLE TO RECEIVE DME I–BB
Advise ATC. Equivalent radar ranges will be provided when established on the LLZ approaching the nominal FAF and 4NM points.

NOTES 1 Aircraft will normally be radar vectored from the STAR Holding/Initial Approach Fixes.
2 Ranging information is provided by ILS–dedicated DME facilities. DME values derived to the nearest 0.1NM from VOR DME LON are also provided for the FAF and 4NM check fixes.

INSTRUMENT APPROACH CHART–ICAO

CHANGE RESTRICTED AREA R151 ADDED. FINAL APPROACH TRACK.

BEARINGS ARE MAGNETIC

AERO INFO DATE 18 MAY 99 (V1)

EFFECTIVE DATE 15 JUL 99

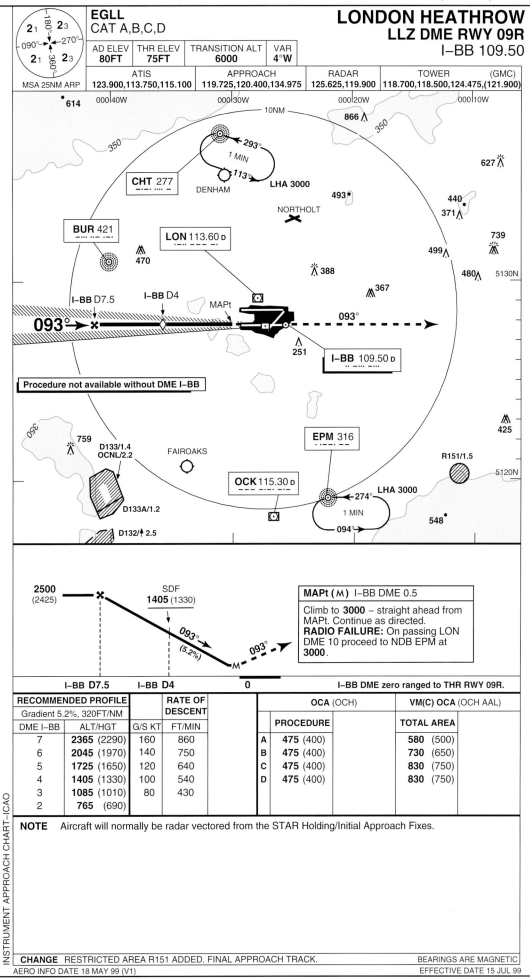

EGLL
CAT A,B,C,D

LONDON HEATHROW
LLZ DME RWY 09R
I–BB 109.50

AD ELEV	THR ELEV	TRANSITION ALT	VAR
80FT	**75FT**	**6000**	**4°W**

MSA 25NM ARP

ATIS	APPROACH	RADAR	TOWER (GMC)
123.900,113.750,115.100	**119.725,120.400,134.975**	**125.625,119.900**	**118.700,118.500,124.475,(121.900)**

Procedure not available without DME I–BB

I–BB DME zero ranged to THR RWY 09R.

MAPt (M) I–BB DME 0.5

Climb to **3000** – straight ahead from MAPt. Continue as directed.
RADIO FAILURE: On passing LON DME 10 proceed to NDB EPM at **3000**.

2500 (2425)   SDF 1405 (1330)   093° (5.2%)   093°

I–BB D7.5   I–BB D4   0

RECOMMENDED PROFILE		RATE OF DESCENT	
Gradient 5.2%, 320FT/NM			
DME I–BB	ALT/HGT	G/S KT	FT/MIN
7	**2365** (2290)	160	860
6	**2045** (1970)	140	750
5	**1725** (1650)	120	640
4	**1405** (1330)	100	540
3	**1085** (1010)	80	430
2	**765** (690)		

	OCA (OCH)		VM(C) OCA (OCH AAL)	
	PROCEDURE		TOTAL AREA	
A	**475** (400)		**580** (500)	
B	**475** (400)		**730** (650)	
C	**475** (400)		**830** (750)	
D	**475** (400)		**830** (750)	

NOTE Aircraft will normally be radar vectored from the STAR Holding/Initial Approach Fixes.

INSTRUMENT APPROACH CHART–ICAO

CHANGE RESTRICTED AREA R151 ADDED. FINAL APPROACH TRACK.

AERO INFO DATE 18 MAY 99 (V1)

BEARINGS ARE MAGNETIC

EFFECTIVE DATE 15 JUL 99

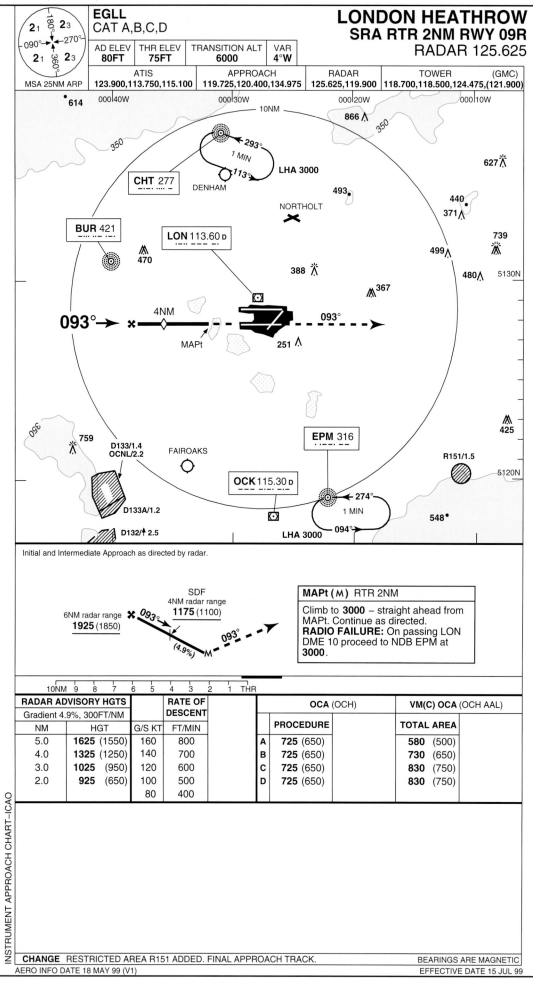

EGLL
CAT A,B,C,D

LONDON HEATHROW
SRA RTR 2NM RWY 09R
RADAR 125.625

AD ELEV	THR ELEV	TRANSITION ALT	VAR
80FT	**75FT**	**6000**	**4°W**

MSA 25NM ARP

ATIS	APPROACH	RADAR	TOWER	(GMC)
123.900,113.750,115.100	**119.725,120.400,134.975**	**125.625,119.900**	**118.700,118.500,124.475,**	**(121.900)**

Initial and Intermediate Approach as directed by radar.

MAPt (M) RTR 2NM

Climb to **3000** – straight ahead from MAPt. Continue as directed.
RADIO FAILURE: On passing LON DME 10 proceed to NDB EPM at **3000**.

SDF
4NM radar range
1175 (1100)

6NM radar range
1925 (1850)

093°
093°
(4.9%)
M

RADAR ADVISORY HGTS		RATE OF DESCENT			OCA (OCH)			VM(C) OCA (OCH AAL)	
Gradient 4.9%, 300FT/NM									
NM	HGT	G/S KT	FT/MIN		PROCEDURE			TOTAL AREA	
5.0	**1625** (1550)	160	800	A	**725** (650)			**580** (500)	
4.0	**1325** (1250)	140	700	B	**725** (650)			**730** (650)	
3.0	**1025** (950)	120	600	C	**725** (650)			**830** (750)	
2.0	**925** (650)	100	500	D	**725** (650)			**830** (750)	
		80	400						

CHANGE RESTRICTED AREA R151 ADDED. FINAL APPROACH TRACK.

BEARINGS ARE MAGNETIC

AERO INFO DATE 18 MAY 99 (V1)

EFFECTIVE DATE 15 JUL 99

INSTRUMENT APPROACH CHART–ICAO

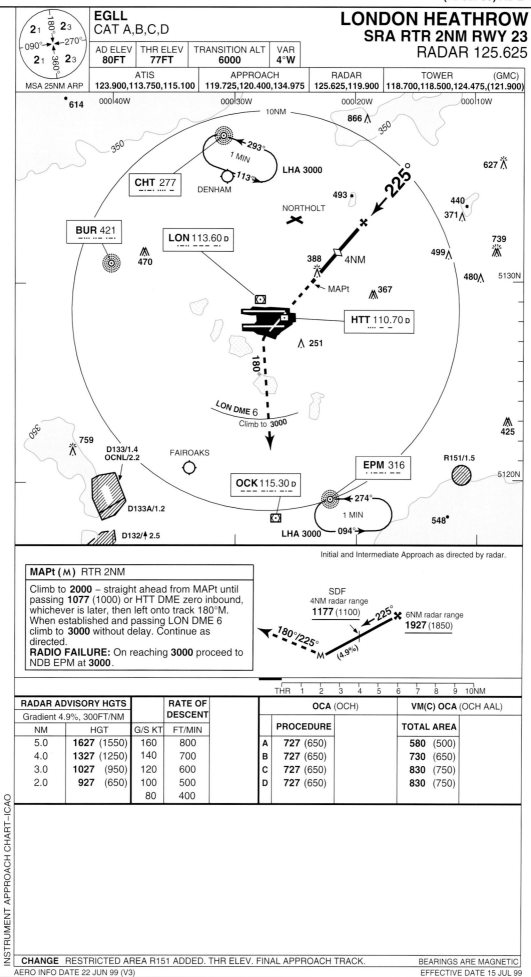

EGLL
CAT A,B,C,D

LONDON HEATHROW
SRA RTR 2NM RWY 23
RADAR 125.625

MSA 25NM ARP

AD ELEV	THR ELEV	TRANSITION ALT	VAR
80FT	**77FT**	**6000**	**4°W**

ATIS	APPROACH	RADAR	TOWER	(GMC)
123.900,113.750,115.100	119.725,120.400,134.975	125.625,119.900	118.700,118.500,124.475,	(121.900)

Initial and Intermediate Approach as directed by radar.

SDF
4NM radar range
1177 (1100)

6NM radar range
1927 (1850)

MAPt (M) RTR 2NM

Climb to **2000** – straight ahead from MAPt until passing **1077** (1000) or HTT DME zero inbound, whichever is later, then left onto track 180°M. When established and passing LON DME 6 climb to **3000** without delay. Continue as directed.
RADIO FAILURE: On reaching **3000** proceed to NDB EPM at **3000**.

THR	1	2	3	4	5	6	7	8	9	10NM

RADAR ADVISORY HGTS		RATE OF DESCENT			OCA (OCH)		VM(C) OCA (OCH AAL)	
Gradient 4.9%, 300FT/NM								
NM	HGT	G/S KT	FT/MIN			PROCEDURE		TOTAL AREA
5.0	**1627** (1550)	160	800	A		**727** (650)		**580** (500)
4.0	**1327** (1250)	140	700	B		**727** (650)		**730** (650)
3.0	**1027** (950)	120	600	C		**727** (650)		**830** (750)
2.0	**927** (650)	100	500	D		**727** (650)		**830** (750)
		80	400					

INSTRUMENT APPROACH CHART–ICAO

CHANGE RESTRICTED AREA R151 ADDED. THR ELEV. FINAL APPROACH TRACK.
AERO INFO DATE 22 JUN 99 (V3)

BEARINGS ARE MAGNETIC
EFFECTIVE DATE 15 JUL 99

LONDON HEATHROW
ILS DME RWY 27L
I–LL 109.50

EGLL
CAT A,B,C,D

AD ELEV	THR ELEV	TRANSITION ALT	VAR
80FT	**77FT**	**6000**	**4°W**

	ATIS	APPROACH	RADAR	TOWER	(GMC)
MSA 25NM ARP	123.900,113.750,115.100	119.725,120.400,134.975	125.625,119.900	118.700,118.500,124.475,	(121.900)

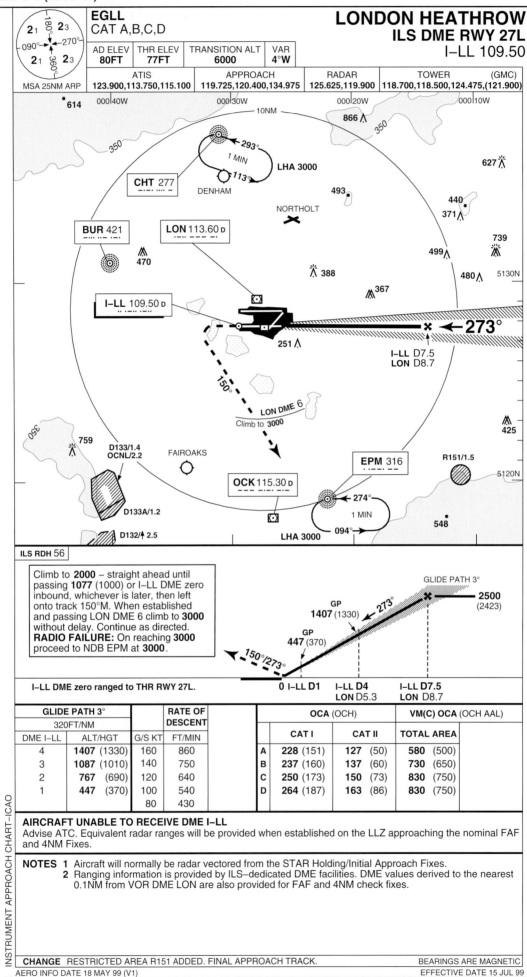

ILS RDH 56

Climb to **2000** – straight ahead until passing **1077** (1000) or I–LL DME zero inbound, whichever is later, then left onto track 150°M. When established and passing LON DME 6 climb to **3000** without delay. Continue as directed.
RADIO FAILURE: On reaching **3000** proceed to NDB EPM at **3000**.

I–LL DME zero ranged to THR RWY 27L.

GLIDE PATH 3°			RATE OF DESCENT			OCA (OCH)			VM(C) OCA (OCH AAL)	
320FT/NM							CAT I	CAT II	TOTAL AREA	
DME I–LL	ALT/HGT		G/S KT	FT/MIN						
4	**1407**	(1330)	160	860		A	**228** (151)	**127** (50)	**580** (500)	
3	**1087**	(1010)	140	750		B	**237** (160)	**137** (60)	**730** (650)	
2	**767**	(690)	120	640		C	**250** (173)	**150** (73)	**830** (750)	
1	**447**	(370)	100	540		D	**264** (187)	**163** (86)	**830** (750)	
			80	430						

AIRCRAFT UNABLE TO RECEIVE DME I–LL
Advise ATC. Equivalent radar ranges will be provided when established on the LLZ approaching the nominal FAF and 4NM Fixes.

NOTES   **1** Aircraft will normally be radar vectored from the STAR Holding/Initial Approach Fixes.
        2 Ranging information is provided by ILS–dedicated DME facilities. DME values derived to the nearest 0.1NM from VOR DME LON are also provided for FAF and 4NM check fixes.

INSTRUMENT APPROACH CHART–ICAO

CHANGE RESTRICTED AREA R151 ADDED. FINAL APPROACH TRACK.   BEARINGS ARE MAGNETIC

AERO INFO DATE 18 MAY 99 (V1)   EFFECTIVE DATE 15 JUL 99

LONDON HEATHROW
LLZ DME RWY 27L
I–LL 109.50

EGLL
CAT A,B,C,D

AD ELEV	THR ELEV	TRANSITION ALT	VAR
80FT	**77FT**	**6000**	**4°W**

MSA 25NM ARP

ATIS	APPROACH	RADAR	TOWER	(GMC)
123.900,113.750,115.100	119.725,120.400,134.975	125.625,119.900	118.700,118.500,124.475,	(121.900)

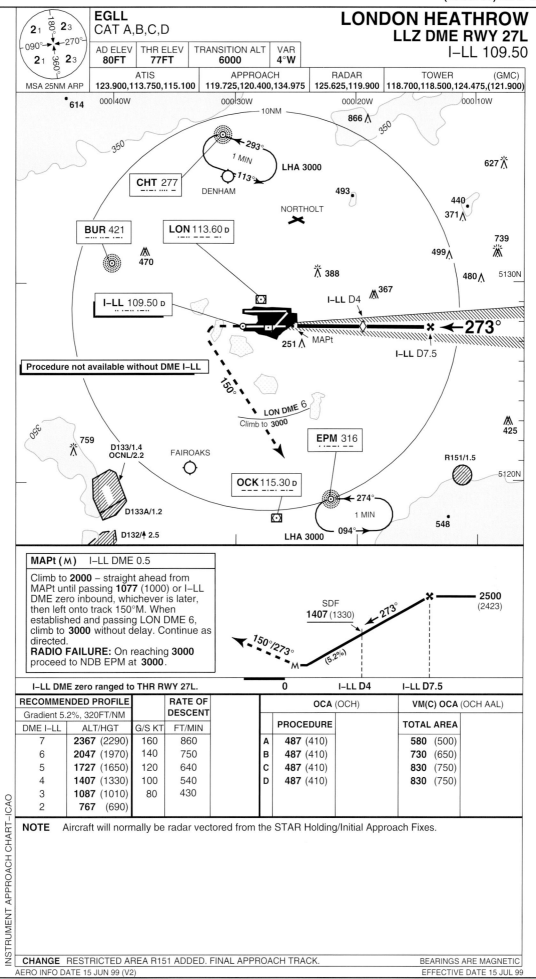

Procedure not available without DME I–LL

MAPt (M)   I–LL DME 0.5

Climb to **2000** – straight ahead from MAPt until passing **1077** (1000) or I–LL DME zero inbound, whichever is later, then left onto track 150°M. When established and passing LON DME 6, climb to **3000** without delay. Continue as directed.
RADIO FAILURE: On reaching **3000** proceed to NDB EPM at **3000**.

I–LL DME zero ranged to THR RWY 27L.

RECOMMENDED PROFILE		RATE OF DESCENT	
Gradient 5.2%, 320FT/NM			
DME I–LL	ALT/HGT	G/S KT	FT/MIN
7	**2367** (2290)	160	860
6	**2047** (1970)	140	750
5	**1727** (1650)	120	640
4	**1407** (1330)	100	540
3	**1087** (1010)	80	430
2	**767** (690)		

	OCA (OCH)		VM(C) OCA (OCH AAL)	
	PROCEDURE		TOTAL AREA	
A	**487** (410)		**580** (500)	
B	**487** (410)		**730** (650)	
C	**487** (410)		**830** (750)	
D	**487** (410)		**830** (750)	

NOTE   Aircraft will normally be radar vectored from the STAR Holding/Initial Approach Fixes.

INSTRUMENT APPROACH CHART–ICAO

CHANGE   RESTRICTED AREA R151 ADDED. FINAL APPROACH TRACK.   BEARINGS ARE MAGNETIC

AERO INFO DATE 15 JUN 99 (V2)   EFFECTIVE DATE 15 JUL 99

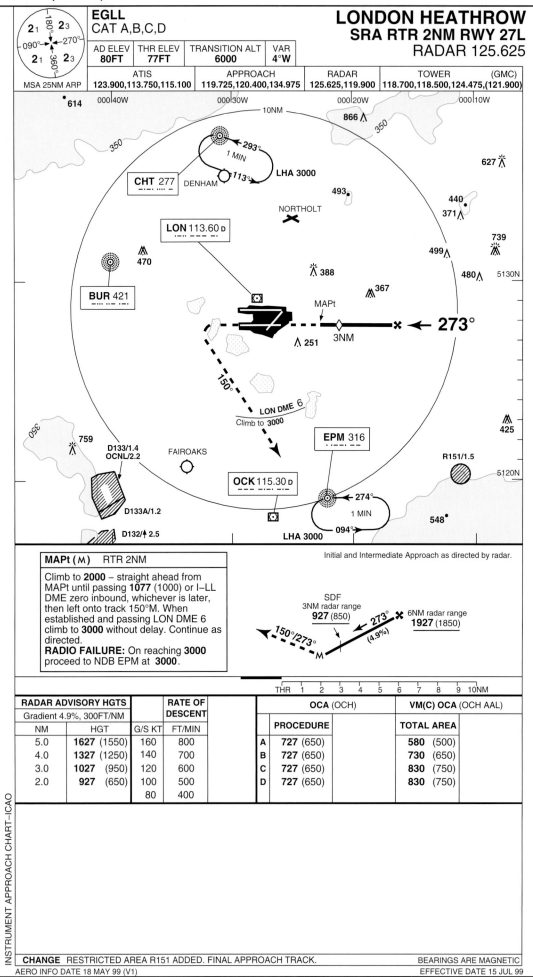

EGLL
CAT A,B,C,D

LONDON HEATHROW
SRA RTR 2NM RWY 27L
RADAR 125.625

AD ELEV	THR ELEV	TRANSITION ALT	VAR
80FT	**77FT**	**6000**	**4°W**

MSA 25NM ARP

ATIS	APPROACH	RADAR	TOWER	(GMC)
123.900,113.750,115.100	**119.725,120.400,134.975**	**125.625,119.900**	**118.700,118.500,124.475,**	**(121.900)**

Initial and Intermediate Approach as directed by radar.

MAPt (M)   RTR 2NM

Climb to **2000** – straight ahead from MAPt until passing **1077** (1000) or I–LL DME zero inbound, whichever is later, then left onto track 150°M. When established and passing LON DME 6 climb to **3000** without delay. Continue as directed.
RADIO FAILURE: On reaching **3000** proceed to NDB EPM at **3000**.

SDF
3NM radar range
927 (850)

273°
(4.9%)

6NM radar range
1927 (1850)

150°/273°

M

THR 1 2 3 4 5 6 7 8 9 10NM

RADAR ADVISORY HGTS				OCA (OCH)		VM(C) OCA (OCH AAL)	
Gradient 4.9%, 300FT/NM		RATE OF DESCENT					
NM	HGT	G/S KT	FT/MIN		PROCEDURE		TOTAL AREA
5.0	**1627** (1550)	160	800	A	**727** (650)		**580** (500)
4.0	**1327** (1250)	140	700	B	**727** (650)		**730** (650)
3.0	**1027** (950)	120	600	C	**727** (650)		**830** (750)
2.0	**927** (650)	100	500	D	**727** (650)		**830** (750)
		80	400				

INSTRUMENT APPROACH CHART–ICAO

CHANGE RESTRICTED AREA R151 ADDED. FINAL APPROACH TRACK.   BEARINGS ARE MAGNETIC

AERO INFO DATE 18 MAY 99 (V1)   EFFECTIVE DATE 15 JUL 99

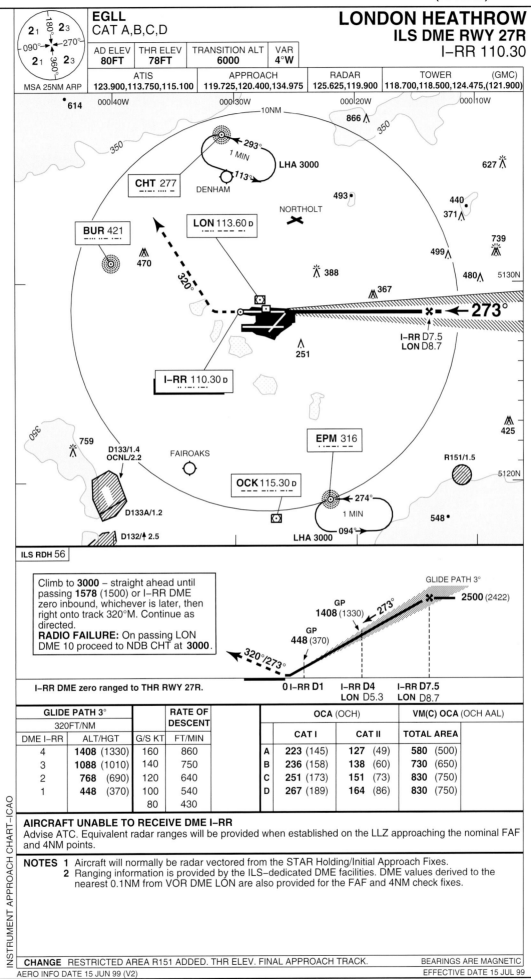

EGLL
CAT A,B,C,D

MSA 25NM ARP

AD ELEV	THR ELEV	TRANSITION ALT	VAR
80FT	**78FT**	**6000**	**4°W**

LONDON HEATHROW
ILS DME RWY 27R
I–RR 110.30

ATIS	APPROACH	RADAR	TOWER (GMC)
123.900,113.750,115.100	119.725,120.400,134.975	125.625,119.900	118.700,118.500,124.475,(121.900)

ILS RDH 56

Climb to **3000** – straight ahead until passing **1578** (1500) or I–RR DME zero inbound, whichever is later, then right onto track 320°M. Continue as directed.
RADIO FAILURE: On passing LON DME 10 proceed to NDB CHT at **3000**.

GLIDE PATH 3°

GP **1408** (1330) 273°   **2500** (2422)

GP **448** (370)

320°/273°

0 I–RR D1   I–RR D4 / LON D5.3   I–RR D7.5 / LON D8.7

I–RR DME zero ranged to THR RWY 27R.

GLIDE PATH 3°			RATE OF DESCENT	
320FT/NM				
DME I–RR	ALT/HGT		G/S KT	FT/MIN
4	**1408**	(1330)	160	860
3	**1088**	(1010)	140	750
2	**768**	(690)	120	640
1	**448**	(370)	100	540
			80	430

	OCA (OCH)		VM(C) OCA (OCH AAL)	
	CAT I	CAT II	TOTAL AREA	
A	**223** (145)	**127** (49)	**580** (500)	
B	**236** (158)	**138** (60)	**730** (650)	
C	**251** (173)	**151** (73)	**830** (750)	
D	**267** (189)	**164** (86)	**830** (750)	

AIRCRAFT UNABLE TO RECEIVE DME I–RR
Advise ATC. Equivalent radar ranges will be provided when established on the LLZ approaching the nominal FAF and 4NM points.

NOTES 1 Aircraft will normally be radar vectored from the STAR Holding/Initial Approach Fixes.
2 Ranging information is provided by the ILS–dedicated DME facilities. DME values derived to the nearest 0.1NM from VOR DME LON are also provided for the FAF and 4NM check fixes.

CHANGE RESTRICTED AREA R151 ADDED. THR ELEV. FINAL APPROACH TRACK.   BEARINGS ARE MAGNETIC

AERO INFO DATE 15 JUN 99 (V2)   EFFECTIVE DATE 15 JUL 99

INSTRUMENT APPROACH CHART–ICAO

EGLL
CAT A,B,C,D

LONDON HEATHROW
LLZ DME RWY 27R
I–RR 110.30

AD ELEV **80FT**	THR ELEV **78FT**	TRANSITION ALT **6000**	VAR **4°W**

MSA 25NM ARP

ATIS	APPROACH	RADAR	TOWER (GMC)
123.900,113.750,115.100	119.725,120.400,134.975	125.625,119.900	118.700,118.500,124.475,(121.900)

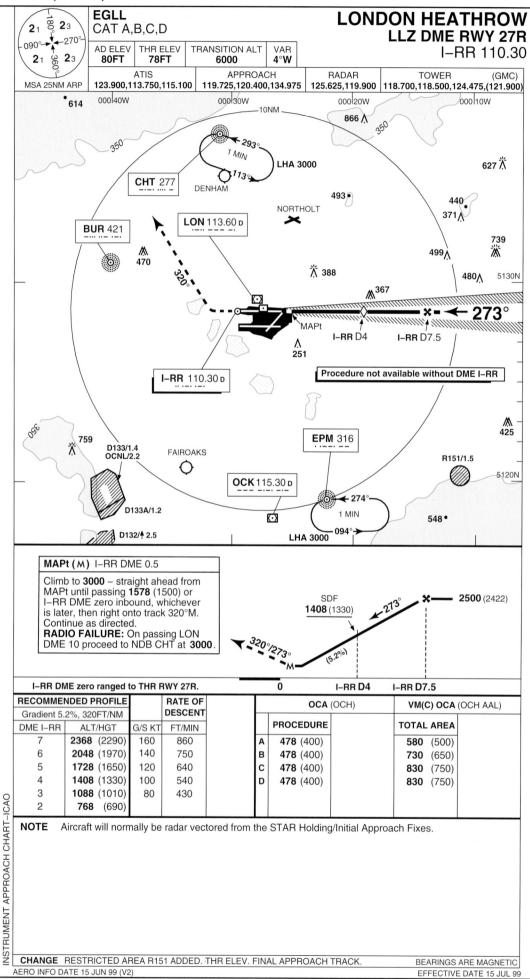

MAPt (M) I–RR DME 0.5

Climb to **3000** – straight ahead from MAPt until passing **1578** (1500) or I–RR DME zero inbound, whichever is later, then right onto track 320°M. Continue as directed.
RADIO FAILURE: On passing LON DME 10 proceed to NDB CHT at **3000**.

I–RR DME zero ranged to THR RWY 27R.

RECOMMENDED PROFILE		RATE OF DESCENT	
Gradient 5.2%, 320FT/NM			
DME I–RR	ALT/HGT	G/S KT	FT/MIN
7	**2368** (2290)	160	860
6	**2048** (1970)	140	750
5	**1728** (1650)	120	640
4	**1408** (1330)	100	540
3	**1088** (1010)	80	430
2	**768** (690)		

OCA (OCH)			VM(C) OCA (OCH AAL)	
	PROCEDURE		TOTAL AREA	
A	**478** (400)		**580** (500)	
B	**478** (400)		**730** (650)	
C	**478** (400)		**830** (750)	
D	**478** (400)		**830** (750)	

NOTE Aircraft will normally be radar vectored from the STAR Holding/Initial Approach Fixes.

INSTRUMENT APPROACH CHART–ICAO

CHANGE RESTRICTED AREA R151 ADDED. THR ELEV. FINAL APPROACH TRACK.
AERO INFO DATE 15 JUN 99 (V2)

BEARINGS ARE MAGNETIC
EFFECTIVE DATE 15 JUL 99

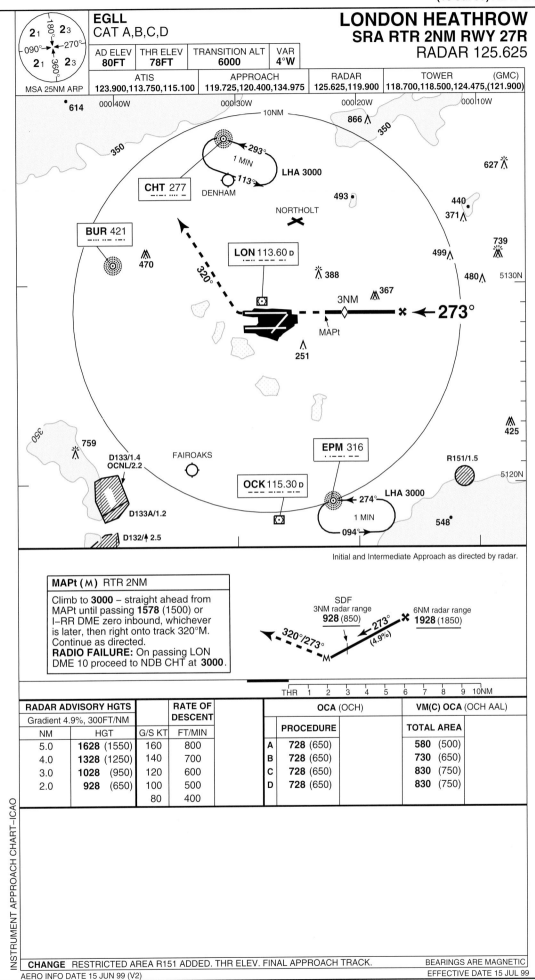

LONDON HEATHROW
SRA RTR 2NM RWY 27R
RADAR 125.625

EGLL
CAT A,B,C,D

AD ELEV	THR ELEV	TRANSITION ALT	VAR
80FT	**78FT**	**6000**	**4°W**

ATIS	APPROACH	RADAR	TOWER (GMC)
123.900,113.750,115.100	**119.725,120.400,134.975**	**125.625,119.900**	**118.700,118.500,124.475,(121.900)**

MSA 25NM ARP

Initial and Intermediate Approach as directed by radar.

MAPt (M) RTR 2NM

Climb to **3000** – straight ahead from MAPt until passing **1578** (1500) or I–RR DME zero inbound, whichever is later, then right onto track 320°M. Continue as directed.
RADIO FAILURE: On passing LON DME 10 proceed to NDB CHT at **3000**.

SDF
3NM radar range **928** (850)
6NM radar range **1928** (1850)
273°
(4.9%)
320°/273°

RADAR ADVISORY HGTS			RATE OF DESCENT	
Gradient 4.9%, 300FT/NM				
NM	HGT	G/S KT	FT/MIN	
5.0	**1628** (1550)	160	800	
4.0	**1328** (1250)	140	700	
3.0	**1028** (950)	120	600	
2.0	**928** (650)	100	500	
		80	400	

	OCA (OCH)		VM(C) OCA (OCH AAL)	
	PROCEDURE		TOTAL AREA	
A	**728** (650)		**580** (500)	
B	**728** (650)		**730** (650)	
C	**728** (650)		**830** (750)	
D	**728** (650)		**830** (750)	

INSTRUMENT APPROACH CHART–ICAO

CHANGE RESTRICTED AREA R151 ADDED. THR ELEV. FINAL APPROACH TRACK.
AERO INFO DATE 15 JUN 99 (V2)

BEARINGS ARE MAGNETIC
EFFECTIVE DATE 15 JUL 99

INTENTIONALLY BLANK

EGTK AD 2.1	EGTK	OXFORD/KIDLINGTON

See AD 2.24 for Chart Listing.

EGTK AD 2.2 – AERODROME GEOGRAPHICAL AND ADMINISTRATIVE DATA		
2	**Direction and distance from city:**	6 nm NW by N of Oxford.
5	**AD Administration:** **Address:**	CSE Aviation Ltd. Oxford Airport, Kidlington, Oxford, OX5 1RA.
	Telephone: **Fax:**	01865-844272 (ATC); 01865-844267 (Operations). 01865-841807.

EGTK AD 2.3 – OPERATIONAL HOURS		
1	**AD:**	**Winter:** Mon-Fri 0800-1730; Sat, Sun and PH 0830-1700; and by arrangement. **Summer:** Mon-Fri 0700-1630; Sat, Sun and PH 0730-1600; and by arrangement.
2	**Customs and Immigration:**	See remarks.
7	**ATS:**	As AD hours. **ATC** Mon-Fri; **AFIS/ATIS** Sat, Sun & PH.
8	**Fuelling:**	As AD hours.
12	**Remarks:**	This aerodrome is **strictly PPR.** At all times all requests must be made via Operations on the numbers listed at AD 2.2. **Customs/Immigration.** All inbound flights must be approved by Operations (on the numbers listed at AD 2.2) and are subject to the following prior notice: (a) All flights with EU crew or PAX 6 hours notice; (b) all flights with non-EU crew or PAX 12 hours notice; (c) all flights requiring Special Branch 24 hours notice (36 hours for weekends). Contact Tel: 01865-844267; or Fax: 01865-841807.

EGTK AD 2.4 – HANDLING SERVICES AND FACILITIES		
2	**Fuel/oil types:**	Fuel: AVTUR JET A-1, AVGAS 100LL. Oil: W80, W100, 80, 100.

EGTK AD 2.6 – RESCUE AND FIRE FIGHTING SERVICES	
1 AD Category for fire fighting:	**Winter:** RFF Category 2 Mon-Fri 0800-1730. Category 1 Sat, Sun 0830-1700 (other times by arrangement). 1 hr earlier in Summer.

EGTK AD 2.10 – AERODROME OBSTACLES

In Approach/Take-off areas			In circling area and at aerodrome		
1			2		
Runway/Area affected	Obstacle type Elevation Markings/lighting	Co-ordinates	Obstacle type Elevation Markings/lighting		Co-ordinates
a	b	c	a		b
		ft amsl		ft amsl	
02/Approach 20/Take-off	Trees Trees Tree Sub-station Vehicles on road	328 315 282 249 259	514932.63N 0011927.95W 514931.93N 0011916.51W 514944.78N 0011917.70W 514945.35N 0011919.49W 514944.66N 0011920.26W	Anemometer (Lgtd) Radio mast (Lgtd) Radio mast (Lgtd)	285 302 272	515018.39N 0011902.46W 514956.93N 0011923.66W 515017.03N 0011900.04W
20/Approach 02/Take-off	Vehicles on road	272	515039.31N 0011903.66W			
03/Approach 21/Take-off	Vehicles on road Trees Trees Tree Tree	262 328 315 282 279	514945.52N 0011916.28W 514932.63N 0011927.95W 514931.93N 0011916.51W 514944.78N 0011917.70W 514945.52N 0011913.51W			
21/Approach 03/Take-off	Chimney	495	515116.93N 0011807.43W			
09/Approach 27/Take-off	Vehicles on road Trees Tree	285 325 308	515005.51N 0011947.45W 515001.84N 0011942.08W 515007.88N 0011955.13W			
27/Approach 09/Take-off	Tree Tree Tree	292 305 289	515005.52N 0011836.05W 515004.15N 0011841.72W 515005.76N 0011846.30W			
12/Approach 30/Take-off	Trees Trees Trees	338 341 322	515031.52N 0011948.48W 515033.82N 0011945.87W 515031.36N 0011943.50W			
3 Remarks:					

EGTK AD 2.13 – DECLARED DISTANCES

Runway Designator	TORA (m)	TODA (m)	ASDA (m)	LDA (m)	Remarks
1	2	3	4	5	6
02 20	1200 1200	1200 1200	1200 1200	1200 1200	
03 21	902 902	902 902	902 902	902 902	
12 30	760 760	760 760	760 760	760 760	
09 27	860 860	860 860	860 860	764 860	
09 (Relief Runway) 27 (Relief Runway)	860 860	860 860	860 860	860 860	

EGTK AD 2.17 – ATS AIRSPACE

Designation and lateral limits	Vertical limits	Airspace Classification
1	2	3
Oxford/Kidlington Aerodrome Traffic Zone (ATZ) Circle radius 2 nm centred on longest notified runway (02/20) 515013N 0011912W.	2000 ft aal SFC	G †

4	**ATS unit callsign:** **Language:**	Oxford Approach/Information. English.
6	**Remarks:**	ATZ hours: **Winter:** **ATC** Mon-Fri 0800-1730; **AFIS** Sat, Sun & PH 0830-1700. **Summer:** **ATC** Mon-Fri 0700-1630; **AFIS** Sat, Sun & PH 0730-1600. † Refer to Section ENR 1.4 for Notifications.

EGTK AD 2.18 – ATS COMMUNICATION FACILITIES

Service Designation	Callsign	Frequency (MHz)	Hours of Operation	Remarks
1	2	3	4	5
APP	Oxford Approach	125.325	Mon-Fri 0800-1730 and by arrangement (Winter). Mon-Fri 0700-1630 and by arrangement (Summer).	
TWR	Oxford Tower Oxford Ground	118.875 121.950	Mon-Fri 0800-1730 and by arrangement (Winter). Mon-Fri 0700-1630 and by arrangement (Summer).	
AFIS	Oxford Information	118.875	Sat, Sun & PH 0830-1700 (Winter). Sat, Sun & PH 0730-1600 (Summer).	
ATIS	Oxford Departure Information	121.750	Sat, Sun & PH 0830-1700 (Winter). Sat, Sun & PH 0730-1600 (Summer).	

EGTK AD 2.19 – RADIO NAVIGATION AND LANDING AIDS

Type Category (Variation)	IDENT	Frequency	Hours of Operation Winter	Summer # and by arrangement	Antenna site co-ordinates	Elevation of DME transmitting antenna	Remarks
1	2	3	4		5	6	7
DME	OX	Ch 124X	Mon-Fri 0800-1730 #	Mon-Fri 0700-1630 #	514956.76N 0011922.14W	279 ft amsl	Zero range is indicated at THR of Runway 02. Related freq 117.70 MHz. Any VOR indications should be ignored.
L	OX	367.5 kHz	Mon-Fri 0800-1730 #	Mon-Fri 0700-1630 #	514956.93N 0011923.66W		On AD. Range 25 nm.
MKR	K	75 MHz	Mon-Fri 0800-1730 #	Mon-Fri 0700-1630 #	515018.65N 0012428.15W		
VDF	Oxford Approach	125.325 MHz	O/R	O/R	515017.03N 0011900.04W		

EGTK AD 2.23 – ADDITIONAL INFORMATION

Aerodrome Regulations. Not applicable.

Warnings

(a) Helicopter training in designated areas takes place on the aerodrome.

(b) Runway 20. Pilots are advised that there is a downslope of 2.25% over the last 200 m at the Southwest end.

(c) EG D129 is located 4.5 nm northeast of the aerodrome.

(d) A jet fuel installation positioned by the eastern taxiway north of the VCR infringes the taxiway strip-width. Aircraft with a wingspan in excess of 15 m should exercise caution when taxiing past the installation.

EGTK AD 2.23 – ADDITIONAL INFORMATION

Noise Abatement Procedures

(a) Pilots are to avoid, where there is no overriding training or Flight Safety requirement, overflying the residential areas, including Blenheim Palace, surrounding Oxford aerodrome.

(b) After departing from Runway 02, climb ahead to 750 ft QFE (1000 ft QNH) and 1.5 DME OX, before turning on course. Pilots carrying out visual departures should endeavour to complete this turn before reaching the Mercury Satellite Station (at 1.5 nm). When turning right, pilots are to avoid overflying the village of Shipton-on-Cherwell.

(c) After departing from Runway 20, climb straight ahead to 750 ft QFE (1000 ft QNH) or 1 DME OX, whichever is earlier, before turning right. Aircraft intending to turn left, climb ahead to 1.5 DME OX (IFR) or until south of Yarnton Village (VFR), remaining clear, in all cases, of the Brize Norton CTR.

(d) After take-off from all other Runways, circuit and departing traffic must climb straight ahead to 750 ft QFE (1000 ft QNH) before turning on course. Circuit height for fixed-wing aircraft is 1200 ft QFE (1450 ft QNH).

(e) Whenever possible aircraft joining the circuit should, subject to ATC approval, plan to join on base leg or via a straight-in approach, giving way to traffic already established in the circuit.

(f) Oxford Airport operates a noise amelioration scheme. All pilots are to comply with the requirements of the scheme, a copy of which is available from airport operations.

Flight Procedures

(a) Circuits variable.

(b) To provide separation between fixed-wing and rotary-wing traffic, the circuit height for fixed-wing aircraft is 1200 ft QFE. All departing fixed-wing aircraft are to climb straight ahead to 750 ft QFE (1000 ft QNH) before turning crosswind.

(c) Visiting pilots should familiarise themselves with the instrument departure routes operated by the Oxford Air Training School. Instrument arrivals should contact Brize Norton ATC as soon as possible after leaving airways on 134.300 MHz or as directed. Arrivals from Daventry: Track to DTY fix 215°/17 nm then turn left to intercept 180° inbound to OX.

(d) Pilots may be asked to use Relief Runway 09/27 which is 40 m wide and marked with white corners.

(e) **Aerodrome Operating Minima - Non-Public Transport Flights:** Refer to AD 1.1.2 before application.

| | | | OCH (ft) | Minima | |
Approach Lighting Category	Runway	Approach Aid	**Acft CAT A** See IAC for other CATs	DH/MDH (ft) **Caution** See AD 1.1.2	RVR (m)
1	2	3	4	5	6
Runway 02 Basic	02	NDB(L) OX/DME OX	420	420	1300
Runway 09 Nil	09	NDB(L) OX/DME OX or MKR K (also without DME)	520	520	1500
	09	NDB(L) OX (MKR inoperative)	950	950	1500
Runway 20 Nil	20	NDB(L) OX/DME OX	500	500	1500

(f) Instrument Approach Procedures (IAP) for this aerodrome are established outside controlled airspace. See ENR 1.5.

EGTK AD 2.24 – CHARTS RELATED TO THE AERODROME

AERODROME CHART – ICAO

BEARINGS ARE MAGNETIC
ELEVATIONS IN FEET AMSL 270

515013N 0011912W   ELEV 270FT   **OXFORD KIDLINGTON**
EGTK

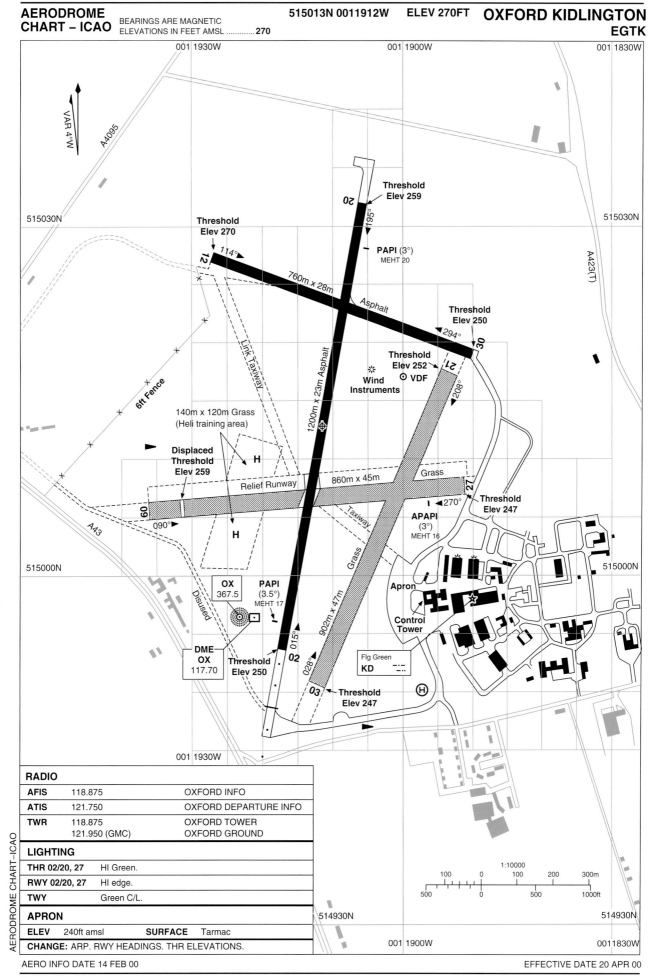

RADIO

AFIS	118.875	OXFORD INFO
ATIS	121.750	OXFORD DEPARTURE INFO
TWR	118.875	OXFORD TOWER
	121.950 (GMC)	OXFORD GROUND

LIGHTING

THR 02/20, 27	HI Green.	
RWY 02/20, 27	HI edge.	
TWY	Green C/L.	

APRON

ELEV	240ft amsl	**SURFACE** Tarmac
CHANGE: ARP. RWY HEADINGS. THR ELEVATIONS.		

AERODROME CHART–ICAO

AERO INFO DATE 14 FEB 00

EFFECTIVE DATE 20 APR 00

Civil Aviation Authority

AMDT 4/00

INTENTIONALLY   BLANK

EGTK
CAT A,B,C

OXFORD/KIDLINGTON
NDB(L) DME RWY 02
OX 367.5/OX 117.70

AD ELEV	THR ELEV	TRANSITION ALT	VAR
270FT	250FT	3000	4°W

MSA 25NM ARP

DEP ATIS	APPROACH	TOWER	AFIS	BRIZE RADAR
121.750	125.325	118.875, 121.950 (GMC)	118.875	134.300

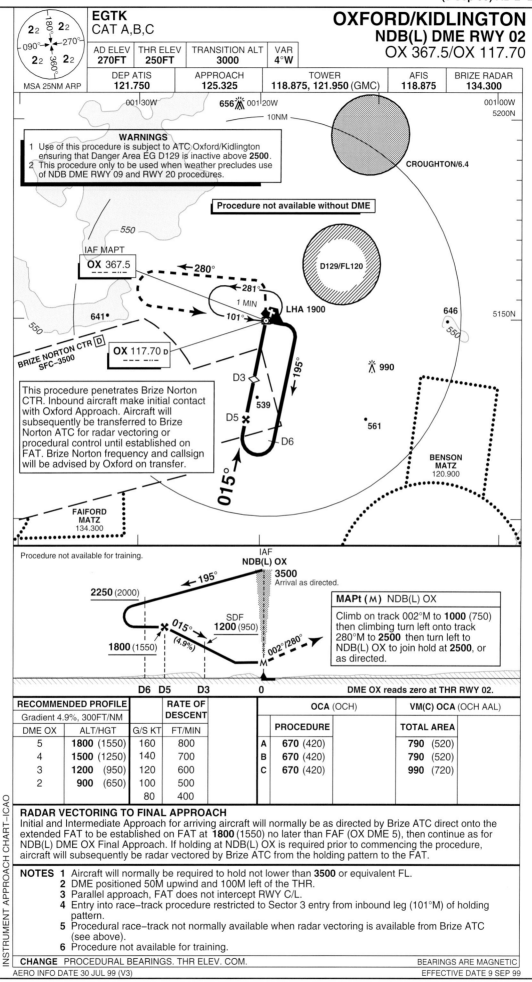

WARNINGS
1 Use of this procedure is subject to ATC Oxford/Kidlington ensuring that Danger Area EG D129 is inactive above **2500**.
2 This procedure only to be used when weather precludes use of NDB DME RWY 09 and RWY 20 procedures.

Procedure not available without DME

CROUGHTON/6.4

D129/FL120

OX 367.5

280°
281°
1 MIN
101°
LHA 1900

646
550

OX 117.70 D

BRIZE NORTON CTR D
SFC–3500

641

550

195°

D3
539
D5
D6

990

561

015°

This procedure penetrates Brize Norton CTR. Inbound aircraft make initial contact with Oxford Approach. Aircraft will subsequently be transferred to Brize Norton ATC for radar vectoring or procedural control until established on FAT. Brize Norton frequency and callsign will be advised by Oxford on transfer.

BENSON
MATZ
120.900

FAIFORD
MATZ
134.300

Procedure not available for training.

IAF
NDB(L) OX

3500
Arrival as directed.

2250 (2000)

195°

015°
(4.9%)

SDF
1200 (950)

002°/280°

1800 (1550)

M

MAPt (M) NDB(L) OX

Climb on track 002°M to **1000** (750) then climbing turn left onto track 280°M to **2500** then turn left to NDB(L) OX to join hold at **2500**, or as directed.

DME OX reads zero at THR RWY 02.

	D6	D5	D3	0

RECOMMENDED PROFILE			RATE OF DESCENT		OCA (OCH)		VM(C) OCA (OCH AAL)	
Gradient 4.9%, 300FT/NM								
DME OX	ALT/HGT	G/S KT	FT/MIN		**PROCEDURE**		**TOTAL AREA**	
5	**1800** (1550)	160	800	A	670 (420)		790 (520)	
4	**1500** (1250)	140	700	B	670 (420)		790 (520)	
3	**1200** (950)	120	600	C	670 (420)		990 (720)	
2	**900** (650)	100	500					
		80	400					

RADAR VECTORING TO FINAL APPROACH
Initial and Intermediate Approach for arriving aircraft will normally be as directed by Brize ATC direct onto the extended FAT to be established on FAT at **1800** (1550) no later than FAF (OX DME 5), then continue as for NDB(L) DME OX Final Approach. If holding at NDB(L) DME OX is required prior to commencing the procedure, aircraft will subsequently be radar vectored by Brize ATC from the holding pattern to the FAT.

NOTES 1 Aircraft will normally be required to hold not lower than **3500** or equivalent FL.
2 DME positioned 50M upwind and 100M left of the THR.
3 Parallel approach, FAT does not intercept RWY C/L.
4 Entry into race–track procedure restricted to Sector 3 entry from inbound leg (101°M) of holding pattern.
5 Procedural race–track not normally available when radar vectoring is available from Brize ATC (see above).
6 Procedure not available for training.

CHANGE PROCEDURAL BEARINGS. THR ELEV. COM.	BEARINGS ARE MAGNETIC
AERO INFO DATE 30 JUL 99 (V3)	EFFECTIVE DATE 9 SEP 99

INSTRUMENT APPROACH CHART–ICAO

EGTK
CAT A,B

OXFORD/KIDLINGTON
NDB(L) DME RWY 09
OX 367.5/OX 117.70

MSA 25NM ARP

AD ELEV	THR ELEV	TRANSITION ALT	VAR
270FT	**259FT**	**3000**	**4°W**

DEP ATIS	APPROACH	TOWER	AFIS	BRIZE RADAR
121.750	**125.325**	**118.875, 121.950** (GMC)	**118.875**	**134.300**

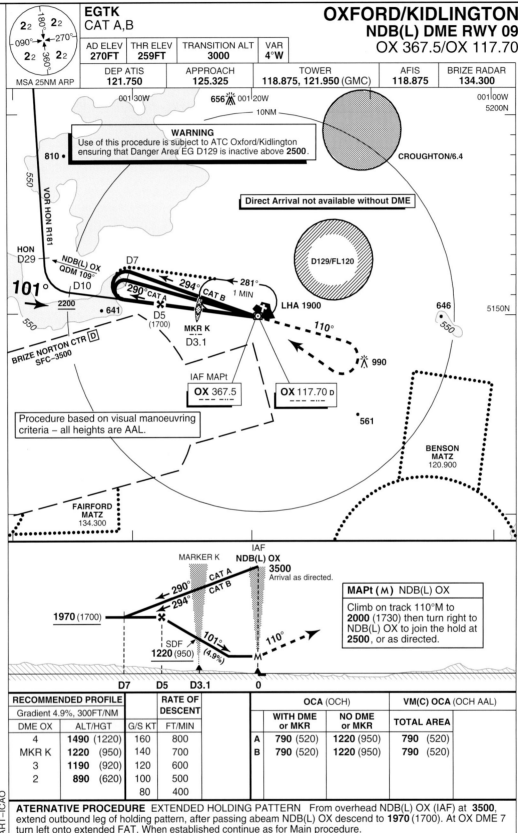

WARNING
Use of this procedure is subject to ATC Oxford/Kidlington ensuring that Danger Area EG D129 is inactive above **2500**.

CROUGHTON/6.4

Direct Arrival not available without DME

D129/FL120

Procedure based on visual manoeuvring criteria – all heights are AAL.

BENSON MATZ 120.900

FAIRFORD MATZ 134.300

IAF MAPt
OX 367.5
OX 117.70 D

BRIZE NORTON CTR D SFC–3500

MAPt (M) NDB(L) OX

Climb on track 110°M to **2000** (1730) then turn right to NDB(L) OX to join the hold at **2500**, or as directed.

MARKER K
IAF NDB(L) OX **3500** Arrival as directed.

1970 (1700)

SDF **1220** (950) (4.9%)

	D7	D5	D3.1	0

RECOMMENDED PROFILE		RATE OF DESCENT		OCA (OCH)			VM(C) OCA (OCH AAL)	
Gradient 4.9%, 300FT/NM					WITH DME or MKR	NO DME or MKR		
DME OX	ALT/HGT	G/S KT	FT/MIN				TOTAL AREA	
4	**1490** (1220)	160	800	A	**790** (520)	**1220** (950)	**790** (520)	
MKR K	**1220** (950)	140	700	B	**790** (520)	**1220** (950)	**790** (520)	
3	**1190** (920)	120	600					
2	**890** (620)	100	500					
		80	400					

ATERNATIVE PROCEDURE EXTENDED HOLDING PATTERN From overhead NDB(L) OX (IAF) at **3500**, extend outbound leg of holding pattern, after passing abeam NDB(L) OX descend to **1970** (1700). At OX DME 7 turn left onto extended FAT. When established continue as for Main procedure.

AIRCRAFT UNABLE TO RECEIVE DME OX As for Main and Alternative procedures, except time outbound 3.5 MIN (CAT A); 3 MIN (CAT B). When established on FAT, descend to cross MKR K **not below 1220** (950), then continue to MDH.

DIRECT ARRIVAL VIA VOR HON R181 Subject to prior approval from Brize Norton/Oxford ATC. Intercept and follow VOR HON R181 **not below** MSA. At HON DME 29 (NDB(L) OX lead QDM 109°) turn left to intercept extended FAT. When established, from OX DME 10 descend to **1970** (1700) and continue as for Main procedure.

NOTES 1 Aircraft will normally be required to hold not lower than **3500** or equivalent FL.
　　　　　2 FAT off-set 11° from RWY C/L.
　　　　　3 DME OX becomes near tangential to FAT after DME 0.5 (approx 0.3NM before THR).

CHANGE COM.　　　　　　　　　　　　　　　　　　　　BEARINGS ARE MAGNETIC

INSTRUMENT APPROACH CHART–ICAO

AERO INFO DATE 30 JUL 99 (V3)　　　　　　　　　　　　　EFFECTIVE DATE 9 SEP 99

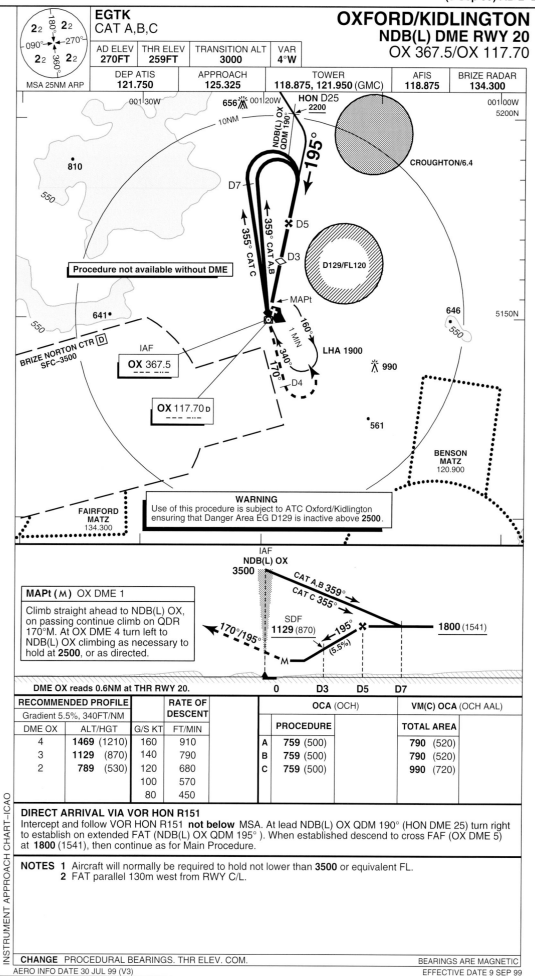

EGTK
CAT A,B,C

OXFORD/KIDLINGTON
NDB(L) DME RWY 20
OX 367.5/OX 117.70

MSA 25NM ARP

AD ELEV	THR ELEV	TRANSITION ALT	VAR
270FT	**259FT**	**3000**	**4°W**

DEP ATIS	APPROACH	TOWER	AFIS	BRIZE RADAR
121.750	**125.325**	**118.875, 121.950** (GMC)	**118.875**	**134.300**

Procedure not available without DME

HON D25 2200
CROUGHTON/6.4
D129/FL120
D7
D5
D3
MAPt
LHA 1900
IAF OX 367.5
OX 117.70 D
D4
BRIZE NORTON CTR D SFC–3500
FAIRFORD MATZ 134.300
BENSON MATZ 120.900

QDM 190° / NDB(L) OX
195°
359° CAT A,B
355° CAT C
160° 1 MIN
340°
170°
810
641
646
561
990
656
5200N
5150N
001°30W 001°20W 001°00W
10NM

WARNING
Use of this procedure is subject to ATC Oxford/Kidlington ensuring that Danger Area EG D129 is inactive above **2500**.

IAF NDB(L) OX 3500
CAT A,B 359°
CAT C 355°
SDF 1129 (870)
195° (5.5%)
1800 (1541)
170°/195°
M
D3 D5 D7
0

MAPt (M) OX DME 1

Climb straight ahead to NDB(L) OX, on passing continue climb on QDR 170°M. At OX DME 4 turn left to NDB(L) OX climbing as necessary to hold at **2500**, or as directed.

DME OX reads 0.6NM at THR RWY 20.

RECOMMENDED PROFILE			RATE OF DESCENT			OCA (OCH)		VM(C) OCA (OCH AAL)	
Gradient 5.5%, 340FT/NM									
DME OX	ALT/HGT		G/S KT	FT/MIN		PROCEDURE		TOTAL AREA	
4	**1469**	(1210)	160	910	A	**759**	(500)	**790**	(520)
3	**1129**	(870)	140	790	B	**759**	(500)	**790**	(520)
2	**789**	(530)	120	680	C	**759**	(500)	**990**	(720)
			100	570					
			80	450					

DIRECT ARRIVAL VIA VOR HON R151
Intercept and follow VOR HON R151 **not below** MSA. At lead NDB(L) OX QDM 190° (HON DME 25) turn right to establish on extended FAT (NDB(L) OX QDM 195°). When established descend to cross FAF (OX DME 5) at **1800** (1541), then continue as for Main Procedure.

NOTES 1 Aircraft will normally be required to hold not lower than **3500** or equivalent FL.
        2 FAT parallel 130m west from RWY C/L.

INSTRUMENT APPROACH CHART–ICAO

CHANGE PROCEDURAL BEARINGS. THR ELEV. COM.   BEARINGS ARE MAGNETIC
AERO INFO DATE 30 JUL 99 (V3)   EFFECTIVE DATE 9 SEP 99

INTENTIONALLY   BLANK

GEN 3.5 — METEOROLOGICAL SERVICES

GEN 3.5.1 — RESPONSIBLE AUTHORITY

1   The Civil Aviation Authority is the Meteorological Authority for the United Kingdom (UK). This authority is derived from Directions issued under section 66(1) of the Transport Act 2000 relating to the Civil Aviation Authority's performance of air navigation functions. The policy of the UK Met Authority is to discharge its responsibilities for the provision of meteorological services to UK based national and international civil aviation operations in accordance with ICAO Annex 3 and other national and international requirements as may be promulgated from time to time.

Postal address:	Head of Met Authority Civil Aviation Authority Directorate of Airspace Policy CAA House, K6 45-59 Kingsway London, WC2B 6TE.
Tel: Fax: AFS: e-mail:	020-7453 6526 020-7453 6565 EGGAYFYM metauthority@dap.caa.co.uk

2   Meteorological forecasting and climatological services for civil aviation in the United Kingdom are provided by The Met Office acting as the agent for the Civil Aviation Authority.

Postal address:	The Met Office Manager Civil Aviation Branch London Road, Bracknell, RG12 2SZ.
Tel: Fax: AFS:	01344-854226 01344-854156 (Administrative) EGRRYTYH (Administrative) or; EGRRYMYX (National Meteorological Centre (NMC))

3   Applicable ICAO Documents

3.1   The Standards, Recommended Practices and, when applicable, the procedures contained in the following ICAO documents are applied:

Annex 3	- Meteorological Service for International Air Navigation.
Doc 7754	- Air Navigation Plan - EUR Region.
Doc 8400	- PANS - ICAO Abbreviations and Codes.
Doc 8755	- Air Navigation Plan - NAT Region.
Doc 8896	- Manual of Aeronautical Meteorological Practice.
Doc 9328	- Manual of Runway Visual Range Observing and Reporting Practices.

3.2   Differences from ICAO Standards, Recommended Practices and Procedures are listed in GEN 1.7.

GEN 3.5.2 — AREAS OF RESPONSIBILITY

1   The United Kingdom provides area Meteorological Watch for the London and Scottish FIR/UIR and for the Shanwick FIR/OCA. The National Meteorological Centre (NMC) at Bracknell acts as the Meteorological Watch Office (MWO) for these areas.

GEN 3.5.3 — OBSERVATIONS AND REPORTS

1   Observing Systems and Operating Procedures

1.1   Surface wind sensors on aerodromes are positioned to give the best practical indication of the winds which an aircraft will encounter during take-off and landing within the layer between 6 and 10 m above the runway(s). The surface wind reported for take-off and landing by ATS Units at aerodromes supporting operations by aircraft whose maximum total weight authorised is below 5700 kg is usually an instantaneous wind measurement with direction referenced to Magnetic North. However, at other designated aerodromes the wind reports for take-off and landing are averaged over the previous 2 minutes. Variations in the wind direction are given when the total variation is 60° or more and the mean speed above 3 kt, the directional variations are expressed as the two extreme directions between which the wind has varied in the past 10 minutes. In reports for take-off, surface winds of 3 kt or less include a range of wind directions whenever possible if the total variation is 60° or more. Variations from the mean wind speed (gust and lulls) during the past 10 minutes are only reported when the variation from the mean speed has exceeded 10 kt. Such variations are expressed as the maximum and minimum speeds attained.

GEN 3.5.3 — OBSERVATIONS AND REPORTS

1.1.1 At aerodromes which normally report surface wind averaged over the previous 2 minutes, the instantaneous wind velocity is available on request. Where an instantaneous wind velocity has been requested the word 'instant' will be inserted in the report (eg. 'G-CD Runway 34 cleared to land instant surface wind 270 7' or 'G-CD Runway 34 cleared to land instant 270 7'). An indication of the wind velocity normally reported at particular aerodromes is included in Table 3.5.3.2. Aerodromes not featuring in the Table use instant reports with the exception of Wycombe Air Park where an averaged value is normally used

1.1.2 Surface wind measurements contained in METAR and SPECI reports are referenced to True North and are averaged over the previous 10 minutes, except when during the 10 minute period there is an abrupt and sustained change in wind direction of 30° or more, with a wind speed of at least 10 kt both before and after the change, or a change in wind speed of 10 kt or more lasting more than 2 minutes. In this case only data occurring since the abrupt change will be used to obtain the mean values. METAR and SPECI reports may give variations in wind direction if during the 10 minute period preceding the time of observation, the total variation in wind direction is 60° or more and the speed greater than 3 kt. The maximum speed is only given if it exceeds the mean speed by 10 kt or more. At aerodromes with wind sensors at two or more sites, METAR surface wind reports are always obtained from one designated 'aerodrome system' irrespective of the system currently in use by the ATS Unit for take-off and landing reports.

1.2   Information on cloud height is obtained by the use of ceilometers (nodding beam or laser), cloud searchlights and alidades, balloons, pilot reports and observer estimation. At some aerodromes an additional, cloud ceilometer may be installed on the approach. The cloud heights reported from an approach ceilometer are:

(a)   The most frequently occurring value during the past 10 minutes if this value is 1000 ft or less;

(b)   if cloud is being indicated at heights 100 ft or more below that indicated at (a) above then the height of the lowest cloud is also reported, prefaced by 'OCNL';

(c)   if the most frequently occurring value is above 1000 ft but the lowest value is 1000 ft or below then only the lowest is reported; for example: most frequent 1200 ft, lowest 900 ft 'CBR 24 OCNL 900 FEET' (24 refers to the nearest runway).

1.3   Temperature is reported in whole degrees from liquid-in-glass or electrical resistance thermometers located in a ventilated screen.

1.4   Horizontal surface visibility is assessed by human observer. Visibility is reported in increments of 50 m up to 500 m, and then increments of 100 m up to 5000 m and in units of kilometres above 5000 m. In METAR, SPECI and TAF the maximum value is '10 km or more'. When the visibility varies in different directions the lowest visibility is reported, however, if there is a marked directional variation, that is the highest visibility exceeds the lowest by 50% or more, and the minimum visibility is less than 5000 m, then the compass direction of the lowest visibility relative to the observer is appended. When the lowest visibility is less than 1500 m and visibility in another direction is more than 5000 m, additionally, the maximum visibility and the direction in which it occurs is reported.

1.4.1 Pilots are reminded that surface visibility forecast in a TAF, TREND or Area Forecast might be subject to marked deterioration caused by smoke at any time. Such deteriorations in surface visibility will be reported as they occur in Routine or Special Aerodrome Meteorological Reports, and forecasts might consequently be amended. It is not possible to forecast the onset or cessation of the smoke, or the precise amount of the visibility deterioration. Turbulence and breathing difficulty might also be encountered.

2   **Accuracy of Meteorological Measurement or Observation**

Element	Accuracy of Measurement or Observation (1994)
Mean surface wind	Direction: ± 5° Speed: ± 1 kt up to 20 kt, ± 5% above 20 kt
Variations from the mean surface wind speed	± 1 kt
Visibility	± 50 m up to 500 m ± 10 % between 500 m and 2000 m ± 20 % above 2000 m up to 10 km
RVR	± 25 m up to 150 m ± 50 m between 150 m and 500 m ± 10 % above 500 m up to 2000 m
Cloud amount	± 1 okta in daylight, worse in darkness and during atmospheric obscuration
Cloud height	± 33 ft up to 3300 ft ± 100 ft above 3300 ft up to 10000 ft
Air temperature and dew point temperature	± 0.2°C
Pressure value (QNH, QFE)	± 0.3 mb

Note: The accuracy stated refers to assessment by instruments (except cloud amount); it is not usually attainable in observations made without the aid of instruments.

3   Details of Meteorological Observations and Reports for UK aerodromes are listed in table 3.5.3.2.

4 Aerodrome Warnings

4.1 Aerodrome Warnings are issued as appropriate when one or more of the following phenomena occurs or is expected to occur:

(a) Gales (when the mean surface wind is expected to exceed 33 kt, or if gusts are expected to exceed 42 kt) or strong winds according to locally-agreed criteria;

(b) squalls, hail or thunderstorms;

(c) snow, including the expected time of beginning, duration and intensity of fall; the expected depth of accumulated snow, and the time of expected thaw. Amendments or cancellations are issued as necessary;

(d) frost warnings are issued when any of the following conditions are expected:

   (i) A ground frost with air temperatures not below freezing point;

   (ii) the air temperature above the surface is below freezing point (air frost);

   (iii) hoar frost, rime or glaze deposited on parked aircraft.

(e) fog (normally when the visibility is expected to fall below 600 m);

(f) rising dust or sand;

(g) freezing precipitation.

4.2 Aerodrome operators requiring notification of the above warnings should apply to the MET Authority (see GEN 3.5.1).

4.3 The normal method of notifying Aerodrome Warnings is by a single AFS, telephone message or Fax by special arrangement to the aerodrome, with local dissemination of the warning being the responsibility of the aerodrome operator.

5 Special Facilities

5.1 Marked Temperature Inversion

5.1.1 At certain aerodromes annotated at Table 3.5.3.2, a Warning of Marked Temperature Inversion is issued whenever a temperature difference of 10°C or more exists between the surface and any point up to 1000 ft above the aerodrome. This warning is broadcast on departure and arrival ATIS at aerodromes so equipped, or in the absence of ATIS passed by radio to departing aircraft before take-off, and to arriving aircraft as part of the report of aerodrome meteorological conditions.

5.2 Windshear Alerting Service - London Heathrow and Belfast Aldergrove Airports.

5.2.2 Forecasters for London Heathrow and Belfast Aldergrove airports regularly review the weather conditions at the two airports and monitor aircraft reports of windshear experienced on the approach or climb out. Where a potential low level (below 1600 ft) windshear condition exists, an Alert is issued; this will be based on one or more of the following criteria:

(a) Mean surface wind speed at least 20 kt;

(b) the magnitude of the vector difference between the mean surface wind and the gradient wind (an estimate of the 2000 ft wind) at least 40 kt;

(c) thunderstorm(s) or heavy shower(s) within approximately 5 nm of the airport.

   Note: Alerts are also issued based on recent pilot reports of windshear on the approach or climb-out.

5.2.3 The Alert message is given in the arrival and departure ATIS broadcasts at Heathrow and by RTF to arriving and departing aircraft at Aldergrove in one of three formats:

(a) 'WINDSHEAR FORECAST' (WSF) - when the meteorological conditions indicate that low level windshear on the approach or climb-out (below 2000 ft) might be encountered;

(b) 'WINDSHEAR FORECAST AND REPORTED' (WSFR) - as above, supported by a report from at least one aircraft of windshear on the approach or climb-out within the last hour;

(c) 'WINDSHEAR REPORTED' (WSR) - when an aircraft has reported windshear on the approach or climb-out within the last hour, but insufficient meteorological evidence exists for the issue of a forecast of windshear.

5.2.4 Pilot reports of windshear on approach or climb-out can greatly enhance the operational efficiency of this service. In addition, they also serve in the continuous evaluation of the criteria upon which Alerts are forecast. Thus pilots who experience windshear on the approach or climb-out are requested to report the occurrence to ATC, as soon as it is operationally possible to do so, even if an Alert has been issued. Windshear reporting criteria are shown at GEN 3.5.6 paragraph 3.2. Pilots who experience windshear at any UK aerodrome are requested to report it in the same way.

5.3 Runway Visual Range (RVR)

5.3.1 RVR assessment is made by either a Human Observer or an Instrumented RVR system (IRVR). Most IRVR systems have an upper limit of 1500 m; the upper limit of the Human Observer system is normally less than this. The system in use at particular aerodromes is indicated in Table 3.5.3.2, and explained in Table 3.5.3.3.

GEN 3.5.3 — OBSERVATIONS AND REPORTS

5.3.2 The United Kingdom standard RVR reporting incremental scale is 25 m between 0 and 400 m, 50 m between 400 and 800 m, and 100 m above 800 m. Some IRVR systems are unable to report every incremental point in the scale. Known limitations in IRVR systems are as follows:

Isle of Man (EGNS):	Due to profile of Main Runway, the maximum reportable value of RVR is 900 m.
Jersey (EGJJ):	System cannot show incremental steps of less than 50 m. RVR Values will be given in 50 m steps from 0 to 800m.
Liverpool (EGGP):	IRVR 50 - 1500 m (175 m not reported).
London/City (EGLC):	IRVR 50 - 1100 m IRVR remains serviceable if TDZ fails.
London Gatwick (EGKK):	RWY 08L IRVR remains serviceable if TDZ fails. RWY 26R TDZ IRVR is considerably displaced from Start-to-Roll position.

5.3.3 The assessment and reporting of RVR begins whenever the horizontal visibility or the RVR is observed to be less than 1500 m. At those aerodromes where IRVR is available, RVR may also be reported when the observed value is at or below the maximum reportable value or when shallow fog is forecast or reported.

5.3.4 RVR is passed to aircraft before take-off and during the approach to landing. Changes in the RVR are passed to aircraft throughout the approach. Additionally, information from pilot reports or ATC observation that the visibility on the runway is worse than that indicated by the RVR report, for example patches of thick fog, are passed.

5.3.5 The table at Table 3.5.3.2 shows which IRVR system is provided at an aerodrome. Aerodromes using AGIVIS systems suppress mid-point and/or stop-end values when:

(a) They are equal to or higher than the touchdown zone value unless they are less than 400 m; or

(b) they are 800 m or more.

Aerodromes using MET-1 systems suppress mid-point and/or stopend values unless they are 550 m or less.

5.3.6 At those aerodromes having multi-site IRVR the standard UK procedure is that the touchdown zone RVR is always given first, followed by any values for the mid-point RVR and/or stopend RVR which have not been suppressed. With the above in mind, the co-operation of the pilots is sought in avoiding unnecessary radio requests for mid-point and/or stopend values when they have not been given. When all three values are given they are passed as a series of numbers, for example, 'RVR 600, 500, 550' relates to touchdown zone, mid-point and stopend respectively. When touchdown zone and only one other value is given, the latter is prefixed with its appropriate position, for example, 'RVR 700, stop-end 650'.

5.3.7 If a single transmissometer fails, and the remainder of the IRVR system is still serviceable, RVR readings are not suppressed for the remaining sites and these values are passed to pilots. For example, if the touchdown zone transmissometer is unserviceable, 'RVR: touchdown missing, mid-point 600, stopend 400'. If two transmissometers fail in a three-site IRVR system, the remaining value is passed and identified provided that it is not the stopend value, in which event the system is considered unserviceable for that runway direction. In a two-site IRVR system, giving touchdown zone and stopend values, if the touchdown zone transmissometer fails, the system is considered unserviceable for that runway direction. Exceptionally, if the Civil Aviation Authority determines in a particular case that the distance between the two transmissometers is sufficiently small, the system may be considered serviceable for that runway direction.

5.3.8 When RVR information is not available, or when the RVR element of Aerodrome Operating Minima falls outside the range of reportable RVR, pilots should use meteorological visibility in the manner specified in their operations manuals, or at AD 1.1.2 (Aerodrome Operating Minima – Non-public transport flights by aircraft).

5.3.9 Information on specific types and locations of observation systems at particular aerodromes is available on request from the Meteorological Authority or the aerodrome operator.

6   Climatological information for certain UK aerodromes is available for civil aviation purposes from the Met Office, according to the following criteria:

	Table 3.5.3.1 — Climatological Information
A	Climatological statistics readily available based on at least 10 years' of three-hourly (usually hourly) data.
B	Climatological statistics readily available based on less than 10 years' data and/or some gaps in night-time data.
C	Limited climatological statistics available based on available METAR reports starting July 1983 or later.
D	No data available or insufficient data available to provide climatological statistics.

The aerodromes are so classified at Table 3.5.3.2.

6.1   Climatological statistics for routes and areas in the United Kingdom are not available. However, global climatology of upper wind and temperature data is held by the Met Office.

GEN 3.5.3 — OBSERVATIONS AND REPORTS

Table 3.5.3.2 — Meteorological Observations at Aerodromes

Aerodrome/ Location Indicator	Observations			Surface Wind	RVR		Obs Hours	Climatological Data
	Type	Freq	Warnings		Sites	Eqpt		
Aberdeen/Dyce EGPD	METAR SYNOP	h H	AW MTI	Average	16 TDZ/END 34 TDZ/END	AGIVIS AGIVIS	HO+ H24	A
Alderney EGJA	METAR	h	AW	Instant	08/26 TDZ	OBS	HO	C
Barra EGPR	METAR SYNOP	I H	AW	Average			HO HO	B
Barrow/Walney Island EGNL	SYNOP §	H		Instant			H24	D
Belfast Aldergrove EGAA	METAR SYNOP	h H	AW MTI Windshear	Average	07 TDZ/MID/END 25 TDZ/MID/END 07 TDZ 17 TDZ 25 TDZ/MID 35 TDZ	AGIVIS AGIVIS OBS OBS OBS OBS	H24 H24	A
Belfast/City EGAC	METAR	h	AW MTI	Average	04/22 TDZ	OBS	HO	C
Benbecula EGPL	METAR	h	AW	Average	06/24 TDZ	OBS	HO	A
Biggin Hill EGKB	METAR	h	AW	Average	21 TDZ	OBS	HO	C
Birmingham EGBB	METAR	h	AW MTI	Average	15 TDZ/MID/END 33 TDZ/MID/END	AGIVIS AGIVIS	H24 H24	C
Blackpool EGNH	METAR	h	AW MTI	Average	10/28 TDZ	OBS	HO	A, C
Bournemouth EGHH	METAR SYNOP §	h	AW	Average	08/26 TDZ	OBS	H24	A
Bristol EGGD	METAR	h	AW MTI	Average	09/27 TDZ/END	Vaisala	H24	C
Cambridge EGSC	METAR	h	AW MTI	Average	05/23 TDZ	OBS	HO	C
Campbeltown EGEC	METAR	H	AW	Average			HO	B
Cardiff EGFF	METAR	h	AW MTI	Average	12/30 TDZ/END	AGVIS	H24	C
Carlisle EGNC	METAR	H	AW	Instant	07/25 TDZ	OBS	HO	C
Coventry EGBE	METAR	h	AW	Average	05/23 TDZ	OBS	HO	C
Cranfield EGTC	METAR	H	AW	Average			HO	C
Dundee EGPN	METAR	h	AW	Average			HO	D
Dunkeswell EGTU	SYNOP §	H	AW	Instant			H24	B
East Midlands EGNX	METAR	h	AW MTI	Average	09 TDZ/MID/END 27 TDZ/MID/END	MET-1 MET-1	H24	C
Edinburgh EGPH	METAR	h	AW MTI	Average	06 TDZ/MID/END 24 TDZ/MID/END 06/24 TDZ	AGIVIS AGIVIS OBS	H24	A
Exeter EGTE	METAR	h	AW MTI	Average	08/26 TDZ	OBS OBS	HO	A, C

Aerodrome/ Location Indicator	Observations			Surface Wind	RVR		Obs Hours	Climatological Data
	Type	Freq	Warnings		Sites	Eqpt		
Fairoaks EGTF	METAR ‡	I	AW	Instant			HO	D
Farnborough EGLF	METAR	h	AW	Instant	06/24 TDZ	OBS	HO	B
Filton EGTG	METAR	h	AW	Average			HO	A, C
Glasgow EGPF	METAR	h	AW MTI	Average	05 TDZ/MID/END 23 TDZ/MID/END	AGIVIS AGIVIS	H24	A
Gloucestershire EGBJ	METAR	h	AW	Average	09/27 TDZ	OBS	HO	C
Guernsey EGJB	METAR SYNOP	h H	AW	Average	09/27 TDZ	OBS	HO H24	A
Hawarden EGNR	METAR ‡ SYNOP §	I H	AW	Average			HO H24	D
Humberside EGNJ	METAR	h	AW	Average	03/21 TDZ	OBS	HO	C
Inverness EGPE	METAR	h	AW	Average	06/24 TDZ	OBS	HO	B
Islay EGPI	METAR	H	AW	Average			HO	D
Isle of Man EGNS	METAR SYNOP	h H	AW MTI	Average	26 TDZ	OBS	HO+ H24	A
Jersey EGJJ	METAR SYNOP	h H	AW	Instant	09/27 TDZ/MID/END	Elecma	HO+ H24	A
Kirkwall EGPA	METAR	h	AW	Average			HO+	A
Leeds Bradford EGNM	METAR	h	AW MTI	Average	14 TDZ 32 TDZ/MID 14 TDZ/MID 32 TDZ/MID	OBS OBS MET-1 MET-1	HO	C
Liverpool EGGP	METAR	h	AW MTI	Average	09 TDZ/MID/END 27 TDZ/MID/END	MET-1 MET-1	H24	A, C
London/City EGLC	METAR	h	AW MTI	Average	10 TDZ/END 28 TDZ/END	MET-1 MET-1	HO+	C
London Gatwick EGKK	METAR	h	AW MTI	Average	08R TDZ/MID/END 26L TDZ/MID/END 08L TDZ/MID 26R TDZ/END	AGIVIS AGIVIS AGIVIS AGIVIS	H24 H24	A
London Heathrow EGLL	METAR SYNOP	h H	AW MTI Windshear	Average	09L TDZ/MID/END 09R TDZ/MID/END 27L TDZ/MID/END 27R TDZ/MID/END	AGIVIS AGIVIS AGIVIS AGIVIS	H24 H24	A
London Luton EGGW	METAR	h	AW MTI	Average	08 TDZ/MID/END 26 TDZ/MID/END	AGIVIS AGIVIS	H24	C
London Stansted EGSS	METAR	h	AW MTI	Average	05 TDZ/MID/END 23 TDZ/MID/END	AGIVIS AGIVIS	H24	A
London/Heliport EGLW	METAR ‡	I	AW	Instant			HO	D
Londonderry/Eglinton EGAE	METAR	h	AW	Average	08/26 TDZ	OBS	HO	C
Lydd EGMD	METAR	I	AW	Instant	04/22 TDZ	OBS	HO	C

GEN 3.5.3 — OBSERVATIONS AND REPORTS								
Aerodrome/ Location Indicator	Observations			Surface Wind	RVR		Obs Hours	Climatological Data
	Type	Freq	Warnings		Sites	Eqpt		
Manchester EGCC	METAR SYNOP	h H	AW MTI	Average	06 TDZ/MID/END 24 TDZ/MID/END	AGIVIS AGIVIS	H24 H24	A
Manchester Woodford EGCD	METAR ‡	H	AW	Average	07/25 TDZ 25 Alt TDZ/DEP	OBS	HO	D
Manston EGMH	METAR	h	AW	Average			HO	A
Newcastle EGNT	METAR	h	AW MTI	Average	07 TDZ/MID/END 25 TDZ/MID/END 07/25 TDZ/MID	Elecma Elecma OBS	H24	C
Norwich EGSH	METAR	h	AW MTI	Average	09/27 TDZ	OBS	HO	C
Oxford/Kidlington EGTK	METAR ‡	H	AW	Instant			HO	D
Penzance Heliport EGHK	METAR ‡	I	AW	Instant			HO	D
Plymouth EGHD	METAR	h	AW	Average	31 TDZ	OBS	HO	C
Prestwick EGPK	METAR SYNOP	h I	AW MTI	Average	13/31 TDZ	OBS	H24 H24	A
Scatsta EGPM	METAR	h	AW	Average			HO	C
Scilly Isles/St Mary's EGHE	METAR SYNOP §	h H	AW	Instant			HO H24	A, C
Sheffield City EGSY	METAR	h	AW	Average	10/28 TDZ	OBS	HO	A
Shobdon EGBS	SYNOP §	H	AW	Instant			H24	D
Shoreham EGKA	METAR SYNOP	h I	AW	Average			HO HO	C
Southampton EGHI	METAR	h	AW	Average	02/20 TDZ	OBS	HO	C
Southend EGMC	METAR	h	AW MTI	Average	06/24 TDZ	OBS	H24	C
Stornoway EGPO	METAR SYNOP	h H	AW	Average	18/36 TDZ	OBS	HO H24	A
Sumburgh EGPB	METAR	h	AW MTI	Average	09/27 TDZ	OBS	HO+	A
Swansea EGFH	METAR	I	AW	Instant	04/22 TDZ	OBS	HO	C
Teesside EGNV	METAR	h	AW MTI	Average	05/23 TDZ	OBS	HO	C
Tiree EGPU	METAR	h	AW	Average			HO	A
Unst EGPW	METAR SYNOP	H h	AW	Instant			HO H24	C
Warton EGNO	METAR	h	AW	Average			HO	D
Wick EGPC	METAR SYNOP §	h H	AW	Average			HO H24	A
Yeovil/Westland EGHG	METAR ‡	H	AW	Instant			HO	D

GEN 3.5.3 — OBSERVATIONS AND REPORTS

Table 3.5.3.3 — Explanation of Terms used in the table 3.5.3.2.

Observation Type (column 2)	METAR METAR ‡ SYNOP SYNOP §	Aviation Routine Weather Report (actual) METAR not distributed routinely meteorological synoptic partial SYNOP via remote automated observing system
Observation Frequency (column 3)	h H I	half-hourly hourly irregular
Warnings (column 4)	AW MTI	Aerodrome warning Marked Temperature Inversion
RVR (columns 6 and 7)	TDZ MID END OBS AGIVIS, MET-1, Elecma & Vaisala	touchdown zone mid-point stop end human observer types of IRVR system
Observing hours (column 8)	HO HO+ H24	available to meet operational requirements (ie during aerodrome opening hours) more than HO but not H24 24 hours
Climatological data (column 9)	A, B, C, D	see Table 3.5.3.1

Note: Table 3.5.3.2 lists only those meteorological facilities at aerodromes at which Observations are formally carried out.

GEN 3.5.4— TYPES OF SERVICE

1 Forecast Offices providing a service to Civil Aviation

1.1 The designated Forecast Office(s) for principal aerodromes as listed at AD 2 is indicated at item AD 2.11 (Meteorological Information Provided) in Part 3 of the AIP. Their telephone numbers are listed at Table 3.5.4.1. Where one provides only a local or limited service, the associated Regional Forecast Office (RFO) is entered above it in AD 2.11. All RFOs, except Sella Ness operate H24.

1.2 Some Military and Government aerodromes provide a forecast and briefing service for Civil Aviation, but only for departures from those aerodromes.

1.3 The designated Forecast Office for Intercontinental flights departing any UK Civil aerodrome is Bracknell NMC.

2 Pre-flight Briefing

2.1 The primary method of meteorological briefing for flight crew in the UK is by self-briefing, using information and documentation routinely displayed in aerodrome briefing areas. Alternatively, flight crew and operators may obtain information direct by using the Met Office's PC Service MIST (Meteorological Information Self-briefing Terminal), the Fax services jointly provided by the Met Office and the CAA, and the CAA's AIRMET telephone service, see GEN 3.5.9 for fuller details of the automated services available. English is the language used for all UK documentation and forecast clarification. **The primary method of briefing does not require prior notification to a Forecast Office.**

2.2 Where this primary method is not available, or is inadequate for the intended flight, Special Forecasts, as described at GEN 3.5.5, may be provided.

2.3 When necessary the personal advice of a forecaster, or other meteorological information, can be obtained from the designated Forecast Office for the departure aerodrome forecast office for the departure aerodrome as given at AD 2.11. Pilots at aerodromes without an AD 2.11 entry should contact Bracknell NMC or the nearest Forecast Office (listed at Table 3.5.4.1) appropriate to the service required, except in the case of departures from Military or Government aerodromes where an on-site Forecaster or Briefing Service may be available. **Forecaster advice or other information for safety related clarification/amplification will only be given from a Forecast Office on the understanding that full use has already been made of all meteorological briefing material available at the departure aerodrome or, where appropriate, by telephone recording.** Forecaster clarification/amplification of conditions en-route is not provided for flights departing from locations outside UK.

GEN 3.5.4— TYPES OF SERVICE

Table 3.5.4.1 — Forecast Offices providing service to Civil Aviation

Forecast Office	Services Available	Telephone
Met Office Aberdeen	A, B, C [1]	01224-407650
Met Office Belfast	A, B, C, D [2]	028-9442 3275
Met Office Birmingham	A, B	0121-717 0580
Met Office Bracknell	A, B, C, D, E	01344-856267
Met Office Cardiff	A, B	029-2039 0492
Met Office Glasgow	A, B, C, D	0141-221 6116
Isle of Man Airport	A, B, C [3]	01624-821641
Jersey Airport	A, B, C [4]	01534-492229
JERSEY AIRMET [4]	–	09006-650033
Met Office Manchester	A, B, C, D	0161-429 0927
Met Office Sella Ness [5]	A	01806-242069

Key to Services Available:

A   – Provision of TAF, warnings and take-off data for the assigned principal aerodrome. Amplification/ clarification of these aerodrome forecasts and warnings only.

B   – Dictation of TAFs and METARs unobtainable from the automated services (usually limited to four aerodromes).

C   – Amplification/clarification of AIRMET Regional/Area Forecasts and Metforms 214/215, and requests for Special Forecasts, including specialised supplementary information for Ballooning/Gliding.

D   – Dictation of AIRMET amendments and Regional Forecasts.

E   – Amplification/clarification and amendments for Metforms 414/415 and EUR Charts.

Note 1:   Aberdeen services B, C available only to North Sea operators.

Note 2:   Belfast services B, C, D available only to departures from within Northern Ireland.

Note 3:   Isle of Man services B, C available only to departures from within the Isle of Man.

Note 4:   Jersey services B, C available only to departures from the Channel Islands. JERSEY AIRMET available 0600-2200, and only to callers in the Channel Islands and UK.

Note 5:   Operational hours 0630-1800; outside these hours Met Office Glasgow provides service.

2.4   Forecast Offices and self-briefing facilities are under no obligation to prepare briefing documentation packages.

2.5   Meteorological observations and forecasts have certain expected tolerances of accuracy. Pilots interpreting observations and forecasts should be aware that information could vary within these tolerances, which are shown at GEN 3.5.3 paragraph 2 and Table 3.5.4.3. respectively. Additionally, observations and forecasts are not normally amended until certain criteria for change are exceeded. These are shown at GEN 3.5.4 paragraph 6.

2.6   The specific value of any of the elements given in a forecast shall be understood to be the most probable value which the element in likely to assume during the period of the forecast. Similarly, when the time occurrence or change of an element is given in a forecast, this time shall be understood to be the most probable time.

2.7   The issue of a new forecast, such as an aerodrome forecast, shall be understood to automatically cancel any forecast of the same type previously issued for the same place and for the same period of validity or part thereof.

3   UK Low Level Weather and Spot Wind Forecast Charts – Metform 215/214

3.1   **The UK Low Level Forecast (Metform 215)** is a forecast of in-flight conditions from the surface to around 15000 ft, covering the UK and near Continent. The form comprises:

(a)   A fixed time forecast weather chart and associated descriptive text covering the period of validity from 3 hours before the fixed time, to 3 hours after;

(b)   a separate outlook chart shows the expected position of the principle synoptic features at the end of the outlook period of 6 hours, with separate text describing the main weather developments during this period.

3.2   Information on Form. The following sub-paragraphs summarise the contents of Metform 216 (Explanatory Notes for Form 215), available in A4 or larger size on application to the address at GEN 3.5.1 paragraph 2 or from METFAX by dialling 09060-700-505.

GEN 3.5.4— TYPES OF SERVICE

3.2.1 Main Forecast Weather Chart and Text

(a) The fixed time weather chart at the top left of the form shows the forecast position, direction and speed of movement of surface fronts and pressure centres for the fixed time shown in the chart legend. The position of highs (H) and lows (L), with pressure values in millibars is shown by the symbols O and X. The direction and speed of movement (in knots) of fronts and other features is given by arrows and figures. Speeds of less than 5 knots are shown as 'SLOW'. All features are given identifying letters to enable their subsequent movements to be followed on the outlook chart.

(b) Freezing levels (0°C) are shown in boxes as thousands of feet at appropriate places on the chart.

(c) Zones of distinct weather patterns are enclosed by continuous scalloped lines, each zone being identified by a number within a circle. The forecast weather conditions (visibility, weather, cloud) during the period of validity, together with warnings and any remarks are given in the text underneath the charts, each zone being dealt with separately and completely.

(d) In the text;
   (i) surface visibility is expressed in metres (m) or kilometres (km), with the change over at 5 km;
   (ii) weather is described in plain language, using well known and self evident abbreviations;
   (iii) cloud amount (in oktas) and type, with the height of base and top, is given, with all heights in feet (ft) amsl;
   (iv) warnings and significant changes and the expected occurrence of icing and turbulence are given in plain language, using standard abbreviations where possible (See GEN 2.2);
   (v) the height of any sub-zero layer below the main layer is given in the text;
   (vi) a forecast of CB or TS implies Hail and moderate or severe turbulence and icing;
   (vii) hill fog is not used but 'cloud covering hills' is thought to be more informative and implies visibility < 200 m;
   (viii) single numerical values given for any element represent the most probable mean in a range of values, covering approximately ±25%.

3.2.2 Outlook Chart and Text

(a) At the top right-hand side of the form an outlook chart shows the expected position of the main synoptic features at the end of the outlook period. No weather zones are given on the outlook chart but the pattern of surface isobars is shown.

(b) The outlook text following the main forecast text describes the principle weather changes expected during the 6 hour outlook period.

3.3 **The UK Low Level Spot Wind Forecast (Metform 214)** is fixed time, and suitable for use for a period three hours before or after the validity time.

(a) The data provided is for Latitude/Longitude positions shown at the top of each box;

(b) Wind Speed and temperature information is provided for a selected range of attitudes and are shown in thousands of feet above mean sea level and Degrees Celsius).

3.4 **Weather Forecast Chart Issues**

3.4.1 The routine issue of weather charts is detailed below. The time and the date when the Metform is issued by the Met Office are shown at the bottom of the form.

Time when Forecast Chart becomes available	Validity time	Period of validity	Outlook to
0500Z 1100Z 1700Z 2300Z	0900Z 1500Z 2100Z 0300Z	0600 – 1200Z 1200 – 1800Z 1800 – 0000Z 0000 – 0600Z	1800Z 0000Z 0600Z 1200Z

3.4.2 Amendments

(a) Amendments may appear as complete re-issues of the Metform in which case the validity start time may be different form the routine issue.

(b) An amended Metform 215 is indicated by the word AMENDED at the top left hand corner of the form.

4   Northwest Europe Low Level Weather and Spot Wind Forecast Charts - Metforms 415/414

(a) These charts are similar in format to Metforms 214/215 and extend the low-level flight forecast coverage more into continental Europe.

(b) They are available only from METFAX (09060 700 542) and are issued daily at 0500 valid for flights between 0600 and 1200 and at 1100 for flights between 1200 and 1800.

(c) Amendments cover the F215 area only, and users are advised to check with NMC Bracknell for any updates over Europe before departure.

5   European Medium / High Level Spot Wind / Temperature Forecast Chart - Metform 614

(a) This chart is similar in format to Metforms 214 and 414 but extends the coverage to the most of Europe and western parts of the Mediterranean and North Africa.

(b) It is available only from METFAX (09060 700 541) and provides a single sheet alternative to part of the area covered by the six standard EUR wind/temperature charts between FL 050 and FL 340 to accompany the EUR significant weather chart.

GEN 3.5.4— TYPES OF SERVICE

Table 3.5.4.2 — Meteorological Forecast Charts — Coverage and Validity Times

Area	Chart	Levels	Coverage	Projection	Issue Times	Validity Times*5
1	2	3	4	5	6	7
UK (F215)	Weather*3	Surface - 15000 ft amsl	British Isles and Near Continent	—	2300*2 0500 1100 1700	0300 0900 1500 2100
UK (F214)	Spot Wind/ Temperature	24000, 18000, 10000, 5000, 2000 ft, 1000ft*4 amsl				
Europe (EUR)	Sig Wx/Tropopause/ Max Wind	FL 100 - FL 450	N53 E065 N25 E034 N26 W018 N54 W050	Polar Stereographic	0315 and approx every 6 hours	1200 and every 6 hours
	Wind/Temperature	FL 390, 340, 300, 240, 180, 100, 050 FL 450*1			0410*2 1630*2 1635*2 0400	0000 0600 1200 1800
North Atlantic (NAT)	Sig Wx/Tropopause/ Max Wind	FL 250 - FL 630	N24 E056 N02 W004 N03 W083 N28 W148	Polar Stereographic	0215 and approx every 6 hours	1200 and every 6 hours
	Wind/Temperature	FL 390, 340, 300, 180 FL 530*1, 450*1, 240*1			0410*2 1630*2 1635*2 0400	0000 0600 1200 1800
	Isobaric and frontal analysis (ASXX)	Surface	N37 E050 N68 W105 N34 W055 N20 E010	Polar Stereographic	0415 0950 1615 2140	0000 0600 1200 1800
	T + 24 hour isobaric prognosis (FSXX)				0430*2 1035*2 1640*2 2230*2	0000 0600 1200 1800
Mid/Far East (MID)	Sig Wx/Tropopause/ Max Wind	FL 250 - FL 450	N23 E150 S06 E102 S03 E033 N30 W020	Polar Stereographic	0230 and approx every 6 hours	1200 and every 6 hours
	Wind/Temperature	FL 390, 340, 300 FL 530*1, 450*1, 240*1, 180*1			0410*2 1630*2 1635*2 0405	0000 0600 1200 1800
Africa (EURAFI)	Sig Wx/Tropopause/ Max Wind	FL 250 and above	N70 W032 N70 E065 S38 E065 S38 W032	Mercator	0230 and approx every 6 hours	1200 and every 6 hours
Caribbean/ South America (EURSAM)	Sig Wx/Tropopause/ Max Wind*1	FL 250 and above	N23 W113 N72 W003 N33 E059 S48 W052	Rotated Mercator	—	1200 and every 6 hours
Africa/ Caribbean/ South America	Wind/Temperature	FL 390, 340, 300 FL 530*1, 450*1, 240*1, 180*1	N65 E022 S10 E078 S60 W050 N14 W105	Rotated Mercator	0410*2 1635*2 1640*2 0405	0000 0600 1200 1800

Note *1 All charts are routinely available on the Broadcast Fax except those so marked.
Note *2 Previous day.
Note *3 This chart includes an outlook to the end of the next forecast period.
Note *4 Where terrain permits.
Note *5 Charts cover the period within 3 hours either side of the quoted fixed time, except the surface isobaric charts, which are valid for the time specified.

GEN 3.5.4— TYPES OF SERVICE

Table 3.5.4.3 — Accuracy of Meteorological Forecasts

The percentages in this table are ICAO minimum standards.

Aerodrome Forecast (TAF)

Element	Operationally desirable accuracy of forecast	Minimum percentage of cases within range
Wind direction	± 30°	80
Wind speed	± 5 kt up to 25 kt ± 20% above 25 kt	80
Visibility	± 200 m up to 700 m ± 30% between 700 m and 10 km	80
Precipitation	Occurrence or non-occurrence	80
Cloud amount	± 2 oktas	70
Cloud height	± 100 ft up to 400 ft ± 30% between 400 ft and 10000 ft	70
Air temperature (if forecast)	± 1°C	70

Landing Forecast (TREND)

Element	Operationally desirable accuracy of forecast	Minimum percentage of cases within range
Wind direction	± 30°	90
Wind speed	± 5 kt up to 25 kt ± 20% above 25 kt	90
Visibility	± 200 m up to 700 m ± 30% between 700 m and 10 km	90
Precipitation	Occurrence or non-occurrence	90
Cloud amount	± 2 oktas	90
Cloud height	± 100 ft up to 400 ft ± 30% between 400 ft and 10000 ft	90

Take-Off Forecast

Element	Operationally desirable accuracy of forecast	Minimum percentage of cases within range
Wind direction	± 30°	90
Wind speed	± 5 kt up to 25 kt ± 20% above 25 kt	90
Air temperature	± 1°C	90
Pressure value (QNH)	± 1 mb	90

Area, Flight and Route Forecast

Element	Operationally desirable accuracy of forecast	Minimum percentage of cases within range
Upper air temperature	± 3°C (mean for 500 nm)	90
Upper wind	± 15 kt up to FL 250 ± 20 kt above FL 250 (modulus of vector difference for 500 nm)	90
Significant en-route WX phenomena and cloud	Occurrence or non-occurrence Location: ± 60 nm Vertical extent: ±2000 ft	80 70 70

GEN 3.5.4— TYPES OF SERVICE

6   Aerodrome Forecast (TAF)

6.1   The Aerodrome Forecast (TAF) is the primary method of providing the forecast weather information that pilots require about an airfield in an abbreviated format. The TAF consists of a concise statement of the mean or average meteorological conditions expected at an aerodrome or heliport during the specified period of validity.

6.2   UK civil TAFs are prepared to cover the notified hours of operation of those principal civil aerodromes that have accredited meterorological observers, who produce regular aerodrome weather reports. Being site-specific, to provide an aerodrome forecast in TAF form requires the forecaster to be confident in the knowledge of the weather conditions prevailing at that aerodrome. In the interests of flight safety, continuity of regular reports, and ideally special reports when significant changes occur (particularly if the deterioration or improvement has not been forecast or is mis-timed), are essential for the routine updates and an adequate amendment service to be provided by the forecast office.

6.3   Therefore, where an aerodrome is not open H24, the issue of a TAF will be delayed until at least two consecutive METARs have been received and accepted by the forecaster at the forecast office responsible for its preparation.

6.4   The METARs will be produced by an accredited observer and separated by an interval of not less than 20 minutes and not more than 1 hour. In practice, when METARs are prepared every 30 minutes, a TAF will be drafted by the forecaster once the first METAR has been seen, and when the second METAR is received 30 minutes later and confirms the prevailing weather over the aerodrome the forecaster will issue the TAF.

6.5   However, in the event that an automatic or semi-automatic observing system located on the aerodrome regularly issues information on wind speed and direction, visibility, cloud amount and height, temperature and dewpoint to the forecast office when an aerodrome is closed, the forecaster will draft the TAF on the basis of the automatic observation. The TAF will not be issued until a METAR has been produced by an accredited observer and received in the forecast office, in order to confirm the weather conditions. Note that automatic observations only should be sent to the forecast office responsible for the TAF.

6.6   If a gap of more that one hour between METAR reports occurs, or if a key element is missing from the report, the TAF will be withdrawn. The TAF will not be re-issued until two complete METARs have been received.

6.7.   Accredited observers at some H24 aerodromes take a duty break overnight, of maximum two hours duration. A supply of automatic observations will be provided to the forecast office during this period, but amendments to the TAF will be made only at the discretion of the forecaster. If the duty observer has not recommenced observations after two hours, the TAF will be withdrawn.

6.8   If a TAF needs to be amended due to a deterioration or improvement that has not been forecast or is mis-timed, such amendments shall be issued within 15 minutes of receipt of the observation at the forecast office.

7   Criteria For Special Meteorological Reports and Forecasts

7.1   The following are the criteria for the issue of Special Aerodrome Meteorological Reports, TRENDS, TAF Variants/Amendments and Amended Route/Area Forecasts:

(a)   **Special Report**

    (i)   **Surface Wind**. Issued only when no serviceable wind indicator in ATC; criteria to be agreed locally, based on changes of operational significance at aerodrome; for example:

        (1)   A change in mean direction of 60° or more, mean speed before or after change being 10 kt or more; and/or a change of 30°, the speed 20 kt or more;

        (2)   a change in mean speed of 10 kt or more;

        (3)   a change in gust speed of 10 kt or more, the mean speed before or after the change being 15 kt or more.

    (ii)   **Visibility**

        (1)   A change from one of the following ranges to another:

            10 km or more

            5000 m to 9 km

            3000 m to 4900 m

            2000 m to 2900 m

            1500 m to 1900 m

            800 m to 1400 m

            750 m or less

        (2)   Additional change groups of 100 m or less, 150 to 300 m, 350 to 550 m and 600 to 750 m are used where an RVR is not available, either permanently or during temporary unservicablilty. These criteria will apply by local arrangement.

        (3)   Additional change groups of 3000 to 3900 m and 4000 to 4900 m apply at Aberdeen/Dyce airport.

GEN 3.5.4— TYPES OF SERVICE

    (iii) **Runway Visual Range (RVR)**

        (1) A change from one of the following ranges to another:

            800 m or more

            600 m to 750 m

            350 m to 550 m

            150 m to 325 m

            125 m or less

        (2) Note that special reports for RVR are only made by local arrangement.

    (iv) **Weather**

        (1) The onset or cessation of:
- moderate or heavy: precipitation, including showers;
- freezing fog and freezing precipitation;
- thunderstorm, squall, funnel cloud;
- low drifting or blowing: snow, sand or dust.

        (2) A change in the intensity of the precipitation and blowing snow from slight to moderate/heavy and vice versa.

            Note: The phenomenon associated with a significant change in visibility or cloud shall be reported, whatever the intensity.

    (v) **Cloud**

        (1) When the base of the lowest cloud of over 4 oktas (BKN or OVC) changes from one of the following ranges to another:

            2000 ft or more

            1500 to 1900 ft

            1000 to 1400 ft

            700 to 900 ft

            500 to 600 ft

            300 to 400 ft

            200 ft

            100 ft

            Less than 100 ft*

        (2) *This includes state of sky obscured.

        (3) When the amount of cloud below 1500 ft changes from 4 oktas or less (nil, FEW, SCT) to more than 4 oktas (BKN or OVC), and vice versa.

    (vi) **QFE/QNH.** When the QNH or QFE changes by 1.0mb or more.

    (vii) **Severe/Icing/Turbulence.** After confirmation by the duty forecaster, pilot reports of severe icing or severe turbulence, either on the approach to, or climb out from, the aerodrome.

  (b) **Trend**

    (i) **Surface Wind**

        (1) A change in mean direction of 30° or more, the mean speed before or after the change being 20 kt or more; a change in mean direction of 60° or more, the mean speed before or after the change being 10 kt or more.

        (2) A change in mean speed of 10 kt or more.

        (3) A change in gust speed or 10 kt or more, the mean speed before or after the change being 15 kt or more.

(ii) **Surface Visibility.**

(1) A change from one of the following ranges to another:

5000 m or more

3000 m to 4900 m

1500 m to 2900 m

800 to 1400 m

600 m to 750 m

350 m to 550 m

150 m to 300 m

100 m or less

(iii) **Weather**

(1) Onset, cessation or change in intensity of:
   - freezing precipitation;
   - freezing fog;
   - moderate or heavy: precipitation, including showers;
   - low drifting: sand, dust or snow;
     blowing: sand, dust or snow;
   - thunderstorm;
   - squall, funnel cloud;
   - other phenomena if associated with a significant change in visibility or cloud, whatever the intensity.

(iv) **Cloud**

(1) When the base of the lowest cloud over 4 oktas (BKN or OVC) changes from one of the following ranges to another:

1500 ft or more

1000 ft to 1400 ft

500 ft to 900 ft

300 ft to 400 ft

200 ft

100 ft

Less than 100 ft*

(2) *This includes state of sky obscured.

(3 Additional change groups of 500 to 600 feet and 700 to 900 feet apply at aerodromes serving oil rig helicopter operations; Aberdeen, Blackpool, Humberside, Inverness, Kirkwall, Liverpool, Norwich, Scatsca, Sumburgh and Wick.

(4) When the amount of the lowest cloud below 1500 ft changes from half or less (nil, FEW or SCT) to more than half (BKN or OVC) and vice versa. A change to 'sky clear' should by shown as 'SKC' and a change to no cloud below 5000 ft and no CB, to 'NSC', unless in either case CAVOK applies.

(c) **TAF Variants/Amendments**

(i) **Surface Wind**

(1) A change in mean direction of 30° or more, the mean speed before or after the change being 20 kt or more; a change in mean direction of 60°, the mean speed before or after the change being 10 kt or more.

(2) a change in mean speed of 10 kt or more;

(3) a change in gust speed of 10 kt or more, the mean speed before or after the change being 15 kt or more.

GEN 3.5 4— TYPES OF SERVICE

(ii) **Surface Visibility**

   (1) A change from one of the following ranges to another

      10 km or more

      5000 m to 9 km

      1500 to 4900 m

      800 to 1400 m

      350 to 750 m

      300 m or less

   (2) Additional change groups of 1500 to 2900 m and 3000 to 4900 m as well as 5000 m to 6 km and 7 to 9 km apply at aerodromes serving oil rig helicopter operations; Aberdeen, Blackpool, Humberside, Inverness, Kirkwall, Liverpool, Norwich, Scatsta, Sumburgh and Wick.

(iii) **Weather**

   (1) Onset, cessation or change in intensity of:
      – freezing precipitation;

      – freezing fog;

      – moderate or heavy: precipitation, including showers;

      – low drifting: sand, dust or snow;

      – blowing: sand, dust or snow;

      – thunderstorm;

      – squall, funnel cloud;

      – other phenomena if associated with a significant change in visibility or cloud, whatever the intensity.

      – CAVOK conditions.

(iv) **Cloud**

   (1) When the base of the lowest cloud of over 4 oktas (BKN or OVC) changes from one of the following ranges to another:

      5000 ft or more

      1500 to 4900 ft

      1000 to 1400 ft

      500 to 900 ft

      200 to 400 ft

      100 feet or less*

   (2) *This includes state of sky obscured.

   (3) Additional change groups of 500 to 600 feet and 700 to 900 ft apply at aerodromes serving oil rig helicopter operations; Aberdeen, Blackpool, Humberside, Inverness, Kirkwall, Liverpool, Norwich, Scatsta, Sumburgh and Wick.

   (4) When the amount of the lowest cloud below 1500 ft changes from half or less (nil, FEW or SCT) to more than half (BKN or OVC) and vice versa. A change to 'sky clear' should by shown as 'SKC' and a change to no cloud below 5000 ft and no CB, to 'NSC', unless in either case CAVOK applies.

GEN 3.5 4— TYPES OF SERVICE

(d) **Amended Route/Area Forecast (Advisory Criteria)**

(i) **Surface Wind.** A change in direction of 30° or more, the speed before and/or after the change being at least 30 kt. A change of speed of 20 kt or more.

(ii) **Cloud.** Changes liable to have an appreciable affect on aircraft operations; eg in AIRMET and low level forecasts: changes in the lowest layer from 4 oktas or less to more than 4 oktas, or more than 4 oktas to 4 oktas or less and changes of any layer of ±25% or 500 ft, whichever is the larger.

(iii) **Temperature/Dew Point.** 5°C or more.

(iv) **Surface Visibility.** Changes liable to have an appreciable affect on aircraft operations; eg in AIRMET and low level forecasts: Changing through: 10 km, 5000 m, 1500 m, 800 m.

(v) **Weather**

	Element	Original Forecast	Revised Opinion
(1)	TS, SQ, GR, SA, freezing precipitation	Not included Included	Now expected Not now expected
(2)	Turbulence	NIL or FBL SEV MOD	MOD or SEV FBL or NIL NIL
(3)	Airframe icing	NIL FBL or MOD SEV MOD	Any intensity A higher intensity FBL or NIL NIL
(4)	Fronts, tropical disturbances	Significant front or disturbances not included Significant front or disturbance included	Now expected Not now expected or position materially different from forecast

GEN 3.5.5 — NOTIFICATION REQUIRED FROM OPERATORS

1 Special Forecasts and Specialised Information

1.1 For departures where the standard pre-flight meteorological self-briefing material cannot be obtained or is inadequate for the intended flight, a Special Forecast may be issued on request to the appropriate Forecast Office for a specific period for a designated route, or an area which includes the route. Normally, a Special Flight Forecast will be supplied from the last UK departure point to the first transit aerodrome outside the coverage of standard documentation, at which point pilots should re-brief. However, by prior arrangement, a forecast may be prepared for other legs, provided initial ETD to final ETA does not exceed 6 hours and no stops longer than 60 minutes are planned.

1.2 The usual method of issuing Special Flight Forecasts is by AFS, Telex or Fax to the aerodrome of departure, but if the Flight Briefing Unit is not so equipped or will not be open, pilots may telephone the Forecasting Office for a dictation of the forecast. Similarly, Aerodrome Forecasts and reports for the destination and up to four alternates will be provided with the forecast, if not otherwise available.

2 Forecast Offices normally require prior notification for Special Forecasts as follows:

(a) For flights up to 500 nm, at least two hours before the time of collection;

(b) for flights of over 500 nm, at least four hours before the time of collection.

2.1 Request for Special Forecasts must include details of the route, the period of the flight and where appropriate the ETD/ETA of each leg, the height to be flown and the time at which the forecast is required. Ideally a forecast should be collected no earlier than 90 minutes before departure.

2.2 It is in the interest of all concerned that the maximum possible period of notice is given. The Forecast Office will give priority to emergencies, in-flight forecast and to forecast requirements which have been properly notified. Other requests could be delayed at busy periods and might not comprise full forecasts. A Forecast collected a long time in advance of departure will be less specific and might be less accurate than one prepared nearer departure time.

2.3 Forecast Offices providing Special Forecasts are shown at Table 3.5.4.1. They are not provided for flights inbound to the UK.

2.4 Take-off forecasts containing information on expected conditions over the runway complex in respect of surface wind, temperature and pressure can be made available from Forecast Offices. Prior notification is not normally required, and can be supplied up to three hours before the expected time of departure.

2.5 Meteorological information for specialised aviation use, as defined below, is not included in the AIRMET service or given as Special Forecasts but arrangements can be made for its provision on prior request:

(a) To enable glider, hang glider, microlight and balloon organisations to obtain surface wind and temperature, lee-wave, QNH and thermal activity forecasts;

(b) to provide meteorological information for special aviation events for which routine forecasts are not adequate;

(c) to provide helicopter operators in off-shore areas with forecast winds and temperatures at 1000 ft amsl, information on airframe icing, and sea state and temperature.

2.6 Appropriate forecasts for (a) and (b) above will be made available up to twice in any 24 hour period. For (a), the initial request should be made to the nearest forecasting office designated as providing service 'C' at Table 3.5.4.1 at least 2 hours in advance of the forecast being required. For (b) and (c), application must be made to the Meteorological Authority (GEN 3.5.1 paragraph 1) for approval, giving at least 6 weeks notice of the requirement. The application must specify the nature of the aviation activity, the location(s) involved, the meteorological information required and the associated time periods. If appropriate an AFTN, Telex or Facsimile address should also be included. Applicants will be advised of the time at which the information will be available and of the means of collection/delivery.

3 Additional Meteorological Services

3.1 When specialist, non-standard, aviation meteorological services additional to those given above are required (eg forecaster briefings for aerial photography, test flying, crop spraying and for outlooks for over a day ahead), they may be obtained on a repayment basis by prior arrangement with The Met Office. Enquiries should be directed to The Met Office address at GEN 3.5.1 paragraph 2, or to one of the Forecast Offices listed at Table 3.5.4.1.

GEN 3.5.6 — AIRCRAFT REPORTS

1   Routine Aircraft Observations

1.1   Routine Aircraft Observations are not required in the London or Scottish FIR/UIR or the Shanwick FIR, but in the Shanwick OCA, aircraft are to conform with the requirements for meteorological observations indicated in the ENR section or applicable NOTAM.

2   Special Aircraft Observations

2.1   Special Aircraft Observations are required in any UK FIR/UIR/OCA whenever:

(a) Severe turbulence or severe icing is encountered; or

(b) moderate turbulence, hail or cumulo-nimbus clouds are encountered during transonic or supersonic flight; or

(c) other meteorological conditions are encountered which, in the opinion of the pilot-in-command, might affect the safety or markedly affect the efficiency of other aircraft operations, for example, other en-route weather phenomena specified for SIGMET messages, or adverse conditions during the climb-out or approach not previously forecast or reported to the pilot-in-command. Observations are required if volcanic ash cloud is observed or encountered, or if pre-eruption volcanic activity or a volcanic eruption is observed; or

(d) exceptionally, they are requested by the meteorological office providing meteorological service for the flight; in which event the observation should be specifically addressed to that meteorological office; or

e) exceptionally, there is an agreement to do so between the Meteorological Authority and the aircraft operator.

3   Turbulence and Icing Reporting Criteria

3.1   Turbulence (TURB)

3.1.1 TURB remains an important operational factor at all levels but particularly above FL 150. The best information on TURB is obtained from pilots' Special Aircraft Observations; all pilots encountering TURB are requested to report time, location, level, intensity and aircraft type to the ATS Unit with whom they are in radio contact. High level turbulence (normally above FL 150 not associated with cumuliform cloud, including thunderstorms) should be reported as TURB, preceded by the appropriate intensity or preceded by Light or Moderate Chop.

Table 3.5.6.1 — TURB and other Turbulence Criteria Table	
Incidence: Occasional — less than 1/3 of the time Intermittent — 1/3 to 2/3 Continuous — more than 2/3	
Intensity	Aircraft Reaction (transport size aircraft)	Reaction Inside Aircraft

Intensity	Aircraft Reaction (transport size aircraft)	Reaction Inside Aircraft
Light	Turbulence that momentarily causes slight, erratic changes in altitude and/or attitude (pitch, roll, yaw). IAS fluctuates 5 - 15 kt. (<0.5 g at the aircraft's centre of gravity) Report as **'Light Turbulence'**. or; turbulence that causes slight, rapid and somewhat rhythmic bumpiness without appreciable changes in altitude or attitude. No IAS fluctuations. Report as 'Light Chop'.	Occupants may feel a slight strain against seat belts or shoulder straps. Unsecured objects may be displaced slightly. Food service may be conducted and little or no difficulty is encountered in walking.
Moderate	Turbulence that is similar to Light Turbulence but of greater intensity. Changes in altitude and/or attitude occur but the aircraft remains in positive control at all times. IAS fluctuates 15 - 25 kt. (0.5-1.0g at the aircraft's centre of gravity). Report as **'Moderate Turbulence'**. or; turbulence that is similar to Light Chop but of greater intensity. It causes rapid bumps or jolts without appreciable changes in altitude or attitude. IAS may fluctuate slightly. Report as 'Moderate Chop'.	Occupants feel definite strains against seat belts or shoulder straps. Unsecured objects are dislodged. Food service and walking are difficult.
Severe	Turbulence that causes large, abrupt changes in altitude and/or attitude. Aircraft may be momentarily out of control. IAS fluctuates more than 25 kt. (>1.0 g at the aircraft's centre of gravity). Report as **'Severe Turbulence'**.	Occupants are forced violently against seat belts or shoulder straps. Unsecured objects are tossed about. Food service and walking impossible.

Note 1: Pilots should report location(s), time(s) (UTC), incidence, intensity, whether in or near clouds, altitude(s) and type of aircraft. All locations should be readily identifiable. Turbulence reports should be made on request, or in accordance with paragraph 2.

Example:
(a) Over Pole Hill 1230 intermittent Severe Turbulence in cloud, FL 310, B747.
(b) From 50 miles north of Glasgow to 30 miles west of Heathrow 1210 to 1250, occasional Moderate Chop TURB, FL 330, MD80.

Note 2: The UK does not use the term 'Extreme' in relation to turbulence.

GEN 3.5.6 — AIRCRAFT REPORTS

3.2 Windshear Reporting Criteria

3.2.1   Pilots using navigation systems providing direct wind velocity readout should report the wind and altitude/height above and below the shear layer, and its location. Other pilots should report the loss or gain of airspeed and/or the presence of up-or-down draughts or a significant change in crosswind effect, the altitude/height and location, their phase of flight and aircraft type. Pilots not able to report windshear in these specific terms should do so in terms of its effect on the aircraft, the altitude/height and location and aircraft type, for example, 'Abrupt windshear at 500 feet QFE on finals, maximum thrust required, B747'. Pilots encountering windshear are requested to make a report even if windshear has previously been forecast or reported.

3.3   Airframe Icing

3.3.1 All pilots encountering unforecast icing are requested to report time, location, level, intensity, icing type* and aircraft type to the ATS Unit with whom they are in radio contact. It should be noted that the following icing intensity criteria are reporting definitions; they are not necessarily the same as forecasting definitions because reporting definitions are related to aircraft type and to the ice protection equipment installed, and do not involve cloud characteristics. For similar reasons, aircraft icing certification criteria might differ from reporting and/or forecasting criteria.

<table>
<tr><td colspan="2" align="center">Table 3.5.6.2 — Airframe Icing Intensity Criteria</td></tr>
<tr><td>Intensity</td><td>Ice Accumulation</td></tr>
<tr><td>Trace</td><td>Ice becomes perceptible. Rate of accumulation slightly greater than rate of sublimation. It is not hazardous even though de-icing/anti-icing equipment is not utilised, unless encountered for more than one hour.</td></tr>
<tr><td>Light</td><td>The rate of accumulation might create a problem if flight in this environment exceeds 1 hour. Occasional use of de-icing/anti-icing equipment removes/prevents accumulation. It does not present a problem if de-icing/anti-icing equipment is used.</td></tr>
<tr><td>Moderate</td><td>The rate of accumulation is such that even short encounters become potentially hazardous and use of de-icing/anti-icing equipment, or diversion, is necessary.</td></tr>
<tr><td>Severe</td><td>The rate of accumulation is such that de-icing/anti-icing equipment fails to reduce or control the hazard. Immediate diversion is necessary.</td></tr>
<tr><td colspan="2">*Rime Ice: Rough, milky, opaque ice formed by the instantaneous freezing of small supercooled water droplets.
*Clear Ice: A glossy, clear, or translucent ice formed by the relatively slow freezing of large supercooled water droplets.</td></tr>
</table>

4   In-flight Procedures

4.1   Information to aircraft in flight is usually supplied in accordance with area Meteorological Watch procedures, supplemented when necessary by an En-route Forecast service. Information is also available from the appropriate ATS Unit at the commanders request, or from meteorological broadcasts.

4.2   An in-flight en-route service is available in exceptional circumstances by prior arrangement with the Meteorological Authority (GEN 3.5.1, paragraph 1). A meteorological office is designated to provide the aircraft in flight with the winds and temperatures for a specific route sector. Applications for this service should be made in advance, stating:

(a)   The flight level(s) and the route sector required;

(b)   the period of validity necessary;

(c)   the approximate time and position in flight at which the request will be made;

(d)   the ATS Unit with whom the aircraft is expected to be in contact.

4.3   Aircraft can obtain aerodrome weather information from any of the following:

(a)   VOLMET broadcasts (see GEN 3.5.7);

(b)   Automatic Terminal Information Service (ATIS) broadcasts;

(c)   by request to an ATS Unit but whenever possible only if the information required is not available from a broadcast.

4.4   When an aircraft diverts, or proposes to divert, to an aerodrome along a route for which no forecast has been provided, the commander may request the relevant information from the ATS Unit serving the aircraft at the time, and the necessary forecasts will be provided by the associated Forecast Office.

GEN 3.5.7 — VOLMET SERVICES

Table 3.5.7.1 — Meteorological Radio Broadcasts (VOLMET)

Call Sign/ID	EM	Frequency MHz	Operating Hours	Stations	Contents	Remarks
1	2	3	4	5	6	7
London Volmet (Main)	A3E	135.375	H24 continuous	Amsterdam Brussels Dublin Glasgow London Gatwick London Heathrow London Stansted Manchester Paris/Charles de Gaulle	(1) Half hourly reports (METAR) (2) The elements of each report broadcast in the following order: (a) Surface wind (b) Visibility (or CAVOK) (c) RVR if applicable (d) Weather (e) Cloud (or CAVOK) (f) Temperature (g) Dewpoint (h) QNH (i) Recent Weather if applicable (j) Windshear if applicable (k) TREND if applicable (l) Runway Contamination Warning if applicable	The spoken word 'SNOCLO' will be added to the end of the aerodrome report when that aerodrome is unusable for take-offs and landings due to heavy snow on runways or runway snow clearance.
London Volmet (South)	A3E	128.600	H24 continuous	Birmingham Bournemouth Bristol Cardiff Jersey London Luton Norwich Southampton Southend		
London Volmet (North) (Note 1)	A3E	126.600	H24 continuous	Blackpool East Midlands Isle of Man Leeds Bradford Liverpool London Gatwick Manchester Newcastle Teesside	(3) Non-essential words such as 'surface wind', 'visibility' etc are not spoken. (4) Except for 'SNOCLO' (see Column 7), the Runway State Group is not broadcast. (5) All broadcasts are in English.	
Scottish Volmet	A3E	125.725	H24 continuous	Aberdeen/Dyce Belfast Aldergrove Edinburgh Glasgow Inverness London/Heathrow Prestwick Stornoway Sumburgh		

Note 1: Broadcasting range extended to cover Southeast England and English Channel.

Note 2: An HF VOLMET broadcast for North Atlantic flights (Shannon VOLMET) is operated by the Republic of Ireland.

GEN 3.5.8 — SIGMET SERVICES

1   MWOs are responsible for the preparation and dissemination of SIGMETs to appropriate ACC/FIC within their own and agreed adjacent FIRs. Aircraft in flight should be warned by the ACC/FIC of the occurrence or expected occurrence of one or more of the following SIGMET phenomena for the route ahead for up to 500 nm or 2 hours flying time:

(a)   At subsonic cruising levels (SIGMET):
- thunderstorm (see **Note 2**);
- heavy hail (see **Note 2**);
- tropical cyclone;
- freezing rain;
- severe turbulence (not associated with convective cloud);
- severe icing (not associated with convective cloud);
- severe mountain waves;
- heavy sand/dust storm;
- volcanic ash cloud.

(b)   At transonic and supersonic cruising levels (see **Note 3**) (SIGMET SST):
- moderate or severe turbulence;
- cumulo-nimbus cloud;
- hail;
- volcanic ash cloud.

Note 1:   In general, SIGMET messages are identified by the letters WS at the beginning of the header line, but those referring to tropical cyclones and volcanic ash will be identified by WC and WV respectively. SIGMETs are usually valid for 4 hours (but exceptionally for up to 6 hours and for volcanic ash cloud and tropical cyclones, a further outlook for up to 12 hours may be included) and are re-issued if they are to remain valid after the original period expires. They can be cancelled or amended within the period of validity. SIGMETs are numbered sequentially from 0001 UTC each day.

Note 2:   This refers only to thunderstorms (including if necessary, cumulo-nimbus cloud which is not accompanied by a thunderstorm) widespread within an area with little or no separation (FRQ), along a line with little of no separation (SQL), embedded in cloud layers (EMBD), or concealed in cloud layers or concealed by haze (OBSC), but does not refer to isolated or occasional thunderstorms not embedded in cloud layers or concealed by haze. Thunderstorms and tropical cyclones each imply moderate or severe turbulence, moderate or severe icing and hail. However, heavy hail (HVYGR) may be used as a further description of the thunderstorm as necessary.

Note 3:   FL 250-FL 600 within London and Scottish UIRs, FL 400-FL 600 within Shanwick OCA.

GEN 3.5.9 — OTHER AUTOMATED METEOROLOGICAL SERVICES

1   Meteorological charts are available via facsimile from two automated services, Broadcast Fax and METFAX.

1.1   **Broadcast Fax** is a routine broadcast service available to users requiring a minimum number of charts regularly each week.

    (a)   Charts routinely transmitted over the Broadcast Fax network cover:

        (i)   Low and medium level flights within the UK and Near Continent;

        (ii)   Medium and high level flights to Europe and the Mediterranean;

        (iii)   High level flights to North America;

        (iv)   High level flights to the Middle/Far East;

        (v)   High level flights to Africa.

    (b)   There are additional charts which are not routinely available by Broadcast Fax, for example EURSAM significant weather for high level flights to Caribbean and South America and upper winds/temperatures at other levels. These may be obtained on prior request from Bracknell NMC (a charge to cover handling and transmission costs will be made for this facility, and if Broadcast Fax is not used, an account must be set up in advance by application to the address at GEN 3.5.1 paragraph 2).

    (c)   Table 3.5.4.2 gives the geographical and vertical coverage, the times of issue and validity of charts which are routinely available by Broadcast Fax.

    (d)   It should be noted that forecasts may be amended at any time, which charts received via facsimile may not show. Therefore it is advisable to check there are no changes to the forecast conditions prior to departure. Amendment criteria for forecasts are given in GEN 3.5.4 paragraph 7.

1.2   **METFAX Services**

1.2.1   Meteorological pre-flight briefing information, in addition to that specified in paragraph 1 is available from the joint Met Office/CAA Broadcast Fax and METFAX services. METAR and TAF bulletins and AIRMET Area Forecasts are available on a 24 hour service. The METFAX service also includes planning forecasts, satellite images and tephigrams, the complete schedule can be found on the index page (dial 09060 700 501).

    (a)   METFAX is a dial up premium rate facsimile service designed primarily for the General Aviation sector, and enables low and medium/high level charts for the British Isles and Continental Europe to be obtained as required by the user.

    (b)   Details of the charts available from METFAX are given on the index page (dial 09060 700 501).

1.3   **Broadcast Text Meteorological Information**

    (a)   This information is distributed within the UK through dedicated communication channels (OPMET 1, 2 and 3) by AFS and by autotelex to supply Aerodrome Meteorological Reports (METAR), Aerodrome Forecasts (TAF) and warnings of weather significant to flight safety (SIGMET) including Volcanic Activity Reports. These are broadcast by teleprinter throughout the UK and internationally in text form. Licensed aerodromes receiving text meteorological information bear indication at AD 2 item 2.11.

    (b)   Details of the OPMET networks and their contents are as shown below:

        (i)   Contents of **OPMET 1** Teleprinter Broadcast - METARs, TAFs and SIGMETs for the following areas, plus QFA Alps:

Belgium	Canaries	France	Germany	Iceland	Irish Republic
Italy	Netherlands	Norway	Portugal	Spain	Switzerland
Turkey #	United Kingdom (inc Channel Islands and Isle of Man)				

        # = 9 Hour TAFs Only

        (ii)   Contents of **OPMET 2** Teleprinter Broadcast - METARs, TAFs and SIGMETs for the following areas:

Algeria	Austria	Bahrain	Belarus	Bulgaria	Canaries ‡
Croatia	Cyprus	Czech Republic	Egypt	Georgia	Greece
Hungary	Iran	Iraq	Israel	Jordan	Kazakhstan
Kuwait	Lebanon	Libya	Malta	Morocco	Poland
Portugal	Romania	Russian Federation	Saudi Arabia	Slovenia	Spain
Syria	Tajikistan	Tunisia	Turkey	Ukraine	United Arab Emirates
Yugoslavia					

        ‡ = 18/24 Hour TAFs only

        (iii)   Contents of **OPMET 3** Teleprinter Broadcast - METARs, TAFs and SIGMETs for the following areas, plus QFA Alps:

Algeria §	Austria	Belgium	Bulgaria	Canaries	Croatia §
France	Germany	Greece	Hungary	Iceland	Irish Republic
Italy	Libya	Malta	Morocco §	Netherlands	Norway
Portugal	Romania	Slovenia	Spain	Switzerland	Tunisia §
Turkey	Yugoslavia	United Kingdom (inc Channel Islands and Isle of Man)			

        § = (SIGMET only)

GEN 3.5.9 — OTHER AUTOMATED METEOROLOGICAL SERVICES

(c) Further details of the OPMET networks, including the reporting aerodromes in each area, are available from the Meteorological Authority.

Note 1: METARs are broadcast as routine at half-hourly (exceptionally hourly) intervals during aerodrome opening hours.

Note 2: TAFs valid for periods of less than 12 hours, usually for 9 hours, (FC) and QFA Alps are broadcast every three hours and TAFs valid for periods of 12 to 24 hours (FT) every six hours. Amendments are broadcast between routine times as required.

(d) Additional Information:

    (i) Short term Landing Forecasts valid for two hours (TREND) may be added to METARs issued by those aerodromes so designated at AD 2 item 2.11;

    (ii) Information on runway state is added to the METAR when weather conditions so require and continues until these conditions have ceased;

    (iii) Special Aerodrome Meteorological Reports are issued for operational use locally when conditions change through limits specified at GEN 3.5.4 paragraph 7. Selected Special Reports (SPECI) are defined as Special Reports disseminated beyond the aerodrome of origin; civil aerodromes in the UK are not required to make Selected Special Reports. At aerodromes where only a synoptic weather report (SYNOP) is available, Special Reports are not provided to the ATSU;

    (iv) In general TAFs are provided only for those aerodromes where official meteorological observations are made and recent reports are available. For other aerodromes, Local Area Forecasts can be made on request by arrangement with the MET Authority. Amended TAFs or Local Area Forecasts are issued when forecast conditions change significantly, see GEN 3.5.4 paragraph 7;

    (v) The formats and codes used for METAR, SPECI, TREND, TAF and the METAR Runway State Group are described at GEN 3.5.10;

    (vi) The actual or forecast meteorological conditions for which a SIGMET warning is prepared are detailed at GEN 3.5.8.

    (vii) Area Forecasts for the AIRMET service (see paragraph 1.4) are broadcast in text form by teleprinter. Amended Area Forecasts are issued when forecast conditions change significantly, see GEN 3.5.4 paragraph 7.

1.4 AIRMET Service

(a) AIRMET is a general aviation weather briefing service. The basic service consists of ten routine Forecasts, in plain language, covering the UK and near Continent and a comprehensive selection of TAFs and METARs for aerodromes and the near Continent. Information is provided in spoken form at dictation speed via the public telephone network and in text form via the AFS, Telex and facsimile.

(b) The AIRMET Forecast telephone service is intended for use by pilots who do not have access to meteorological information disseminated by Fax or to the AIRMET Forecasts disseminated in text form by teleprinter to aerodrome flight briefing rooms. It is provided in a standard format to facilitate transcription on to a Pilots Proforma (see sub paragraph (e)). A map, showing the coverage of the ten Forecasts is at GEN 3-5-25 and on the available Proforma. AIRMET Forecasts are obtainable from the telephone numbers given on the map: usually in recorded form but at night from the designated Forecast Offices.

(c) The forecasts will reflect the contents of SIGMETs which are current at the time of issue or amendments of the forecasts. Safety related amplification of an AIRMET forecast may be obtained from a forecaster by telephoning one of the forecast offices listed at paragraph 1 as providing service 'E'. Callers must be able to confirm that they have obtained a current AIRMET forecast on contacting the forecast office, otherwise no additional forecast information will be given.

(d) Special Forecasts in accordance with GEN 3.5.5 paragraph 1 are not provided for flights within the coverage of AIRMET Forecasts. For flights which extend beyond the area of coverage, Special Forecasts will be available on request from selected forecast offices providing a service to civil aviation (paragraph 1 refers).

(e) A copy of the Pilots Proforma (reproduced at GEN 3-5-26 and 3-5-27) is a prerequisite to making optimum use of AIRMET. Supplies of the Proforma may be obtained, free of charge, on request to:

The Technical Secretary, Aircraft Owners and Pilots Association, 50a Cambridge Street, London SW1V 4QQ.

provided a stamped, self-addressed A4 size envelope is included with the request. Photocopying of the Proforma is permitted. An AIRMET explanatory leaflet, 'Users Notes', is also available from AOPA on the same terms.

(f) DIALMET. An Automated METAR and TAF Telephone Service, giving weather reports and forecasts for the UK, near Continent and Eire is available on 09063-800400. The aerodromes available and the three digits to be dialled for each individual aerodrome is published in "GET MET", a pocket sized booklet that can be obtained by writing to the following address: Met Office, Sutton House, London Road, Bracknell, Berkshire RG12 2SZ. When using this service it should be remembered that the three selective digits should not be dialled before the call is answered; however, the <*> key followed by the three digits will provide quick access to the information once the recorded message begins.

(g) JERSEY AIRMET (Available from UK). The States of Jersey Meteorological Office provides a 50 nm radius Channel Islands Low Level telephone recorded forecast service called JERSEY AIRMET, with a format very similar to the UK AIRMET. Amplification or clarification of the current JERSEY AIRMET forecast may be obtained, following receipt of the recording, by consulting the Forecast Office at Jersey Airport. Telephone calls to JERSEY AIRMET and the Forecast Office are charged at a premium rate and are not available outside the Channel Islands and UK. The appropriate telephone numbers are shown at table 3.5.4.1.

GEN 3.5.9 — OTHER AUTOMATED METEOROLOGICAL SERVICES

AIRMET AREAS AND BOUNDARIES

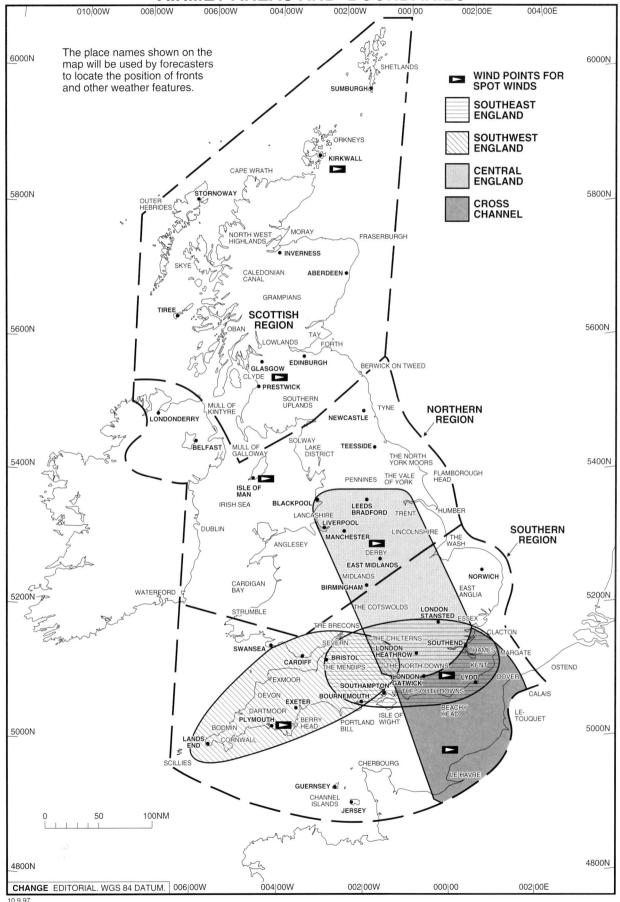

The place names shown on the map will be used by forecasters to locate the position of fronts and other weather features.

WIND POINTS FOR SPOT WINDS

SOUTHEAST ENGLAND

SOUTHWEST ENGLAND

CENTRAL ENGLAND

CROSS CHANNEL

CHANGE	EDITORIAL. WGS 84 DATUM.

10.9.97

GEN 3.5.9 — OTHER AUTOMATED METEOROLOGICAL SERVICES

Format of AIRMET Forecasts

AIRMET COPY FORM
All temperatures degrees Celsius. All heights AMSL. All times UTC.

Forecast Areas
Southern Region - 40 ☐
Northern Region - 41 ☐
Scottish Region - 42 ☐
Southwest England - 46 ☐
Southeast England - 47 ☐
Cross Channel - 48 ☐
Central England - 49 ☐

1   Forecast valid from:

                to:

2   Met Situation:

3   Winds

        (i)   Strong wind warnings (included if the surface wind anywhere in the region is forecast to exceed 20 kt).

        (ii)   Winds at 1000 feet, 3000 feet and 6000 feet;

4   0°C level:

5   Weather conditions:

6   Remarks/Other warnings:

7   Outlook until:

GEN 3.5.9 — OTHER AUTOMATED METEOROLOGICAL SERVICES

Format of AIRMET Forecasts

1   Forecast valid from:

                      to:

Forecast Areas	
UK SigWeather - 43	☐
UK Upper Winds - 44	☐
UK Update/Outlook - 45	☐

2   Met Situation:

3   Winds

Location	Winds and Temperatures		
	1000 Ft (UK sig Wx) or 10000 FT (UK Upper Winds)*	3000 Ft (UK sig Wx) or 18000 FT (UK Upper Winds)*	3000 Ft (UK sig Wx) or 24000 Ft (UK Upper Winds)*
50N 00E			
GATWICK			
PLYMOUTH			
DERBY			
RONALDSWAY			
GLASGOW			
KIRKWALL			

* Delete as applicable

4   Other information

Quick-Reference Telephone Numbers
All Airmet forecasts available on 09068 7713 plus the two digit code listed for each area above
Automated actuals and forecasts for UK and near Continent aerodromes on 09063 800 400
Clarification and amplification on **all** forecasts on 01344-856267 (Bracknell), 0161-429 0927 (Manchester), or 0141-221 6116 (Glasgow)

GEN 3.5.9— OTHER AUTOMATED METEOROLOGICAL SERVICES

Table 3.5.9.1 — Area Forecasts available on the AIRMET Telephone Menu Service

Dial: 09068-77 13 plus appropriate AIRMET code

AIRMET Code	Area Ident	Forecast	Period of Validity (UTC)	Available By (UTC)	Outlook Hours	Outlook To (UTC)
40	Southern Region	FAUK44	0500-1300	*	6	1900
		FAUK45	1100-1900	1030	6	0100 ND
		FAUK46	1700-0100	1630	8	0900
41	Northern Region	FAUK54	0500-1300	*	6	1900
		FAUK55	1100-1900	1030	6	0100 ND
		FAUK56	1700-0100	1630	8	0900
42	Scottish Region	FAUK70	0500-1300	*	6	1900
		FAUK71	1100-1900	1030	6	0100 ND
		FAUK72	1700-0100	1630	8	0900
43	UK-Weather	FAUK10	0600-1200	0530	9	2100
		FAUK11	1200-1800	1130	6	2400
		FAUK12	1800-2400	1730	9	0900 ND
44	UK-Upper Winds	FAUK14	0600-1200	0530		
		FAUK15	1200-1800	1130		
		FAUK16	1800-2400	1730		
45	UK-Update and Outlook	FAUK19	0900-1500	0830	30	2100 ND
		FAUK20	1500-2100	1430	27	2400 ND
46	Southwest England	FAUK30	0800-1700	0730	5	2200
		FAUK31	1300-2100	1230	4	0100 ND
		FAUK32	1700-0100	1630	8	0900
47	Southeast England	FAUK34	0800-1700	0730	5	2200
		FAUK35	1300-2100	1230	4	0100 ND
		FAUK36	1700-0100	1630	8	0900
48	Cross-Channel	FAUK38	0800-1700	0730	5	2200
		FAUK39	1300-2100	1230	4	0100 ND
49	Central England	FAUK50	0800-1700	0730	5	2200
		FAUK51	1300-2100	1230	4	0100 ND
		FAUK52	1700-0100	1630	8	0900

Note 1:   * = Will be recorded as soon as possible after 0500.

Note 2:   ND = Next day.

GEN 3.5.10 — METEOROLOGICAL CODES

1 Aerodrome Weather Report Codes (Actuals)

1.1 The content and format of an actual weather report is as shown in the following table:

Report Type	Location Identifier	Date/Time	Wind	Visibility	RVR
METAR	EGSS	231020Z	31015G30KT 280V350	1400SW 6000N	R24/P1500

Present WX	Cloud	Temp/Dew Pt	QNH	Recent WX	WindShear	TREND	Rwy State
SHRA	FEW005 SCT010CB BKN025	10/03	Q0995	RETS	WS RWY23	NOSIG	88290592

2 Identifier

2.1 The identifier has three components as shown below:

(a) Report type

    (i) METAR - Aviation routine weather report. These are compiled half-hourly or hourly at fixed times while the aeronautical station is open;

    (ii) SPECI - Aviation selected special weather report. Special reports are prepared to supplement routine reports when improvements or deteriorations through certain criteria occur. However, by ICAO Regional Air Navigation agreement, they are not disseminated by either OPMET in UK or MOTNE in Europe.

(b) Location indicator
ICAO four-letter code letters (for UK aerodromes, see GEN 2.4).

(c) Date/Time
The date and time of observation, specified as the day of the month, hours and minutes UTC, followed by the letter Z.

Example: METAR EGSS 231020Z.

Note: In the UK, windshear will normally not appear in the individual reports

3 Wind

3.1 Wind direction is given in degrees True (three digits) rounded to the nearest 10 degrees, followed by the windspeed (two digits, exceptionally three), both usually meaned over the ten minute period immediately preceding the time of observation. These are followed without a space by one of the abbreviations KT, KMH or MPS, to specify the unit used for reporting the windspeed.

Example: 31015KT.

3.2 A further two or three digits preceded by a G gives the maximum gust speed in knots when it exceeds the mean speed by 10 kt or more.

Example: 31015G30KT.

3.3 Calm is indicated by '00000', followed by the units abbreviation, and variable wind direction by the abbreviation 'VRB' followed by the speed and unit.

3.4 If, during the 10 minute period preceding the time of the observation, the total variation in wind direction is 60° or more, the observed two extreme directions between which the wind has varied will be given in clockwise order, separated by the indicator letter V but only when the speed is greater than 3 kt.

Example: 31015G30KT 280V350.

4 Horizontal Visibility

4.1 When there is no marked variation in visibility by direction the minimum is given in metres. When there is a marked directional variation in the visibility, the reported minimum will be followed by one of the eight points of the compass to indicate the direction.

Example: 4000NE.

4.2 When the minimum visibility is less than 1500 metres and the visibility in another direction greater than 5000 metres, additionally the maximum visibility and its direction will be given.

Example: 1400SW 6000N.

Note: 9999 indicates a visibility of 10 km or more; 0000 a visibility of less than 50 metres.

4.3 Visibility is recorded in metres (m) rounded down to:

(a) the nearest 50 m when the visibility is 800 m or less;

(b) the nearest 100 mwhen the visibility is greater than 800 m but less than or equal to 5000 m;

(c) the nearest 1 kilometre when the visibility is greater than 5000 metres, and expressed in kilometres (km).

5 RVR

5.1 An RVR group always includes the prefix R followed by the runway designator and a diagonal, in turn followed by the touch-down zone RVR in metres. If the RVR is assessed on two or more runways simultaneously, the RVR group will be repeated; parallel runways will be distinguished by appending, to the runway designator, L, C or R indicating the left, central or right parallel respectively.

Examples: R24L/1100 R24R/0750.

5.2 When the RVR is greater than the maximum value which can be assessed the group will be preceded by the letter indicator P followed by the highest value which can be assessed. When the RVR is assessed to be more than 1500 metres it will be reported as P1500.

Example: R24/P1500.

5.3 When the RVR is below the minimum value which can be assessed, the RVR will be reported as M followed by the appropriate minimum value assessed.

Example: R24/M0050.

5.4 If it is possible to determine mean values of RVR, the mean value of RVR over the 10 minute period immediately preceding the observation will be reported; trends and significant variations may be reported as follows:

(a) Trends. If RVR values during the 10 minute period preceding the observation show a distinct increasing or decreasing tendency, such that the mean during the first five minutes varies by 100 m or more from the mean during the second five minutes, this will be indicated by subscripts U or D for increasing or decreasing tendencies; otherwise, subscript N will indicate no distinct change during the period.

Example: R24/1100D.

(b) Significant Variations. When the RVR at a runway varies significantly such that, during the 10 minute period preceding the observation, the 1 minute mean extreme values vary from the 10 minute mean value by either more than 50 metres or more than 20% of the 10 minute mean value (whichever is greater), the 1 minute mean minimum and maximum values will be given in that order, separated by V, instead of the 10 minute mean.

Example: R24/0750V1100.

5.5 If the 10-minute period immediately preceding the observation includes a marked discontinuity in runway visual range values, only those values occurring after the discontinuity should be used to obtain mean values.

5.6 A complete RVR group may therefore be of the form:

Example: R24L/0750V1100U.

Note: Until further notice, UK aerodromes will not be required to report RVR trends and significant variations. RVR is reported when the horizontal visibility or RVR is less than 1500 m. For multi-site RVR/IRVR systems, the value quoted is that for the Touch Down Zone (TDZ). If the RVR is assessed for two or more runways simultaneously, the value for each runway is given.

6 Weather

6.1 Each weather group may consist of appropriate intensity indicators and letter abbreviations combined in groups of two to nine characters and drawn from the following table:

Table 3.5.10.1 — Significant Present and Forecast Weather Codes				
Qualifier		Weather Phenomena		
Intensity or Proximity	Descriptor	Precipitation	Obscuration	Other
- Light	MI — Shallow	DZ — Drizzle	BR — Mist	PO — Dust/Sand Whirls (Dust Devils)
Moderate (no qualifier)	BC — Patches	RA — Rain	FG — Fog	
	BL — Blowing	SN — Snow	FU — Smoke	SQ — Squall
+ Heavy ('Well developed' in the case of FC and PO)	SH — Shower(s)	SG — Snow Grains	VA — Volcanic Ash	FC — Funnel Cloud(s) (tornado or water-spout)
	TS — Thunderstorm	IC — Ice Crystals (Diamond Dust)	DU — Widespread Dust	
	FZ — Freezing (Super-Cooled)	PL — Ice-Pellets	SA — Sand	SS — Sandstorm
VC In the vicinity (not at the aerodrome but not further away than approx 8 km from the aerodrome perimeter)	PR — Partial (covering part of aerodrome)	GR — Hail	HZ — Haze	DS — Duststorm
		GS — Small hail (<5 mm diameter) and/or snow pellets		

GEN 3.5.10 — METEOROLOGICAL CODES

6.2   Mixture of precipitation types may be reported in combination as one group, but up to three separate groups may be inserted to indicate the presence of more than one independent weather type.

Examples: MIFG, VCSH, +SHRA, RASN, -DZ HZ.

Note 1:   BR, HZ, FU, IC, DU and SA will not be reported when the visibility is greater than 5000 m.
Note 2:   Some codes are shown that will not be used in UK METARs and TAFs but may be seen in continental reports and when flying in Europe.

7   Cloud

7.1   A six character group will be given under normal circumstances. The first three to indicate cloud amount:

(a)   **FEW** to indicate 1 to 2 oktas;

(b)   **SCT** (scattered) to indicate 3 to 4 oktas;

(c)   **BKN** (broken) to indicate 5 to 7 oktas;

(d)   **OVC** (overcast) to indicate 8 oktas.

and the last three characters indicate the height of the base of the cloud layer in hundreds of feet above aerodrome level.

Example: FEW018.

7.2   Types of cloud other than significant convective clouds are not identified. Significant clouds are.

(a)   **CB** Cumulonimbus;

(b)   **TCU** Towering Cumulus.

Example: SCT018CB.

7.3   Reporting of layers or masses of cloud is made as follows:

(a)   First Group: Lowest individual layer of any amount;

(b)   Second Group: Next individual layer of more than 2 oktas;

(c)   Third Group: Next higher layer of more than 4 oktas;

(d)   Additional Group: Significant convective cloud if not already reported.

The cloud groups are given in ascending order of height.

Example: FEW005 SCT010 SCT018CB BKN025.

7.4   When there is no cloud to report, and **CAVOK** (see paragraph 8) does not apply, then the cloud group is replaced by **SKC** (sky clear).

7.5   Sky obscured is coded by VV followed by the vertical visibility in hundreds of feet. When the vertical visibility cannot be assessed the group will read VV///: (See GEN 1.7)

Example: VV003.

8   CAVOK

8.1   The visibility, RVR, weather and cloud groups are replaced by CAVOK when the following conditions exist:

(a)   Visibility is 10 km or more;

(b)   No cloud below 5000 ft or below the highest Minimum Sector Altitude, whichever is the greater, and no Cumulo-nimbus;

(c)   No significant weather phenomena at or in the vicinity of aerodrome.

9   Air Temperature/Dewpoint

9.1   These are given in Degrees Celsius, M indicates a negative value.

Examples: 10/03, 01/M01.

If the dew point is missing, the temperature would be reported as 10///.

9.2   Temperatures are reported to the nearest whole degree Celsius, with observed values involving 0.5°C rounded up to the next higher degree Celsius, for example +2.5°C is rounded off to +3°C, -2.5°C is rounded off to -2°C.

10   QNH

10.1   QNH is rounded down to the next whole millibar and reported as a four digit group preceded by the letter indicator Q. If the value of QNH is less than 1000 mbs the first digit will be 0.

Example: Q0995.

10.2   Where reported in inches of mercury, the pressure is prefixed by 'A', and the pressure entered in hundredths of inches, viz with the decimal point omitted between the second and third figure.

Example: A3027.

GEN 3.5.10 — METEOROLOGICAL CODES

11   Supplementary Information

(a) **Recent Weather.** Recent Weather will be operationally significant weather observed since the previous observation (or in the last hour whichever period is the shorter), but not now. The appropriate present weather code will be used, preceded by the letter indicator RE; up to three groups may be inserted to indicate the former presence of more than one weather type.

(b) **Windshear.** The Windshear may be inserted if reported along the take-off or approach paths in the lowest 1600 ft with reference to the runway. WS is used to begin the group.

Examples: WS RWY20, WS ALL RWY.

Until further notice, UK aerodromes will not insert windshear groups.

12   TREND

12.1   For selected aerodromes, this is a forecast of significant changes in conditions during the two hours after the observation time:

(a) Change Indicator:

BECMG (becoming) or TEMPO (temporary), which may be followed by a time group (hours and minutes UTC) preceded by one of the letter indicators FM (from), TL (until), AT (at);

(b) Weather:

Standard codes are used. NOSIG replaces the trend group when no significant changes are forecast to occur during the trend forecast period.

Examples: BECMG FM1100 25035G50KT; TEMPO FM0630 TL0830 3000 SHRA.

13   Runway State Group

13.1   An eight-figure Runway State Group may be added to the end of the METAR (or SPECI) (following any TREND) when there is lying precipitation or other runway contamination. It is composed as follows:

(a) **Runway Designator** (First Two Digits)

27	=	Runway 27 or 27L	88	=	All runways
77	=	Runway 27R	99	=	A repetition of the last message received because
		(50 added to the designator for 'right' Runway)			no new information received.

(b) **Runway Deposits** (Third Digit)

0	=	Clear and dry	6	=	Slush
1	=	Damp	7	=	Ice
2	=	Wet or water patches	8	=	Compacted or rolled snow
3	=	Rime or frost covered	9	=	Frozen ruts or ridges
		(depth normally less than 1 mm)	I	=	Type of deposit not reported
4	=	Dry Snow			(eg due to runway clearance in progress).
5	=	Wet Snow			

(c) **Extent of Runway Contamination** (Fourth Digit)

1	=	10% or less	9	=	51% to 100%
2	=	11% to 25%	I	=	Not reported (eg due to runway clearance in progress).
5	=	26% to 50%			

(d) **Depth of Deposit** (Fifth and Sixth Digits)

The quoted depth is the mean number of readings or, if operationally significant, the greatest depth measured.

00	=	less than 1 mm	96	=	30 cm
01	=	1 mm etc	97	=	35 cm
		through to	98	=	40 cm or more
90	=	90 mm	99	=	Runway(s) non-operational due to snow,
91	=	not used			slush, ice, large drifts or runway clearance,
92	=	10 cm			but depth not reported.
93	=	15 cm	II	=	Depth of deposit operationally not significant
94	=	20 cm			or not measurable.
95	=	25 cm			

(e) **Friction Co-efficient or Braking Action** (Seventh and Eighth Digits)

The mean value is transmitted or, if operationally significant, the lowest value. For example:

28	=	Friction co-efficient 0.28		95	=	Braking action: Good
35	=	Friction co-efficient 0.35		99	=	Figures unreliable (eg if equipment has been used which does not measure satisfactorily in slush or loose snow)
	or					
91	=	Braking action: Poor				
92	=	Braking action: Medium/Poor		II	=	Braking action not reported (eg runway not operational; closed; etc)
93	=	Braking action: Medium				
94	=	Braking action: Medium/Good				

Note 1:   **CLRD.** If contamination conditions on all runways cease to exist, a group consisting of the figures 88, the abbreviation CLRD, and the Braking Action, is sent.

Note 2:   It should be noted that runways can only be inspected as frequently as conditions permit, so that a re-issue of a previous half hourly report does not necessarily mean that the runway has been inspected again during this period, but might mean that no significant change is apparent.

Note 3:   It is emphasised that this reporting system is completely independent of the normal NOTAM system and these reports are not used by AIS for amending SNOWTAM received from originators.

14   'AUTO' and 'RMK'

14.1   Where a report contains fully automated observations with no human intervention, it will be indicated by the code word 'AUTO', inserted immediately before the wind group. Such reports are not broadcast from UK civil aerodromes.

14.2   The indicator 'RMK' (remarks) denotes an optional section containing additional meteorological elements. It will be appended to METARs by national decision, and will not be disseminated internationally. In UK, the section will not be inserted without prior authority of the MET Authority.

15   Missing Information

15.1   Information that is missing in a METAR or SPECI may be replaced by diagonals.

16   Examples of METAR:

(a)   METAR EGGX 301220Z 14005KT 0450E R12/1000N DZ BCFG VV/// 08/07 Q1004 NOSIG=

(b)   METAR EGLY 301220Z 24015KT 200V280 8000 -RA SCT010 BKN025 OVC080 18/15 Q0983 TEMPO 3000 RA BKN008 OVC020=

(c)   METAR EGPZ 301220Z 30025G37KT 270V360 1200NE 6000S +SHSNRAGS FEW005 SCT010 BKN020CB 03/M01 Q0999 RETS BECMG AT1300 9999 NSW SCT015 BKN100=

The above METAR for 1220 UTC on the 30th of the month, in plain language:

EGGX: Surface wind: mean 140 Deg True, 5 kt; Minimum vis 450 metres (to east); mean RVR 1000 metres (at threshold Runway 12, no apparent tendency); moderate drizzle with fog patches; Sky obscured, vertical visibility not available; dry bulb temperature Plus 8 C, dew point Plus 7 C; Aerodrome QNH 1004 mb; Trend: no significant change expected next two hours;

EGLY: Surface wind: mean 240 Deg True, 15 kt; varying between 200 and 280 deg; Minimum vis 8 km; Light rain; cloud 3-4 oktas base 1000 ft, 5-7 oktas 2500 ft, 8 oktas 8000 ft; dry bulb: plus 18 C, dew point; plus 15 C; QNH 983 mb; Trend: temporarily 3000 m in moderate rain with 5-7 oktas 800 ft, 8 oktas 2000 ft;

EGPZ: Surface wind: mean 300 Deg True, 25 kt; Gust 37 kt, varying between 270 and 360 deg; Minimum vis 1200 m (to northeast), Maximum vis 6 km (to south); heavy shower of snow, rain and small hail; 1-2 oktas base 500 ft, 3-4 oktas base 1000 ft, 5-7 oktas CB base 2000 ft; dry bulb: plus 3 C, dew point: minus 1 C; QNH 999 mb; thunderstorm since previous report. Trend: improving at 1300 UTC to 1 km or more, nil weather, 3-4 oktas 1500 ft, 5-7oktas 10000 ft.

17   Aerodrome Forecast (TAF) Codes

17.1   TAFs describe the forecast prevailing conditions at an aerodrome and usually cover a period of 9 to 24 hours. The validity periods of many of the latter do not start until 8 hours after the nominal time of origin and the forecast details only cover the last 18 hours. The 9 hour TAFs are updated and re-issued every 3 hours and those valid for 12 to 24 hours, every 6 hours. Amendments are issued as and when necessary. The forecast period of a TAF may be divided into two or more self contained parts by the use of the abbreviation FM followed by a time. TAFs are issued separately from the METAR or SPECI and do not refer to any specific report; however, many of the METAR groups are also used in TAFs and significant differences are detailed below.

GEN 3.5.10 — METEOROLOGICAL CODES

17.2   The content and format of a TAF is as in the following table:

Report Type	Location Identifier	Date/Time of Origin	Validity Time	Wind	Visibility	Weather
TAF	EGZZ	130600Z	130716	31015KT	8000	-SHRA

Cloud	Variant	Validity Times
FEW005 SCT018CB BKN025	TEMPO	1116

Visibility	Weather	Cloud	Probability	Validity Time	Weather
4000	+SHRA	BKN010CB	PROB30	1416	TSRA

17.2   Example of TAF

(a)   9 hr TAF:

FCUK31 EGGY 300900
TAF EGGW 301019 23010KT 9999 SCT010 BKN018 BECMG1114 6000 -RA BKN012 TEMPO 1418 2000 RADZ
    OVC004FM1800 30020G30KT 9999 -SHRA BKN015CB=

(b)   18 hr TAF:

FTUK31 EGGY 302200
TAF EGLL 310624 13010KT 9000 BKN010 BECMG 0608 BKN020 PROB30 TEMPO 0816 17025G40KT 4000 TSRA
    BKN015CB BECMG 1821 3000 BR SKC=

18   Differences from the METAR

(a)   **Identifier**. In the validity period, the first two digits indicate the day on which the period begins, the next two digits indicate the time of commencement of the forecast in whole hours UTC and the last two digits are the time of ending of the forecast in whole hours.

(b)   **Wind**. In the forecast of the surface wind, the expected prevailing direction will be given. When it is not possible to forecast a prevailing surface wind direction due to its expected variability, for example, during light wind conditions (3 kts or less) or thunderstorms, the forecast wind direction will be indicated by the use of the abbreviation 'VRB'.

(c)   **Horizontal Visibility**. As with the METAR code, except that only one value (the minimum) will be forecast. Visibility is reported in steps detailed in para 4.3.

(d)   **Weather**. If no significant weather is expected the group is omitted. However, after a change group, if the weather ceases to be significant, the abbreviation NSW is used for No Significant Weather.

(e)   **Cloud**. When clear sky is forecast the cloud group is replaced by SKC (sky clear). When no cumulo-nimbus or clouds below 5000 ft or below the highest minimum sector altitude, whichever is the greater, are forecast and CAVOK or SKC are not appropriate, then NSC (No Significant Cloud) is used. Only CB cloud will be specified.

(f)   **Significant Changes**. The abbreviation FM followed by the time to the nearest hour and minute UTC is used to indicate the beginning of a self contained part in a forecast. All conditions given before this group are superseded by the conditions indicated after the group.

Example: FM1220 27017KT 4000 BKN010.

The change indicator BECMG followed by a four figure time group, indicates an expected permanent change in the forecast meteorological conditions, at either a regular or irregular rate, occurring at an unspecified time within the period. The time period should not normally exceed 2 hours but in any case should not exceed 4 hours.

Example: BECMG 2124 1500 BR.

The change indicator TEMPO followed by a four figure time group indicates a period of temporary fluctuations to the forecast meteorological conditions which may occur at any time during the period given. The conditions following these groups are expected to last less than one hour in each instance and in aggregate less than half the period indicated. The conditions following these groups are expect to last one hour in each instance and in aggregate less than half the period indicated.

Example: TEMPO 1116 4000 +SHRA BKN010CB.

(g)   **Probability**. The probability of occurrence happening will be given as a percentage, although only 30% and 40% will be used. The abbreviation PROB is used to introduce the group, followed by a time group, or an indicator and a time group.

Examples: (a)   PROB30 0507 0800 FG BKN004:   (b)   PROB40 TEMPO 1416 TSRA.

(h) **Amendments**. When a TAF requires amendment (including correction), the amended forecast may be indicated by inserting AMD after TAF in the identifier and this new forecast covers the remaining validity period of the original TAF.

Example: TAF AMD EGZZ 130820Z 130816 21007KT 9999 BKN020 BECMG 0912 4000 RADZ BKN008=

Any further amendments or corrections to a TAF that has already been amended or corrected will result in the same 'TAF AMD' coding being used.

If an amended TAF **bulletin** is issued, amendment indicators 'AAA', or 'AAB' etc, and the time of origin will appear in the header line beginning FC or FT.

'CCA', 'CCB' may also be seen for corrections and 'RRA', 'RRB' for retards.

(i) **Other Groups**. Three further TAF groups are not used for civil aerodromes in the UK but are shown here to assist in decoding overseas and UK military TAF.

    (i) Forecast Temperature TTFTF/GFGFZ
        T   Group Indicator
        TFTF   Forecast Temperature (M indicates minus)
        GFGF   Time in whole hours to which temperature refers

        Example: TM05/09Z.

    (ii) Airframe Ice Accretion 6IchihihitL

        Example: 680205.

6	Group indicator		6	moderate in precipitation
Ic	Type of airframe ice accretion:		7	severe
0	none		8	severe in cloud
1	light		9	severe in precipitation
2	light in cloud		hihihi	Height above ground level of lowest icing level (hundreds of feet)
3	light in precipitation			
4	moderate		tL	Thickness of icing layer (for decode see turbulence layer below)
5	moderate in cloud			

        Example 680205

    (iii) Turbulence 5B$h_Bh_Bh_B$tL

5	Group indicator		7	severe in clear air, frequent
B	Turbulence:		8	severe in cloud, infrequent
0	none		9	severe in cloud, frequent
1	light		hBhBhB	Height above ground level of lowest level of turbulence (hundreds of feet)
2	moderate in clear air, infrequent			
3	moderate in clear air, frequent		tL	Thickness of turbulent layer:
5	moderate in cloud, frequent		1-9	thickness in thousands of feet
6	severe in clear air, infrequent			

        Example: 520021

19   Reports in Abbreviated Plain Language

19.1   Some reports may be disseminated in abbreviated plain language. These will use:

    (a)   Standard ICAO abbreviations and

    (b)   numerical values of a self explanatory nature.

The abbreviations referred to under (a) are contained in the Procedures for Air Navigation Services - ICAO Abbreviations and Codes (ICAO Doc 8400)

INTENTIONALLY   BLANK